Cell Signalling: Experimental Strategies

METHODOLOGICAL SURVEYS IN BIOCHEMISTRY AND ANALYSIS

Series Editor: Eric Reid

   Guildford Academic Associates
   72 The Chase
   Guildford GU2 5UL, United Kingdom

This series is divided into Subseries A: Analysis, and B: Biochemistry.
Enquiries concerning Volumes 1-18 should be sent to the above address.

Recent Titles

How to obtain future titles on publication

A standing order plan is available for this series. A standing order will bring delivery of each new volume immediately upon publication. For further information, please write to:

The Royal Society of Chemistry
Distribution Centre
Blackhorse Road
Letchworth
Herts. SG6 1HN

Telephone: Letchworth (0462) 672555

Methodological Surveys in Biochemistry and Analysis,
Volume 21

Cell Signalling:
Experimental Strategies

Edited by

Eric Reid
Guildford Academic Associates, Guildford, United Kingdom

G.M.W. Cook and J.P. Luzio
University of Cambridge

Based on proceedings of the Twelfth International Subcellular Methodology Forum entitled Cell Signalling: Experimental Strategies, held September 4–7, 1990 in Guildford, UK.

Special Publication No. 92

ISBN 0-85186-436-8

A catalogue record for this book is available from the British Library

Published by the Royal Society of Chemistry, Thomas Graham House, Science Park, Cambridge CB4 4WF

Printed in Great Britain by Redwood Press Limited, Melksham, Wiltshire

Senior Editor's Preface

This book complements its predecessor *Biochemical Approaches to Cellular Calcium* (Vol. 19; still topical), and likewise comprises material presented at a Methodology Forum (Sept. 1990) and now thoroughly edited and indexed. Strategies and practical guidance feature strongly, along with actual observations. Many authors have also broadly surveyed their own area of interest. This is helpful to readers unfamiliar with signalling concepts and mechanisms, who could usefully read firstly the introductory material in arts. #A-1 to #A-5, #B-3 and #B-6, and others such as #C-3, #D-1 and #E-2.

The 'cellular' (as distinct from the 'bioanalytical') subseries may not continue after the present book, largely because mounting a high-calibre Forum is an increasing drain on the modest funds of the sponsoring Trust, notwithstanding book-royalty input. The Forum's trail-blazing era (from 1969) ended a decade ago when, with the world-wide proliferation of symposia, the number of non-speaking Forum registrants began to shrink. Books in the cellular field have also proliferated. Consideration of methodology has become 'respectable'; hence a decision to cease, whilst regretted by Forum enthusiasts, would represent retiral with honour.

Publication texts have improved in presentation in recent years, to the relief of the Editors. Yet some 'blind spots' persist, notably in respect of references. Mistakes that have escaped shrewd editorial scrutiny at the outset may come to light when already proofed texts are finally re-inspected; so an appeal is now made to authors:- please compare your references as printed with the versions in your files!

Some authors have a statistical blind spot, blithely making error estimates from a mere 3 observations (even 2!). Moreover, for products in the wet state, some authors leave the reader to infer that stated weights are based on protein. A reiterated personal dislike is use of the term 'incubate' where the material is kept cold rather than warmed. To investigators in general, not particularly to our book contributors, the following remark is addressed, especially applicable where it is for a 'molecular' study that subcellular material is needed.- It is mandatory that subcellular fractions be skilfully prepared and properly characterized, as was emphasized in early volumes (still useful and procurable) and in a Forum-generated article aimed at Journal Editors as well as authors [D.J. Morré *et al.* (1979) *Eur. J. Cell Biol. 20*, 195-199 — *Markers for membranous cell components*; reprints requestable from this Editor].

Since the book has been constructed with care to ensure that it is free from the raggedness that is manifested by many 'Conference Proceedings' books, it may seem paradoxical that Forum discussion points have been included (in the 'nc' subsections). They are justifiable as being an asset to the book, as is the miscellaneous supplementary material, mainly comprising literature which this Editor has come across. It serves to reinforce and extend the main articles in respect of topic coverage, and is indexed; even a 'suboptimal' citation may have value as a spur to a follow-up search of literature.

Book layout. - The diverse subject-matter of the supplementary (nc) material and of the articles, often wide-ranging, offered few possibilities for neat 'packaging', posing the challenge of achieving tolerably coherent placement within the book. The section categories adopted, and the assignment strategy, are set down in the Contents list *(opposite)* which opens with an explanatory paragraph that is worth perusal.

Acknowledgements. - Thanks are expressed to our authors for trouble taken to furnish publication texts. Dr. K-J. Andersen (Univ. of Bergen) played a valuable role as Forum Co-Organizer. Generous support came from U.K. pharmaceutical companies – Glaxo, ICI and SmithKline-Beecham. Some Forum contributors attended with little or no support from Forum funds. Certain Fig. legends carry an acknowledgement for permission to reproduce, mostly courtesy of the Biochemical Society's publishing arm, Portland Press.

Conventions and abbreviations. - Throughout the book, ° signifies °C. Divergences in nomenclature have been largely eliminated; thus, an inositol phosphate is denoted InsP, and only in some Figs. has the synonymous term IP lingered; similarly the term Ptd-Ins *(not* PI) has been adopted for a phosphatidylinositol (phosphoinositide), not to be confused with phosphoinositol (InsP). Free Ca^{2+} within cells, usually cytosolic, is generally connoted $[Ca^{2+}]_i$, usually referring to its level; Ca^{2+} outside the cell is termed $[Ca^{2+}]_e$ or sometimes $[Ca^{2+}]_o$. Such terms are generally explained at the start of each article, accompanied where applicable by others such as Ab = antibody, PKC = protein kinase C, PL = phospholipase, p.m. = plasma membrane. For 'GPI', see p. 199. GTPγS is a GTP analogue, guanosine 5'-O-(3-thiotriphosphate); it can arise from ATPγS, which otherwise is metabolically stable. (Some authors use the synonym GTP[S].)

Guildford Academic Associates   ERIC REID
72 The Chase, Guildford,
Surrey GU2 5UL, U.K.   *10 April 1991*

Contents

Layout.- Each **'nc'** subsection includes Forum discussion material, and reinforcing material for which there is some sublisting (on **'ncE'** title page in particular). Even where bioactive agents were investigated for their own sake rather than used as 'tools', generally they are not grouped in a particular section. Exceptionally, a section (**#C**) is devoted to certain endocrine tissues and targets. Other hormonal subject-matter appears elsewhere, particularly in **#B** & **#D**. Material on interactions between nearby cells and on locally acting agents, e.g. eicosanoids, endothelins and growth factors, appears largely in **#E** but also elsewhere (**#ncB** houses nitric oxide). **#E** also houses items on cell proliferation and cognate topics such as tyrosine kinases. **#E** also contains neural subject-matter that would otherwise have been disadvantageously dispersed through other sections.

List of Authors

Primary author

*Co-authors, with relevant name
to be consulted in left column*

*see art. for author's new address

Primary author	*Co-authors, with relevant name to be consulted in left column*
J-P. Mauger - pp. 149-158 INSERM U 274, Univ. Paris-Sud, Orsay, France	P. Meda - Bruzzone R.H. Michell - Barker F.M. Mitchell - Milligan
G. Milligan - pp. 39-48 Univ. Inst. of Biochem., Glasgow	J.O. Møskaug - Madshus I. Mullaney - Milligan
J.T. O'Flaherty - pp. 127-138 Bowman Gray Sch. of Medicine, Wake Forest Univ., Winston-Salem, NC	K. Müller - Knot B.M. Mullock - Luzio M.S. Nenseter - Berg
L.A.J. O'Neill - pp. 31-38 Strangeways Lab., Cambridge*	
C.A. Pasternak - pp. 391-396 as for Austen	D.L. Paul - Bruzzone J.H. Perez - Luzio
J. Pfeilschifter - pp. 3-12 Ciba-Geigy, Basle, Switzerland	S.J. Persaud - Howell F. Pietri - Mauger
D. Scheller - pp. 373-379 Janssen Res. Foundation, Neuss, Germany	J.H. Reinders - Jansen A. Robinson - Austen P. Rorsman - Berggren U.T. Rüegg - Knot
S. Soboll - pp. 257-265 Inst. f. Physiol. Chemie I, Universität, Dusseldorf, Germany	J. Saklatvala - O'Neill G. Sala-Newby - Campbell E. Samols - Marks C. Schmitz-Salue - Joost A. Schürmann - Joost H. Sies - Soboll J. Stagner - Marks E. Stylianou - O'Neill K.I. Swenson - Bruzzone K. Tan - Marks F. Tegtmeier - Scheller & Wermelskirchen
N.T. Thompson - pp. 69-77 (& see Bonser; same address)	G.M.H. Thomas - Birch J. Urenjak - Wermelskirchen
D. Wermelskirchen - pp. 369-372 as for Scheller	M.B. Vallotton - Lang G. Van Dessel - Hilderson O.K. Vintermyr - Døskeland T.M. Weber - Joost O. Westwood - Austen
*see art. for author's new address	B. Wilffert - Wermelskirchen N.S. Wong - Barker

#A

THE SIGNALLING SCENE, AND RESPONSE INITIATION

#A-1

INTERACTING SIGNAL TRANSDUCTION PATHWAYS REGULATE COMPLEX RENAL FUNCTIONS

Josef Pfeilschifter

Ciba-Geigy Ltd., Research Department (R-1056.P.23), Pharmaceuticals Division, CH-4002 Basel, Switzerland

The complex physiological functions of the kidney are regulated by humoral, neuronal and local functions. The responsiveness of renal tissue to a large variety of hormones and autocoids is particularly interesting when considering the marked heterogeneity of cell types existing in this organ. The three major signalling mechanisms employed by a variety of hormones use cAMP*, cGMP and Ca^{2+} as the intracellular messengers. Besides these three second messengers, the role of PKC in the control of physiological processes in the kidney is discussed in this survey [intended to set the scene for the rest of the book - *Ed.*].

Three major renal functions are presented to exemplify interacting signalling cascades: regulation of glomerular filtration by mesangial cells, of renin secretion from juxtaglomerular cells, and of collecting-duct water transport. Furthermore, PTH signalling in proximal tubular cells is considered as an illuminating example of a hormone that exerts its action on kidney via *two different signalling cascades.*

REGULATION OF MESANGIAL CELL CONTRACTION

Mesangial cells are a major determinant of the glomerular filtration rate. Morphologically, mesangial cells resemble vascular smooth muscle cells and are able to contract upon stimulation by vasoactive hormones (review: [1]). Analogously to smooth muscle cells, it is assumed that Ca^{2+} occupies a central role in excitation-contraction coupling in mesangial cells. When appropriate agonists such as ANG-II or VP bind to specific cell surface receptors on mesangial cells, they activate a PL-C that hydrolyses $PtdInsP_2$ with the formation of $InsP_3$ and DAG (Fig. 1, A). $InsP_3$ releases Ca^{2+} from intracellular stores and may also participate in increasing

*Abbreviations (some with variants in Figs.; consult these for others).- cAMP, cyclic AMP (similarly cGMP); ANG-II, angio- and the prefix Ptd denote respectively the 1,4,5-trisphosphate and phosphatidyl (IP_3 and prefix P in Figs.)]; PKC, protein kinase C; PL, as in PL-A_2 (PLA_2 in Figs.), phospholipase; PTH, parathyroid hormone; VP, vasopressin.

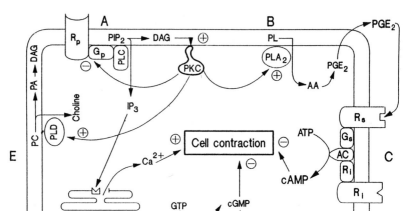

Fig. 1. Signalling pathways in mesangial cells (see text for details). Abbreviations besides those on previous p.: Rp & Gp, receptor and G-protein coupling to PL-C [P(td)I(ns)P$_2$ is the 4,5-bisphosphate]; PL, phospholipids; AA, arachidonic acid; PG, prostaglandin; R$_s$ & G$_s$, receptor and G-protein coupling in a stimulatory mode to adenylate cyclase (AC); similarly R$_i$ & G$_i$, but inhibitory mode; GC$_p$ & GC$_s$, particulate and soluble guanylate cyclase; PC, Ptd-choline; PA, phosphatidic acid. *From ref. [3], by permission.*

Ca^{2+} influx, while DAG activates PKC. Ca^{2+} binds to calmodulin and initiates activation of myosin light-chain kinase with subsequent phosphorylation of 20 kDa myosin light chains, whereby contractile force is regulated (review: [2]).

The delicate regulation of glomerular haemodynamics demands a well tuned system of counteracting regulatory cycles. Three such cycles which exert a strong feedback on the contractile input of hormone-like ANG-II are known in mesangial cells. PKC plays a central role in two of these cycles (as reviewed [3]). Although the bifurcating inositol lipid signalling system looks at first unnecessarily complex, it does provide an elegant mechanism of auto-regulation. Treatment of mesangial cells with PKC activators such as phorbol esters decreases or abolishes ANG-II-, VP- and ATP-stimulated generation of InsP$_3$ and subsequent Ca^{2+} mobilization, suggesting a role for PKC as a negative feedback regulator of cell responsiveness [4-6]. Furthermore, such a modulation of InsP hydrolysis by PKC has been proposed to be involved in

ANG-II-induced homologous desensitization in mesangial cells
[7]. Reinforcing this hypothesis, PKC inhibitors augmented
hormone-stimulated $InsP_3$ generation. Furthermore, in PKC-
depleted mesangial cells a potentiation of hormone-evoked $InsP_3$
formation and Ca^{2+} mobilization was observed [5, 8]. These
data conform well with the idea that blockage or down-regulation
of PKC would result in an increased responsiveness of the
inositol lipid signalling cascade to hormone stimulation. An
important consequence of this PKC-mediated feedback regulation
is that it is limiting for $InsP_3$ formation and Ca^{2+} mobilization
in stimulated cells (Fig. 1, A).

The second feedback cycle triggered by PKC is the activation
of $PL-A_2$ and the subsequent formation of vasodilatory prostaglandins (PG's; Fig. 1, B). Phorbol esters, which activate
PKC, are capable of stimulating PG synthesis in mesangial
cells [6, 7]. Inhibitors of PKC suppress the stimulatory
effect of phorbol esters on PG synthesis, suggesting a role
for PKC in the regulation of $PL-A_2$ [7]. Furthermore, a
synergistic action of Ca^{2+} and phorbol esters on $PL-A_2$ activity
was demonstrated by mesangial-cell arachidonic acid release
and PG synthesis [9]. Recently, a $PL-A_2$ from mesangial
cells has been purified and characterized [10]: it has M_r
~60 kDa and has an absolute Ca^{2+} requirement. It is noteworthy
that its activity can be increased by pre-incubation of
the cells with VP or phorbol ester, indicating that it is
a hormone-sensitive $PL-A_2$ and that PKC can increase its activity.
The observed increase in enzyme activity may be due to
direct phosphorylation of $PL-A_2$ by PKC. Alternatively, PKC
may phosphorylate a $PL-A_2$-modulating protein. Also of interest
is the observation by Troyer and colleagues [6] of an additive
effect of phorbol ester and VP on PG synthesis, which contrasts
with the inhibitory effect of phorbol ester on agonist-stimulated $InsP_3$ generation. The authors suggested the existence
of separate coupling mechanisms for $PL-A_2$ and PL-C activation.

In PKC down-regulation experiments [unpublished work]
we observed that stimulation of PG synthesis and inhibition
of $InsP_3$ generation were abolished with different time-courses,
indicating that different isozymes of PKC with different
sensitivities to down-regulation may be involved in these
two cellular responses. PGE_2 formed in mesangial cells
is released into the extracellular fluid and acts in an
autocrine fashion on the cells to stimulate adenylate cyclase.
The cAMP thus formed initiates mesangial cell relaxation
by protein kinase A-mediated phosphorylation and inactivation
of myosin light-chain kinase (Fig. 1, C). This is associated
with myosin light-chain dephosphorylation and stress fibre
disassembly [2].

The third regulatory cycle is initiated by an increase
in intracellular cGMP (Fig. 1, D). There are two ways
of increasing cGMP in mesangial cells: atrial natriuretic
factor activates a membrane form of guanylate cyclase, whereas
nitro-compounds or endothelium-derived relaxing factor activates a soluble form of guanylate cyclase [11, 12]. Increases
in cGMP levels induced in either way accompany relaxation
of mesangial cells [11]. It has been reported that atrial
natriuretic factor inhibits ANG-II- and VP-induced Ca^{2+} mobilization in mesangial cells [13]. Inhibition of Ca^{2+} influx
from the extracellular compartment, inhibition of hormone-
induced release of Ca^{2+} from intracellular storage sites
or stimulation of calcium sequestration are candidates in
the search for possible mechanisms underlying mesangial cell
relaxation in response to cGMP increases. Very recently
Barnett and colleagues [14] reported that atrial natriuretic
factor and sodium nitroprusside inhibit ANG-II-stimulated $InsP_3$
formation in mesangial cells [14]. Obviously, 'cross-talk'
between different transmembrane signalling pathways proves
to be a pre-requisite for the subtle regulation of glomerular
filtration.

REGULATION OF RENIN SECRETION

The highly specific aspartyl proteinase renin is synthesized
and stored in the granular juxtaglomerular ('JG') cells,
which are modified smooth-muscle cells present in the media
of the afferent arteriole. These cells also show the characteristics of secretory cells, as evidenced by the presence of
a well developed endoplasmic reticulum, a prominent Golgi apparatus and secretory granules. Four basic mechanisms are
responsible for the control of renin release: the sympathetic
nerve activity, the renal vascular or stretch receptor, the
macula densa receptor, and humoral factors. Renin release
from JG cells shows some unusual features. The most striking
one is the so-called 'calcium paradox': in contrast to almost
every other secretory process, increased intracellular Ca^{2+}
leads to inhibition of renin release from JG cells. Stimulation
of Ca^{2+} mobilization by hormones that activate PL-C and
generate $InsP_3$, or inhibition of Ca^{2+} extrusion, which occurs
mainly *via* Ca^{2+}-ATPase and Na^+/Ca^{2+} exchange, causes inhibition
of renin release as shown in Fig. 2. Activation of PKC,
the second branch of the inositol lipid signalling pathway,
by phorbol esters leads to a dose-dependent inhibition of
renin secretion (as reviewed: [15, 16]).

The mechanism of Ca^{2+}- and PKC-mediated inhibition of
renin secretion from JG cells is largely unknown. Recently
it has been reported that 12- and 15-lipoxygenase products
of arachidonic acid metabolism are potent inhibitors of

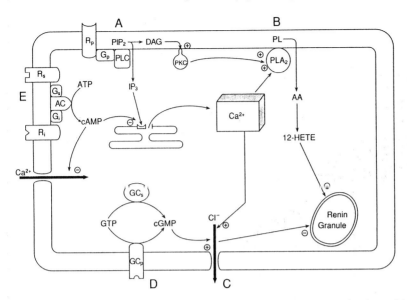

Fig. 2. Signalling pathways in juxtaglomerular (JG) cells
(see text for details). 12-HETE, 12-hydroxyeicosatetraenoic
acid; for other abbreviations see first p. and Fig. 1 legend.

renal release [17]. Furthermore, ANG-II stimulates 12-hydroxy-
eicosatetraenoic acid formation in renal cortical slices,
suggesting that ANG-II inhibition of renin secretion is mediated
by arachidonic acid release and subsequent synthesis of 12-
lipoxygenase products [18]. Ca^{2+} and PKC could therefore
inhibit renin release from JG cells by stimulating PL-A_2
(Fig. 2, B). An alternative explanation for the Ca^{2+}-mediated
inhibition of renin secretion has been suggested [16].

        Increased Ca^{2+} levels were shown to activate, in JG
cells, a large chloride conductance which is assumed to
modulate the volume of JG cells (Fig. 2, C).   The efflux
of Cl^- and the accompanying K^+ would result in cell shrinkage
and impair renin granule swelling, which is a pre-requisite
for renin exocytosis [19].   It has been observed that cGMP
increases the Ca^{2+} sensitivity of the Cl^- channels [16].
This conforms well to the inhibition of renin release by
atrial natriuretic factor or endothelium-derived relaxing
factor, which are thought to exert their effects through
an increase in cGMP (Fig. 2, D) [16, 20].

        The most potent signal that initiates renin secretion
is cAMP. PG's, β-adrenergic antagonists, glucagon and PTH
stimulate renin release by an increase in intracellular cAMP
levels.   cAMP has been shown to inhibit Ca^{2+} mobilization
from intracellular stores in single JG cells [21]. Furthermore,

cAMP decreases the Ca^{2+} permeability of the plasma membrane
in JG cells (Fig. 2, E), thus suggesting that the stimulatory
effect of cAMP on renin secretion is indirect and mediated
by modulation of the intracellular Ca^{2+} concentration. In
summary, Ca^{2+} seems to be the dominant regulator of renin
secretion from JG cells. The exact pathway of Ca^{2+} action,
however, still needs to be unravelled.

REGULATION OF COLLECTING-DUCT WATER TRANSPORT

The intracellular messengers generated by PL-C have been
found to interact with other signalling pathways, especially
with the cAMP-mediated reactions. This cross-talk between
the $InsP_3$/PKC cascade and the cAMP cascade is especially
important for the distal tubule and the collecting duct.
In this part of a nephron, VP regulates water transport
via the cAMP pathway, and activators of PKC were observed
to inhibit the hydroosmotic effect of VP (reviews: [15,
22]). Furthermore, other hormones known to activate PL-C
have been observed to inhibit VP-induced water flow. Ando
and colleagues [22] reported that phorbol esters and the
calcium ionophore A23187 have additive but mechanistically
separate effects on VP action in rabbit collecting tubules.

Elevation of intracellular Ca^{2+}, as well as activation
of PKC, suppresses the hydroosmotic effect of VP. This
inhibition is primarily at a step prior to cAMP formation
and involves, at least in part, a cyclooxygenase metabolite
of arachidonic acid. PGE_2 is especially well known to
inhibit VP-stimulated cAMP formation by interacting in an
autocrine fashion with the G_i-protein of the adenylate cyclase
system (Fig. 3). In contrast, the inhibitory effect of
PKC occurs at a post-cAMP step and is independent of cyclooxygenase products of arachidonic acid [22]. These data are
summarized in Fig. 3 and show that both branches of the
inositol lipid cascade are potent negative modulators of
VP-induced water flow. Although most of the agents that
antagonize the effects of VP on the collecting duct inhibit
cAMP formation *via* the G_i-protein, bradykinin and epidermal
growth factor seem to act *via* InsP hydrolysis [22]. Bradykinin
has been demonstrated to activate PL-C and PKC in cortical
collecting-tubule cells, which then inhibit VP-stimulated cAMP
formation at the level of the hormone receptor or coupling
of the receptor to G_s [23].

PTH SIGNALLING IN PROXIMAL TUBULAR CELLS

The well-known renal effects of PTH include an increase
in Ca^{2+} reabsorption as well as an inhibition of phosphate
reabsorption. Modulation of the Na^+/P_i co-transport rate

Fig. 3. Signalling
pathways in a collecting
duct cell (see text for
details). For abbrevi-
ations see first p. and
Fig. 1 legend.

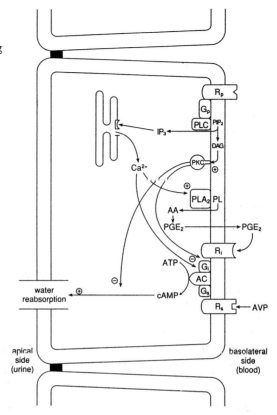

across the apical membrane of the proximal tubule is the
predominant means of regulation of phosphate reabsorption by
the kidney. PTH binding studies demonstrate a single class
of receptors, although Scatchard analysis of the binding
of b-PTH 1-34 revealed a possible heterogeneity of receptors
(as reviewed: [15, 24]). PTH receptors couple to the adenylate
cyclase system, and it has been shown that PTH increases
cAMP levels and thereby mediates inhibition of phosphate
transport. There is recent evidence that PTH inhibits phosphate
transport by triggering InsP hydrolysis [15, 24].

Interestingly, the dose-dependent stimulation of $InsP_3$
and DAG formation is more sensitive than the reported dose
response of cAMP production in opossum kidney ('OK') cells,
a cell line with proximal tubule characteristics. PTH inhibits
Na^+/P_i co-transport with a half-maximal response at ~10 pM.
This hormone concentration accords nicely with the levels
required to stimulate $InsP_3$ and DAG production but not with
those needed for cAMP generation (1-10 nM) [25]. Accordingly,
activation of PL-C may be more important than stimulation
of adenylate cyclase at physiological PTH plasma levels.

Reinforcing this suggestion, down-regulation of PKC leads to a refractory state in which OK cells do not respond to PTH, even though Na^+/P_i co-transport is similar to that in control cells [26].

These results suggest that PKC is the predominant mediator of PTH inhibition of Na^+/P_i co-transport in OK cells. A recent study by Martin and colleagues [27], however, suggests that cAMP is an insensitive indicator for protein kinase A activity. This kinase was activated at PTH concentrations that correlated well with the inhibition of phosphate transport. Previously it had been reported [28, 29] that this kinase can be activated in stimulated cells without concomitant changes in cAMP levels, and it was concluded that dynamic cAMP turnover could be a means of signal amplification.

Fig. 4 depicts two hypothetical models for PTH signalling in target cells. Either there are two PTH receptors, one coupling to adenylate cyclase and the other to the inositol lipid signalling pathway, or a single PTH receptor is coupling to both signalling cascades. The physiological significance of the adenylate cyclase and the $InsP_3$/PKC systems needs further investigation. Maybe the cross-talk between these two signalling cascades is important in mediating the physiological response and in triggering events that may attenuate each other's cellular functions.

References

1. Pfeilschifter, J. (1989) *Eur. J. Clin. Invest. 19*, 347–361.
2. Kamm, E.K. & Stull, J.T. (1989) *Annu. Rev. Physiol. 51*, 299–313.
3. Pfeilschifter, J. (1990) *Klin. Wochenschr. 68*, 1134–1137.
4. Pfeilschifter, J. (1986) *FEBS Lett. 203*, 262–266.
5. Pfeilschifter, J. (1990) *Cell Signalling 2*, 129–138.
6. Troyer, D.A., Gonzalez, O.F., Douglas, J.G. & Kreisberg, J.I. (1988) *Biochem. J. 251*, 907–912.
7. Pfeilschifter, J. (1988) *Biochim. Biophys. Acta 969*, 263–270.
8. Pfeilschifter, J., Fandrey, J., Ochsner, M., Whitebread, S. & de Gasparo, M. (1990) *FEBS Lett. 261*, 307–311.
9. Bonventre, J.V. & Swidler, M. (1988) *J. Clin. Invest. 82*, 168–176.
10. Gronich, J.H., Bonventre, J.V. & Nemenoff, R.A. (1988) *J. Biol. Chem. 263*, 16645–16651.
11. Singhal,P.C., DeCandido, S., Satriano, J.A., Schlondorff, D. & Hays, R.M. (1989) *Am. J. Physiol. 257*, C86–C93.
12. Shultz, P.J., Schorer, A.E. & Eaij, L. (1990) *Am. J. Physiol. 258*, F162–F167.
13. Meyer-Lehnert, H., Tsai, P., Caramelo, C. & Schrier, R.W. (1988) *Am. J. Physiol. 255*, F771–F780.

Fig. 4.
PTH signalling pathways
in proximal tubule cells
(see text for details).
For abbreviations see
first p. and Fig. 1
legend.
Model **A**: two receptors
postulated.
Model **B**: a single PTH
receptor coupling to
both signalling cascades.

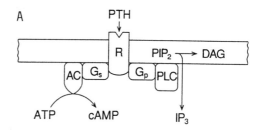

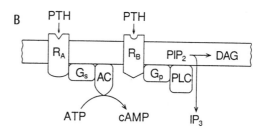

14. Barnett, R., Oritz, P.A., Blaufox, S., Singer, S., Nord, E.P. & Ramsammy, L. (1990) *Am. J. Physiol. 258,* C37-C45.
15. Pfeilschifter, J. (1989) *Renal Physiol. Biochem. 12,* 1-31.
16. Kurtz, A. (1989) *Rev. Physiol. Biochem. Pharmacol. 113,* 1-40.
17. Antonipillai, I., Nadler, J. & Horton, R. (1988) *Endocrinology 122,* 1277-1281.
18. Antonipillai, I., Horton, R., Natarajan, R. & Nadler, J. (1989) *Endocrinology 125,* 2028-2034.
19. Skott, O. & Taugner, R. (1987) *Cell Tissue Res. 249,* 325-329.
20. Vidal, M.J., Romero, J.C. & Vanhoutte, P.M. (1988) *Eur. J. Pharmacol. 149,* 401-402.
21. Kurtz, A. & Penner, R. (1989) *Proc. Nat. Acad. Sci. 86,* 3423-3427.
22. Ando, Y., Jacobson, H.R. & Breyer, M.D. (1988) *News Physiol. Sci. 3,* 235-240.
23. Dixon, B.S., Breckon, R., Fortune, J., Sutherland, E., Simon, F.R. & Anderson, R.J. (1989) *Am. J. Physiol. 257,* F808-F817.
24. Dunlay, R. & Hruska, K. (1990) *Am. J. Physiol. 258,* F223-F231.
25. Quamme, G., Pfeilschifter, J. & Murer, H. (1989) *Biochem. Biophys. Res. Comm. 158,* 951-957.
26. Quamme, G., Pfeilschifter, J. & Murer, H. (1989) *Biochim. Biophys. Acta 1013,* 159-165.
27. Martin, K.J., McConkay, C.L., Garcia, J.C., Montani, D. & Betts, C.R. (1989) *Endocrinology 125,* 295-301.

28. Litvin, Y., Pasmantier, R., Fleischer, N. &
    Erlichman, J. (1984) *J. Biol. Chem.* *259*, 10296–10302.
29. Weiss, A. & Erlichman, J. (1988) *Mol. Endocrinol.* *2*,
    412–419.

#A-2

ADVANCES IN RECEPTOR INVESTIGATION 1980-1990

Harry LeVine, III[⊗]and Stephen C. Brown

Department of Molecular Biology and
  Department of Structural and Biophysical Chemistry,
Glaxo Research Laboratories,
Research Triangle Park, NC 27709, U.S.A.

*Improvements in technology and conceptual developments
have advanced our knowledge of receptors far beyond our mostly
phenomenological knowledge of the early 1980's*. The primary
structures of molecular families of signalling proteins are
now available for nearly all classes of receptor ligand.
The sequences are arranged in common motifs, often within-family
functional domains. There is comparable organization of
other signal-transduction components, ion channels and transducing and effector proteins including DNA-binding regulatory
proteins. Such studies have helped explain response specificities and inter-system interactions. Recombinant technology
allows primary structure and function to be related through
chimaera formation and expression with exchange of functional
domains among receptor subtypes or between families. Convenient
assay systems for expression of rare receptors or single
subtypes facilitate their characterization.*

*Improved sensitivity and resolution of physical techniques
for structure determination, along with the availability of
cloned proteins, offer the prospect of tertiary structural detail.
Computational techniques and software allow rapid processing
and use of detailed structural information to infer binding
properties of prospective ligands and may help towards rational
drug design. The tools now exist to launch the era of increasing
ability to predict ligand interactions through ligand-receptor
tertiary structure determination, bringing in structure-
activity relationships from medicinal chemistry. One could
design molecules with apt pharmacological characteristics to
elicit the desired response from a specific receptor from
a specific tissue with minimal side-effects.*

Explanation of the myriad of biological responses elicited
in different tissues by the same agonist has been aided
by the elucidation of a diversity of receptor subtypes constituting gene families, families of coupling proteins and families
of effector molecules. Similar heterogeneity is found in
voltage- or ligand-gated ion channels, cell-substratum or
cell-cell receptors of the Ig-like superfamily, cytokine receptors,

---

*Note by Ed..- Consult *Investigation of Membrane-located Receptors*
(Vol. 13, this series; ed. E. Reid *et al.*; Plenum, 1984); it includes
a survey by LeVine *et al.*   *[Abbreviations* are listed *overleaf.*
[⊗]now at Warner-Lambert (Parke-Davis), Ann Arbor, MI 48105-1047.

and DNA-binding proteins. So far, the only receptor molecules that lack this gene-family organization, besides possibly the PDGF receptor and a spliced variant (12 amino acids) of the insR* [1], appear to be the insR and receptors for certain other peptides, e.g. EGF, where the receptor and effector constitute domains in a polypeptide chain. General organizational types grouped around subunit structure have been identified, but hitherto no homologies approaching those of the G-protein-linked receptors have been found. Yet nature remains conservative; a basic theme is elaborated upon, substituting domains in the different molecules and altering their properties and with which molecules they interact.

The recognition of heterogeneity at the molecular level resulted from knowledge gained by the cloning and sequencing of cDNA's coding for the receptors. The presence of similar structural elements in related proteins reinforced the concept of protein domains encoding structural and therefore possible functional motifs. Molecular-biology techniques have enabled this hypothesis to be tested directly by experimental manipulation of these domains. Expression of the products in homologous or heterologous cell systems has allowed their analysis, and even led to description of new functional motifs. The extent and interaction of the molecular families of receptors and potential new signalling and control mechanisms is being explored. The next level of understanding is likely to come from the determination of the 3-D molecular structure of receptors, in particular that of the ligand in the binding site. [Vol. 19, this series,† has pertinent molecular-biology arts.-Ed.]

The intent of this article is to briefly illustrate the investigation of receptor domain structure for two important receptor classes: peptide growth factors linked to TK's, and the 7-transmembrane-helix (rhodopsin-like) receptors linked to G-proteins which account for many pharmacologically important biological responses. Following a discussion of uses of receptors in the medicinal chemistry of drug development, advances in physical structure determination and their implications for future pharmaceutics are considered.

RECEPTOR CHIMAERAS

Recombinant DNA technology allows the study of receptor structure/function relationships by random or selective

*Abbreviations.- Ab, antibody; NMR, nuclear magnetic resonance. PKC, protein kinase C; TK, tyrosine kinase - hence RTK, receptor with TK activity. Receptors include: aspR, aspartate; insR, insulin. Growth factors: CSF, myeloid-cell; EGF, epidermal; IGF-I, insulin-like; PDGF, platelet-derived; TGF, transforming. Other agents: NMDA, N-methyl-D-aspartate; CCK, cholecystokinin.

†Biochemical Approaches to Cellular Calcium (eds. & publisher as now).

mutagenesis or domain interchange among subtypes of a given
receptor or between different receptor types. These altered
molecules can then be expressed in different cell systems
depending upon the response to be measured. The types of
information available from this approach are useful in assigning
functional significance to particular residues and domains
within the protein. Such capabilities have been expected to
assist those interested in rational drug design as well as
making it possible to devise a screening system wherein a
single intracellular response is generated by a common intracellular domain coupled to binding domains specific for different ligands. So far these possibilities have remained
primarily of interest to specialty biotechnology companies
and academic laboratories, but the pharmaceutical industry
has shown some interest in the latter system as it is adaptable
to general screening of compounds and has the potential to
distinguish agonists from antagonists, an advantage over ligand
displacement assays.

Selective application of mutagenesis to receptors has
been utilized to identify amino acid residues involved in
the ligand binding event and specificity determination. In
systems where inter-subunit or other protein-protein or protein-
nucleic acid interactions are involved in signal transduction,
such techniques have been useful in delineating regions and
sometimes residues involved in the interaction.

GROWTH FACTOR RECEPTORS

A number of growth factor receptors have been cloned
and expressed in a variety of systems. These are to be
distinguished from an ever-increasing number of cytokine receptors (not discussed here) which are growth or mitogenic factors
for specific populations of cells, notably various leucocyte
subpopulations, and of considerable importance in their own
right. The receptor group studied the most intensively in
terms of domain structure and relationship to function comprises receptors that contain a tyrosine kinase catalytic
activity (RTK's). Interest in these kinase domains has arisen
from their homology to viral oncogene products, such as *src*.
A number of viral oncogenes appear to be mutants of normal
cellular genes that are related to cell proliferation with
enhanced, unregulated TK activity. The list of these
viral oncogenes and their cellular counterparts continues
to grow. Interesting receptor molecules can and have been
cloned on the basis of their homology to oncogenes, e.g,
the CSF-1 receptor (fms-like), c-erbB-2 and c-kit genes.
We focus on the RTK's, on which the most domain structure/
function work has been done (as well reviewed in detail:
[2-4]).

The RTK's have been classified according to conserved characteristics of their extracellular domains. Those of subclass I, which includes the EGF/TGF-α receptor, the HER2/neu cellular oncogene analogue, and the v-erbB gene product, possess two cysteine-rich regions (20-25 cys residues) and appear to be monomeric. Subclass II comprises the insulin and IGF-1 receptors; these receptors contain one cysteine-rich region, but consist of a heterodimer formed by dimers held together by -S-S- bonds with only one type of subunit penetrating the lipid bilayer. Subclass III includes the PDGF receptor, the CSF-1 receptor and the c-kit gene product; they are single-chain receptors with a series of 10 conserved cysteines spaced through the extracellular domain.

Extracellular domain

The extracellular domains of all the subclasses are extensively glycosylated. Their primary structures, as expected from the different specificities, differ between subclasses; yet 40-50% homology exists between the receptors for EGF and HER2/neu (subclass I) and between those for insulin and IGF-I (subclass II) in this region, with less conservation among the subclass III receptors. Insulin and IGF-I are highly homologous peptide hormones with distinct biological functions and different affinities for the insulin, IGF-I and IGF-II receptors, as demonstrated by mutation of residues of IGF-I which alter its ability to bind to the insulin receptor and to the IGF-II receptor (which is unrelated to the RTK's) [5].

Homologies between the insulin and IGF-I receptors [6] have been used by modellers to try to localize the regions of the polypeptide chain involved in hormone binding. The lack of good data (X-ray or NMR) on structure has hampered their efforts. Primary sequence comparison of the insulin and IGF-I receptors suggests some structural determinants that might define the functional differences between them. While the cys's are generally conserved, the C-terminal portion of the α-subunit and the N-terminal portion of the β-subunit show lower homology. These are extracellular domains and are probably involved in ligand binding [6]. The ligand-binding ectodomain has been cloned and expressed in mammalian cells [7]. Insulin binding requires the ectodomain of the α-subunit and causes oligomerization when a ligand is bound.

Within the IGF-IR TK domain (84% overall homology) there is a nonapeptide sequence beginning at residue 1073, whereas in the insulin receptor (insR) this is in the same position as a 70-100 amino acid heterogeneous insertion in subclass III TK regions. The C-terminal cytoplasmic region from 1233 is also of relatively low homology; this is suggested, as for most receptors, to be related to the divergent intracellular actions of the receptors.

Transmembrane region (TM region)

The transmembrane region of the RTK's consists of 23-26 amino acids with little discernable homology other than their common property of hydrophobicity, crossing the membrane just once. Sternberg & Gullick [8] propose that a sequence motif present in the transmembrane α-helices of the TK growth factor receptors is responsible for the dimerization of receptors with concomitant activation of the kinase activity. They suggest a model with close packing of helices, interacting through a 5-residue segment where P0 is a small side-chain (G, A, S, T, P), P3 an aliphatic side-chain (A, V, L, I) and P4 the smallest side-chains (G, A). Sequence differences adjacent to the TM region could lead to different rotational orientation of the helix with respect to other domains of the protein. This might suffice to account for the failure to successfully interchange TM regions among different members of the growth-factor receptor family [9]. For some receptors, notably the T-cell antigen receptor, the specificity of the interaction between subunits is thought to occur through the TM region [10].

Cytoplasmic domain

The cytoplasmic portion of the RTK's contains the TK catalytic and autophosphorylation sequences, separated by 70-100 amino acids. This insertion is characteristic of the particular receptor and has been postulated to govern some of the intracellular interactions of the receptor and to account for the cytosolic specificity of the receptor.

SH2 and SH3 regions (reviewed in [11])

N-terminal to the kinase domain is a region of 100 amino acids with 30% overall homology to the cytosolic protein TK's src, fps and abl, termed 'SH2' (**Src-H**omology-region 2). While not essential for TK activity, mutagenesis suggests a role in governing transformation, perhaps *in cis* and *in trans* to the kinase function and/or *in trans* with a cellular factor(s). Another 50-amino acid domain further N-terminal, SH3, is thought to have an opposing influence to SH2. The putative interactions of SH regions with factors is thought to have regulatory significance in signalling pathways. Deletions in this region in the PDGF receptor resulted in loss of mitogenic activity although other PDGF actions such as TK activity and activation of phosphatidylinositol turnover were normal. This region may be important for interaction with other molecules involved in mitogenesis or may affect the substrate specificity of the TK [12].

Other cellular proteins unrelated to RTK's also contain SH2 and SH3 domains. Two of these, the phospholipase isozyme

PLC-148 which gives rise to the second messengers diacylglycerol and InsP$_3$, and GAP (the GTPase-Activating-Protein) which interacts with p21ras, are RTK signal-transducing elements. In contrast, p47crk, while not itself a TK, appears to induce elevated levels of phosphotyrosyl cellular proteins; it contains SH2 and SH3 regions requisite for display of this activity, and preferentially associates with phosphotyrosine-containing proteins [13]. α-Spectrin, a cytoskeletal scaffolding element, contains a single SH2 region which could directly furnish RTK signal information to the cytoskeleton which is so drastically affected by growth factors. These similarities provide the basis for the hypothesis that proteins having a SH2,3-region may interact with common elements to integrate mitogenic and developmental information at the level of the plasma membrane. This is one of the emerging avenues of inquiry which may allow dissection of the complex phenomena surrounding cell growth.

Tyrosine kinase (TK) region

The ATP-binding site is located 50 amino acids from the transmembrane sequence. The cytoplasmic portion of the receptor C-terminal to the kinase domain contains phosphorylation sites for RTK and also others (serine) which are modified by other protein kinases in the cell and may govern some of the receptor's actions *in vivo*. A 41 kDa insR TK domain has been over-expressed in the SF9-baculovirus system and purified to homogeneity [14]. The availability of a soluble, catalytically active kinase domain will facilitate studies of the activation mechanism and the role of autophosphorylation.

CHIMAERIC RECEPTOR TYROSINE KINASES

The possession in common of one or two hydrophobic transmembrane segments connecting extracellular and intracellular domains has led to the postulate that some of these receptors may share a common mechanism of signal propagation through the membrane despite lack of similarity of other portions or activities of the receptor (growth-factor, LDL and T-cell receptors, and the chemotactic receptors of Gram-negative bacteria). Proposed mechanisms include multimerization of receptors *via* either subunit structure or by multimeric ligands (PDGF and CSF-1), as well as interaction with other proteins [15]. The ligand-binding domain of the aspR including the transmembrane domain was attached to the first hydrophilic residue of the cytoplasmic portion of the insR and expressed in *E. coli*. The purified aspR-insR in the presence or absence of aspartate did not autophosphorylate and demonstrated an aspartate-dependent altered exogenous substrate profile compared with insR. Possible explanations for the specificity alteration include lack of autophosphorylation (non-multimeric receptor), of post-translational modification or of disulphide formation.

Thus, a limited ability exists to transmit an activation signal across the bilayer between recognition and effector domains from two unrelated receptors. It is also apparent that the information transfer between domains of different receptors may not be commutative. An insR ligand-binding domain attached to the aspR cytoplasmic region fails to demonstrate signal transduction [16]. Similar results were found for several other RTK chimaeras including the IL-2 receptor [17] and the v-abl gene product [18].

Experiments done with more homologous receptors have been much more successful. Reciprocal exchanges of the ligand binding and intracellular portions of the insulin and IGF-1 receptors and expression in NIH-3T3 fibroblasts resulted in TK activities regulated by the ligand specified by the binding domain, and all were internalized in response to receptor occupation by ligand. There was similarity in respect of how short-term responses were mediated by the receptors; but their long-term mitogenic potentials were governed by the identity of the cytoplasmic portion [19]. Other RTK's and RTK chimaeras have shown biological activity, e.g. EGF/insR [20], EGF/verbB [21] and insR/v-ros [22, 23].

MUTAGENESIS

Mutagenesis studies of the domains of the RTK's have been highly informative concerning the roles of these regions and mechanisms of activation of the receptor and of intracellular activation. The EGF receptor has been most extensively studied, as will be considered here to exemplify the utility of the approach. The RTK's exist in the membrane in high- and low-affinity forms with respect to the ligand. Low affinity seems to be intrinsic to the extracellular domain while the juxtamembrane and C-terminus cytosolic domains control the high-affinity state. N-terminal deletions lower the affinity for EGF and block the self-dimerization of homomutant receptor.

Receptor dimerization is thought to be the mechanism whereby RTK's of subclasses I and II are activated through an undefined conformational change [24]. Deletions also increase TK activity as if the non-liganded extracellular domain was negatively regulating the kinase. The juxtamembrane region contains a phosphorylation site for the Ca^{2+}/phospholipid-dependent PKC; phosphorylation of this site, T654, decreases the receptor's affinity for EGF. The TK domain activity is thought to be required for all tested biological effects of EGF and insulin, including ligand-induced endocytosis and down-regulation, although conflicting evidence exists [25-27]. Catalytic activity does *not* suffice for signal generation, as a non-catalytic cytoplasmic element is required too. Deletions in the C-terminus of either the insulin or the EGF receptor have no effect on mitogenesis or kinase activity.

Some of the secondary effects of insulin on glucose uptake and glycogen synthesis are altered for that receptor when Y1146-->F [28]. EGF-dependent erythroblast self-renewal and erythropoietin-dependent differentiation by EGF require intact C-terminal sequences [29]. Deletion of the autophosphorylation site is without effect, although replacement with another amino acid decreases both kinase activity and mitogenic potential [28, 30].

Endocytosis of the EGF and insulin receptors follows two pathways inside the cell. A pathway targeting the ligand-receptor complexes to the lysosome for destruction requires TK activity and is independent of the non-catalytic tail of the receptor [31]. Recycling of the complex through endosomes and routing the receptor back to the cell surface is not TK-dependent but requires C-terminal sequences [32].

G-PROTEIN-LINKED RECEPTORS

These receptors have been studied the most intensively, beginning with the prototypical 7-transmembrane helix receptor for photons, rhodopsin. The adrenergic receptors are the most studied hormone receptor of the G-protein-linked class and provide a challenging test of the resolving power and interpretability of mutagenesis results. These proteins come in two major varieties, α and β, which differ in their relative responses to epinephrine and nonepinephrine and in the type of G-protein with which they interact. Moreover, subtypes of each variety, $\beta 1$, $\beta 2$ and, recently, $\beta 3$ β-receptors and $\alpha 1$, $\alpha 2$ (2A, 2B, 2C) and other α-receptors [33] have been defined, genetically, biochemically and pharmacologically. These molecules are distinct entities with characteristic tissue distribution patterns and are of great pharmacological and therapeutic significance, particularly in determining cardiovascular and pulmonary function. Accordingly, there is available an ample set of subtype-specific agonists and antagonists of widely ranging affinities and efficacies as well as high specific-radioactivity ligands to perform the desired studies.

While binding studies can be performed with receptor inserted into the *E. coli* membrane [34], more complex interactions such as those involving receptor/G-protein and desensitization reactions [35, 36] require transient expression in mammalian (COS) cells. All of the following studies, except where indicated, have been performed in mammalian cells. Homology among G-protein-linked receptors is greatest between the 7-α-helical segments which are thought to cross the plasma membrane. Very little homology amongst different receptors is seen at the amino or carboxyl termini or in the cytoplasmic loop between helices 5 and 6, in the case of the α- or β-adrenergic receptors, a substance-K receptor, muscarinic cholinergic receptors, the mas oncogene and the rhodopsins. This also holds for the subtypes of α- and β-receptors or the different opsins. The non-homologous intracellular helix 5-helix 6 loop and C-terminal regions are of variable lengths and amino acid composition, although in all of the receptors they are rich in polar and acidic residues and helix-breaking amino acids. The hydrophobic core of the receptor contains residues involved in ligand binding and represents a structural entity in itself that is stable and self-associating in functional manner as was assessed for the hamster β2-adrenergic receptor after proteolysis [37].

The β-adrenergic receptor has been intensively studied by recombinant techniques. A series of deletion mutations in the hamster β2-adrenergic receptor established that the N- and C-terminal hydrophilic regions, the 2nd and 3rd intracellular loops, and the 1st and 3rd extracellular loops, could all be removed without affecting antagonist binding (^{125}I-CYP) or protein folding and membrane insertion. Insertion of the protein was severely affected by deletion of any of the transmembrane domains or sequences adjacent to the helices, or of intracellular loop 1. Thus, the hydrophobic core of the receptor seems to suffice for ligand binding and receptor stability.

The redox state of 4 cysteine residues in extracellular loops 1 and 2 is involved in modulation of ligand binding and perhaps receptor activation [38]. None of the other cysteines (11/15) have any effect on protein accumulation or ligand binding. Four glycosylation sites on the receptor can be removed without affecting ligand binding or protein accumulation [38].

Photoaffinity-labelling studies have suggested that W330 in helix 7 as well as residues in helices 3 or 4 can interact with ligand, which accords with the conclusion that among all of the conserved acidic counter-ion side-chain residues in helices 2 and 3 for the cationic amine residue of the adrenergic ligands, only replacement of D113-->A abolishes both agonist and antagonist binding. D79-->A lowered the affinity for agonists only. Thus, agonists and antagonists probably occupy overlapping sites in the receptor. Another acidic residue, D130, conserved in adrenergic and muscarinic receptors, is involved in transmission or generation of receptor activation by agonists. When D130-->N, then high affinity guanine nucleotide-modulated agonist binding is obtained, suggesting that the interaction with G-protein is substantially intact, but adenylate cyclase activity is not stimulated [39]. Serine residues (204, 207, 319) in helices 6 and 7 are involved in agonist binding but have little effect on antagonists. Hydrophobic residues F289, 290 and Y326 have major effects on agonist but not antagonist binding.

Interaction with G-protein.- Deletion analysis indicates that both the amino and the carboxyl ends of the 3rd intracellular loop are required for transmission of the agonist signal to adenylate cyclase through Gs [38, 39]. The amino sequences of this region of the G-protein-linked receptors coupled through the same G-protein are not homologous apart from predicted formation of amphipathic helices. These may interact with a corresponding amphipathic helix on the C-terminus of the G-protein α-subunit [37]. The effects of mastoparan, a wasp venom polypeptide mimicking an activated receptor surrogate, appear to be similarly mediated through the C-terminus of the G-protein α-subunit [40]. Competition with antipeptide Ab's for binding to the G-protein segments has also been a powerful technique [41] as an adjunct to mutational analysis. Similar mutational analyses have been performed with the G-proteins [42], but their description is beyond the scope of this receptor update.

Regions involved in signal termination.- The C-terminal cytoplasmic portion of the receptor is the site for regulatory phosphorylation by the β-adrenergic receptor kinase, an analogue of the rhodopsin kinase well studied in the visual system. β-Adrenergic or other adenylate cyclase-coupled stimulation leads to translocation of the kinase from cytosol to membranes [43]. Phosphorylation of the agonist-occupied receptor by this kinase has been shown to correlate with uncoupling (desensitization) of the β-adrenergic receptor as well as with a cardiac muscarinic receptor [44]. The kinase also phosphorylates α2 (adenylate cyclase-coupled) but not α1 (phosphoinositide cycle-coupled) adrenergic receptors [45]. Removal of the serine/threonine-rich C-terminus from residue 365 of the human β2 adrenergic receptor delayed the onset of agonist-promoted desensitization [46]. Replacing the hdroxy-amino acid residues with glycine and alanine was even more effective. Mutant β2-adrenergic receptors lacking either the putative cAMP-dependent protein kinase sites or the β-adrenergic receptor kinase sites have been used to show that the cyclic nucleotide-dependent sites are phosphorylated at low concentrations of agonist and are responsible for the decreased sensitivity to agonist stimulation of adenylate cyclase. Higher agonist concentrations elicit phosphorylation at both types of site, reducing the maximal adenylate cyclase responsiveness [47]. The extent of sequestration was not diminished by the modifications and so must be limited by some other process(es).

It is also possible that the rate of sequestration rather than the extent is affected by phosphorylation.

In the visual system, analogous modification of rhodopsin leads to interaction with arrestin, blocking further G-protein interaction and interrupting the signal. An analogous 'arrestin-like' protein has been suggested to be active on the β-adrenergic receptor [48]. Y350 and Y354 of the C-terminal portion of the human β2-adrenergic receptor are required for agonist-induced down-regulation but not receptor sequestration [49], similarly to the LDL, poly-Ig and mannose-phosphate receptors,

ASPECTS OF RECEPTOR-BASED DRUG DESIGN

In determining receptor structures, much of the effort is aimed at elucidating the receptor-bound conformation or agonists and antagonists - the goal of medicinal chemists, to provide lead compounds for pharmaceutical development. The structure of the receptor-bound ligand rather than that of the ligand in solution or in a ligand crystal is likely to be the relevant conformation for triggering biological activity. While nature produces nearly exclusively agonist-receptor ligands, with the sole exception of an IL-1 antagonist whose sequence was recently reported [50], pharmacological manipulation of biological activity usually involves an antagonist. The structural basis for the difference appears to involve agonist-induced receptor conformation change not inducible by antagonists. The ability to control the intrinsic agonism of the ligand is a valuable pharmacological tool.

Maximal resolution of the receptor-ligand could best come from X-ray crystallography, but as indicated above it is hard to obtain the data. High-resolution information is obtainable, in appropriate circumstances, directly on the receptor-bound ligand with little interference from the bulky receptor [51], using NMR with computer modelling of intra- and inter-molecular interactions.

While many of the ligands being considered are peptides, for practical reasons the object is normally the development of low-mol. wt. non-peptidic compounds of highly constrained conformation. Peptides are notoriously difficult to deliver and in general have too many degrees of conformation freedom to yield high binding affinity. Nature has provided us with the example of the alkaloid morphine, which binds to the μ-enkephalin receptor that normally recognizes a penta-peptide ligand. Limited success has been obtained with penta-peptide structures constrained on the basis on morphine and

morphine analogues.   Given a pentapeptide linear sequence
of amino acids it has not been possible to design an effective
non-peptidic ligand from first principles. Hence more empirical
methods, albeit intellectually less satisfying, have served
in searching for active components binding a particular receptor
in complex plant and microbial products, giving small-molecule
starting points for medicinal chemistry programs.   Thereby
non-peptidic ligands for CCK, gastrin and angiotensin II have
been discovered [52]. The success of this approach depends
on access to diverse compound mixtures (~10,000-100,000 mixtures),
on a selective, high-throughput screen to detect competing
agents, on secondary screens - often biological - for verification of selectivity and evaluation of agonism/antagonism
efficacy, and on rapid compound purification and molecular
structure determination so as to provide the chemists with
a lead compound to be followed up.   Finally, a reliable
whole-animal test system, a disease model, is needed for
evaluation of *in vivo* efficacy, drug disposition and toxicity.

    Assay developments.- High-throughput assays could become
revolutionized through an advance in assay methodology. Receptor-
ligand assays rely on the ability to clearly distinguish
free from bound ligand, usually accomplished by rapidly washing
away free ligand from immobilized (on a filter or bead)
receptor-bound ligand. There are some limitations on interaction affinities amenable to this method.   Some success
had been reported for homogeneous solution assays with fluorescence transfer or polarization, but there was always a signal/
noise problem and ligands had to be derivatized. A method
was developed [53] based on radiative energy transfer of
radioisotope decay to a nearby acceptor scintillant molecule
in a matrix by an immobilized ligand, achievable if a membrane-
associated receptor were bound to the matrix.   Only the
bound ligand was detectable, by scintillant fluorescence,
and it was unnecessary to physically separate free and receptor-
bound ligand.   This principle has been commercialized by
Amersham International (Scintillation Proximity Assay, SPA;
see contribution by R. Heath to this vol.) and is being
extended to diverse receptor and enzyme systems. Moderately
rich sources of receptor-containing membranes are required,
but these can be provided by recombinant expression methods.

STRUCTURAL DETERMINATION OF RECEPTOR-LIGAND INTERACTIONS

    Directly determining membrane-bound receptor structure has
generally been precluded because of several difficulties,
notably poor abundance that can now be alleviated by genetic
over-expression techniques. Receptors are large, amphipathic
molecules, often glycosylated, that are adapted to exist
in the anisotropic environment of the lipid membrane.   They

fail to crystallize, readily forming aggregates even in detergent
micelles.   In lieu of direct structure determination on the
whole receptor, its structure is being assembled, often with
information derived from model systems, as a mosaic of substruc-
tural domains.   These domains, often conserved in protein
families as if they subserve similar functions, can in some
cases be removed from the rest of the protein, retaining
their functional properties, and be studied in isolation.
Ligand-binding domains in the extracellular portions of some
receptors, e.g. insR, have been expressed in soluble form
and shown to undergo ligand-dependent oligomerization in solution.
Since chimaera formation has shown that ligand-binding activation could be transferred to a heterologous effector domain,
conformational changes in the soluble ectodomain may well
reflect relevant transitions.   Further dissection of the
domain could lead to a domain amenable to physical structural
analysis.

    In-depth understanding of receptor signal transduction
will call for a sufficient number of receptor/receptor-ligand
high-resolution (<3 Å) structures to be established. Enzymes
show a complex range of structure changes from <1 Å during
catalysis to >10 Å loop movements upon inhibitor binding [54].
Detailed understanding of enzyme mechanisms has emerged from
multidisciplinary analysis (organic chemistry, spectroscopy,
molecular modelling, mutational manipulation) based on high-
resolution structures. Crystal diffraction is the only physical
method so far capable of such resolution for assemblies of
mol. wt. >20,000.   For smaller entities, solution structures
with comparable resolution are obtainable in favourable cases
by NMR techniques. Progress made with these various techniques
is now considered.

Diffraction analysis

    In tackling the above-mentioned difficulties of directly
determining structures for membrane-bound receptors, limited
success has been obtained with the photoreaction centre of
purple bacteria [55] and bacteriorhodopsin.   Some information
has come from analysis of submolecular domains as identified
in receptor families (see earlier).   Ligand-binding domains
in the extracellular portion of some receptors, e.g. InsR, are
expressible in soluble form [7]. The recent report of a co-crystal
containing IL-2 and a soluble fragment of the IL-2 Tac receptor
(p55) [56] is the first for a structural determination.

    Anti-idiotypic Ab's have been suggested as convenient
models of receptor-ligand binding sites (reviewed in [57]*),
They are more readily obtained than receptors and are soluble
molecules of which much of the structure is known.   Several
*Investigation and Exploitation of Antibody Binding Sites (Vol. 15 in
present series; ed. E. Reid et al.; Plenum, 1985) is pertinent.-Ed.

antigen-Ab complexes have been solved by crystal diffraction to high resolution [e.g. 58, 59], providing insight into protein-protein recognition processes.

Methods are becoming available to ease the difficulties of receptor structure determination. The requisite quantities of receptor can be produced without carbohydrate by 'engineering out' glycosylation sites, provided that the product is metabolically stable and biologically active. Phase reference can be supplied by incorporation of selenocysteine and seleno-methionine for cloned proteins, thereby avoiding the requirement for preparation of heavy-metal derivatives when using data obtained on a synchrotron with anomalous diffraction methodology [60]. Once a high-resolution structure of a complex has been obtained, difference-Fourier methods or Laue diffraction can be used to solve highly related structures. It is even possible to obtain time-resolved structure information on crystals. Laue diffraction data can be collected in 10^{-10}nsec with a synchrotron source. The 'snapshots' obtained at these time-scales are approaching the range of the molecular dynamics simulations performed to calculate molecular conformations. Such an analysis requires that the process can be synchronized over the entire crystal [61].

Solution analysis

New developments in NMR methodology can, for proteins of M_r <20,000, provide structural data with comparable resolution to crystal diffraction, limited primarily by resolution and assignment of individual resonances but also by motional characteristics of the molecule [62]. Stable isotope labelling of protein residues (^{13}C, ^{15}N) [63] in conjunction with 3-D and 4-D NMR experiments [64] have allowed rapid structural determination for calmodulin (M_r 17,000), IL-1β (M_r 19,000) and staphylococcal nuclease (M_r 18,000) in aqueous solution. The attainable upper limit is at present M_r 30,000. Again, submolecular domain analysis by multidimensional NMR techniques for fragments of larger proteins such as fibronectin [65] can be used to solve partial structures. Modelling techniques, with the normal caveats of independent structures, can then be applied to assemble the whole protein. The glucocorticoid-receptor DNA-binding domain is the first such application of this approach to a receptor [66].

Methods for macromolecule-bound ligand structure determination by high-resolution NMR are reviewed in an entire issue [Vol. 40(1), 1990] of *Biochemical Pharmacology*. Stable free-radical probes such as nitroxides covalently attached to haptens or other small ligands have given data on Ab binding-site residues [67]. Various other NMR methods can also determine the bound conformation of small ligands as well as reveal

details of active-site structure [68]. Most noteworthy is
the transferred nuclear Overhauser Enhancement (TRNOE) experiment which gives data on the bound conformation of a small
ligand regardless of the size or complexity of the macromolecular
complex.   The TRNOE method has been successfully applied
to allosteric effector-protein complexes [69], enzyme-inhibitor
complexes [70] and antigen-Ab complexes [71, 72].   A study
with the acetylcholine-acetylcholine receptor complex [73]
showed a structure very similar to that of acetylcholine
bound to phospholipid vesicles and different from the solution
or crystal structures.

New solid-state NMR methods can provide precise but
limited inter-atomic distance information for protein complexes
of M_r >30,000.   Distances of <5-10 Å can be measured   for
isotopically enriched (^2H, ^{13}C, ^{15}N) or labelled (^{19}F) sites.
A rotational resonance (R2) experiment can measure distances
between like isotope labels (^{15}N-^{15}N, ^{13}C-^{13}C) [74], while
REDOR and RETRO experiments can determine distances between
dissimilar labels (^{13}C-^{15}N) [75].   The primary advantage
of these techniques is that there is no upper limit to
the complex's M_r;   but large amounts of material, >100 mg,
are required.

Computational aids to determining and applying crystal structures

The prediction of protein structure from primary sequence
information is in its infancy. Improved computational algorithms and programs are available to help derive approximate
models [76] which can be tested by various spectroscopic
techniques.   If the structure of a related receptor is known,
then a new sequence can be mapped onto the known structure
with methods [77] far more sophisticated than when the adrenergic receptors were mapped onto the bacteriorhodopsin structure.

A model of a receptor-ligand binding site or an active
site can be derived from a series of ligands. New conformational
searching algorithms [78, 79] can efficiently search torsional
angle space to obtain the 'pharmacophore' [80] or consensus
binding mode for the series of structurally related compounds.
An alternative, the ensemble distance geometry approach, first
applied to the nicotinic acetylcholine receptor   and   then to
other systems [81], can be used to gain a more precise
map of receptor binding sites.   Free energy perturbation
methods have been extended to modelling the NMDA-receptor
binding site [82] to calculate relative binding energies
given a hypothesized arrangement of receptor functional groups
in the ligand binding site. These basic protein-engineering
concepts and methods are sufficiently accurate to have allowed
the *de novo* design of several selective ion channels [83].

LIGAND DESIGN

Computational chemistry and protein modelling methods can be used to design better agonists or antagonists using high-resolution structures. Shape-complementarity algorithms help search small-molecule databases, such as the Cambridge crystallographic database, to find potential ligands fitting into pockets or clefts on macromolecular surfaces. Based on the crystal structure of the HIV-protease complexed with a peptide inhibitor [84], the DOCK algorithm [85] was used to search for a potential ligand for a non-active site cleft of the HIV protease, which proved to be a selective inhibitor of the target protease but not other aspartyl proteases. Similar methods are routinely used in the pharmaceutical industry to provide novel lead compounds for drug development.

References (JBC = J. Biol. Chem.; PNAS = Proc. Nat. Acad. Sci.)

1. Mosthaf, L., Graco, K., Dull, T.J., Coussens, L., Ullrich, A. & McClain, D.A. (1990) *EMBO J. 9*, 2409-2413.
2. Schontal, A. (1990) *Cellular Signalling 2*, 215-225.
3. Yarden, Y. & Ullrich, A. (1988) *Annu. Rev. Biochem. 57*, 443-478.
4. Schlessinger, J. (1988) *Biochemistry 27*, 3119-3123.
5. Cascieri, M.A., Chicchi, G.G., Applebaum, J., Hayes, N.S., Green, B.G. & Bayne, M.L. (1988) *Biochemistry 27*, 3229-3233.
6. Ullrich, A., Gray, A., Tam, A.W., Yang-Feng, T., Tsubokawa, M., Collins, C., Henzel,W., LeBon, T., Kathuria, S., Chen, E.,Jacobs, S., Francke, U., Ramachandran, J. & Fujita-Yamaguchi, Y. (1986) *EMBO J. 5*, 2503-2512.
7. Johnson, J.D., Wong, M.L. & Rutter, W.J. (1988) *PNAS 85*,7516-7520.
8. Sternberg, M.J.E. & Gullick, W.J. (1990) *Prot.Eng. 3*, 245-248.
9. Escobedo, J.A., Barr, P.J. & Williams, L.T. (1988) *Mol. Cell. Biol. 8*, 5126-5131.
10. Manolios, N., Bonifacino, J.S. & Klausner, R.D. (1990) *Science 249,*
11. Pawson, T. (1988) *Oncogene 3*, 491-495.   [274-277.
12. Escobedo, J.A. & Williams, L.T. (1988) *Nature 335*, 85-87.
13. Matsuda, M., Mayer, B.J., Fukui, Y. & Hanafusa, H. (1990) *Science*
14. Kallen, R.G., Smith, J.E., Sheng, A. & Tung, L.   [248, 1537-1538. (1990) *Biochem. Biophys. Res. Comm. 168*, 616-624.   [5687.
15. Moe, G.R., Bollag, G.E. & Koshland, Jr.,D.E. (1989) *PNAS 86*, 5683-
16. Ellis, L., Morgan, D.D., Koshland, Jr., D.E., Clauser, E.,Moe, G.R., Bollag,G.,Roth, R.A. & Rutter, W.J. (1986) *PNAS 83*, 8137-8141.
17. Bernard, O., De St. Groth, B.F., Ullrich, A., Green, W. & Schlessinger, J. (1987) *PNAS 84*, 2125-2129.
18. Prywes, R., Livneh, E., Ullrich, A. & Schlessinger, J. (1986) *EMBO J. 5*, 2179-2190.   [8, 1369-1375.
19. Lammers,R.,Gray, A., Schlessinger,J. & Ullrich, A. (1989) *EMBO J.*
20. Riedel, H., Dull, T.J., Schlessinger, J. & Ullrich, A. (1986) *Nature 324*, 68-70.   [200.
21. Riedel, H., Schlessinger, J. & Ullrich, A. (1987) *Science 236*,197-
22. Ellis, L., Morgan, D.D., Jong, S-M., Wang, L-H., Roth, R.A. & Rutter, W.J. (1987) *PNAS 84*, 5101-5105.

23. Berhanu, P., Rohilla, A.M.K. & Rutter, W.J. (1990) *JBC 265*, 9505-9511.

24. Yarden, Y. & Schlessinger, J. (1987) *Biochemistry 26*, 1443-1451.

25. McClain, D.A., Maegawa,H., Lee, J., Dull, T.J., Ullrich, A. & Olefsky, J.M. (1989) *JBC 262*, 14663-14671.

26. Debant, A., Ponzio, G., Clauser, E., Contreras, J.L. & Rossi, B. (1989) *Biochemistry 28*, 14-17.

27. Debant, A., Clauser, E., Ponzio, G., Filloux, C., Auzen, C., Contreras, J.L. & Rossi, B. (1988) *PNAS 85*, 8032-8036.

28. Wilden, P.A., Backer, J.M., Kahn, C.R., Cahill, D.A., Schroeder, G.J. & White, M.F. (1990) *PNAS 87*, 3358-3362.

29. Khazaie, K., Dull, T.J., Graf, T., Schlessinger, J., Ullrich, A., Beug, H. & Vennstrom, B. (1988) *EMBO J. 7*, 3061-3071.

30. Honegger, A.M., Dull, T.J., Bellot, F., Van Obberghen, E., Szapary, D., Schmidt, A., Ullrich, A. & Schlessinger, J. (1988) *EMBO J. 7*, 3045-3052.

31. Felder, S., Miller, K., Moehren, G., Ullrich, A., Schlessinger, J. & Hopkins, C.R. (1990) *Cell 61*, 623-634.

32. Honegger, A.M., Schmidt, A., Ullrich, A. & Schlessinger, J. (1990) *J. Cell Biol. 110*, 1541-1548.

33. Lomasney, J.W., Lorenz, W., Allen, L.F., King, K., Regan, J.W., Yang-Feng, T.L., Caron, M.G. & Lefkowitz, R.J. (1990) *PNAS 87*, 5094-

34. Marullo, S., Emorine, L.J., Strosberg, A.D. &   [5098.
    Delavier-Klutchko, C. (1990) *EMBO J. 9*, 1471-1476.   [2889.

35. Hausdorff, W.P., Caron, M.G. & Lefkowitz, R.J. (1990) *FASEB J. 4*, 2881-

36. Benovic, J.L., Bouvier, M., Caron, M.G. & Lefkowitz, R.J. (1988) *Annu. Rev. Cell Biol. 4*, 405-428.

37. Ross, E.M., Wong, S.K-F., Rubenstein, R.C. & Higashijima, T. (1988) *Cold Spr. Harb. Symp. Quant. Biol. 53*, 499-506.

38. Dixon, R.A.F., Sigal, I.S. & Strader, C.D. (1988) *as for* 37., 487-497.

39. Fraser, C.M., Chung, F.Z., Wand,C.D. & Venter, J.C. (1988) *PNAS 85*, 5478-5482.

40. Weingerten, R., Ransnas, L., Mueller, H., Sklar, L.A. & Bokoch, G.M. (1990) *JBC 265*, 11044-11049.

41. Hamm, H.E., Deretic, D., Mazzoni, M.R., Moore, C.A., Takahashi, J.S. & Resenick, M.M. (1989) *JBC 264*, 11475-11482.

42. Masters, S.B., Landis, C.A. & Bourne, H.R. (1990) *Enzyme Regul. 30*,

43. Strasser, R.H., Benovic, J.L., Caron, M.G. &   [75-87.
    Lefkowitz, R.J. (1986)   *PNAS 83*, 6362-6366.

44. Kwatra, M.M., Leung, E., Maan, A.C., McMahon, K.K., Ptasienski, J., Green, R.D. & Hosey, M.M. (1987) *JBC 262*, 16314-16321 *(& 261*, 12429-12432).

45. Benovic, J.L., Regan, J.W., Matsui, H., Mayor, Jr., F., Cotecchia, S., Leeb-Lundberg, L.M.F., Caron, M.G. & Lefkowitz, R.J. (1987) *JBC 262*,

46. Bouvier, M. (1988) *Nature 333*, 370-372.   [17251-17253.

47. Hausdorff, W.P. (1989) *JBC 264*, 12657-12665.

48. Lohse, M.J., Benovic, J.L., Codina, J., Caron, M.G. & Lefkowitz, R.J. (1990) *Science 248*, 1547-1551.

49. Vallquette, M., Bonin, H., Ilnatowich, M., Caron, M.C., Lefkowitz, R.J. & Bouvier, M. (1990) *PNAS 87*, 5089-5093.

50. Eisenberg, S.P., Evans, R.J., Arend, W.P., Verderber, E., Brewer, M.T., Hunnum, C.H. & Thompson, R.C. (1990) *Nature 343*, 341-346.

51. Hruby, V.J., Al-Obeidi, F. & Kazmierski, W. (1990) *Biochem. J. 268*,

52. Fredinger, R.M. (1989) *Trends Pharmacol. Sci. 10*, 270-274. [249-262.

53. Bosworth, N. & Towers, P. (1989) *Nature 341*, 167-168.

54. Lolis, E. & Petsko,G.A. (1990) *Annu. Rev. Biochem.* 59, 597-630.
55. Deisenhofer, J. & Michel, H. (1989) *Science 245*, 1463-1473.
56. Lambert, G., Stura, E.A. & Wilson, I.A. (1989) *JBC 264*, 12730-12736.
57. Linthicum, D.S., Bolger, M.B., Kussie,P.H., Albright, G.M.,
    Linton, T.A., Combs, S. & Marchetti, D. (1988) *Clin.Chem. 34*, 1676-
58 Colman, P.M., Laver, W.G.,Varghesi, J.N., Baker, A.T.,   [1680.
    Tulloch, P.A., Air, G.M. & Webster, R.G.(1987) *Nature 326*, 358-363.
59. Sheriff, S., Silverton, E.W.,Padlan, E.Q., Cohen, G.H.,
    Smith-Gill, S.J., Finzel, B.C. & Davies, D.R. (1987) *PNAS 84*, 8075-8079.
60. Hendrickson, W.A.,Smith, J.L., Phizackerley, R.P. & Merritt, E.A.
    (1988) *Proteins: Structure, Function and Genetics*, 4, 77-88.
61. Helliwell,J.R., Habash, J., Cruickshank, D.W.J., Harding, M.M.,
    Greenhough, T.J., Campbell, J.W., Clifton, I.J., Elder, M.,   [497.
    Machin, P.A., Papiz, M.Z. & Zurek, S. (1989)*J. Appl. Crystallogr. 22*,483-
62. Wuthrich, K. (1989) *Acc. Chem. Res. 22*, 36-44.   [908-911.
63. Oh, B.H., Westler, W.M., Darba, P. & Markley, J.L. (1988) *Science 240*,
64. Kay, L.E., Clore, G.M., Bax, A. & Gronenborn, A.M. (1990) *Science 249*,
65. Baron, M., Norman, D., Willis, A. & Campbell. I.D.   [411-413.
    (1990) *Nature 345*, 642-646.
66. Hard, T., Kellenbach, E., Boelens, R., Maler, B.A., Dahlman, K.,
    Freedman, L.P., Carlstedt-Duke, J., Yamamoto, K.R., Gustafsson, J-A.
    & Kaptein, R. (1990) *Science 249*,157-160.
67. Anglister, J., Frey, T. & McConnell, H.M. (1985) *Nature 315*, 65-67.
68. Fesik, S.W. (1989) in *Computer-Aided Drug Design: Methods and
    Applications* (Perun, T.J. & Propst, C.L., eds.),M. Dekker, N. York, 133-183.
69. Gronenborn, A.M., Clore, G.M., Brunori, M., Giardina, B.,
    Falconi, G. & Perutz, M.F. (1984)*J. Mol. Biol. 178*, 731-742.
70. Meyer, E.F., Clore, G.M., Gronenborn, A.M. & Hansen, H.A.S. (1988)
    Biochemistry 27, 725-730.
71. Tsang, P., Feiser, T.M., Ostresh, J.M., Lerner, R.A. &
    Wright, P.E. (1988) *Peptide Res. 1*, 87-92.
72. Glasel, J.A. (1989) *J. Mol. Biol. 209*, 747-761.
73. Behling, R.W., Yamane, T., Navone, G. & Jelinski, L.W.
    (1989) *PNAS 85*, 6721-6725.
74. Raleigh, D.P., Creuzet, F., DasGupta, S.K., Levitt, M.H. &
    Griffin, R.G. (1989) *J. Am. Chem. Soc. 111*, 4502-4503.
75. Guillon, T. & Shaefer,J. (1989)*J. Magn. Reson. 81*, 196-200.
76. Cohen, F.E., Kosen, P.A.,Kuntz, I.D., Epstein, L.B., Ciardelli, T.L.
    & Smith, K.A. (1987) *Science 234*, 349-352.
77. Blundell, T., Carney, D., Gardner, S., Hayes, F., Howlin, B.,
    Hubbard, T., Overington, J., Singh, D.A., Sibanda, B.L.& Sutcliffe, M.
    (1988) *Eur. J. Biochem. 172*, 513-520.   [(1986) *as for* 60., 342-362.
78. Fine, R.M., Wang, H., Shenkin, P.S., Yarmush, D.L. & Levinthal, V.
79. Bruccoleri, R. & Karplus, M. (1987)*Biopolymers 26*, 137-154. [5959-5965
80. Cramer, R., Patterson, D. & Bunce, J. (1988) *J. Am. Chem. Soc. 110*,
81. Ghose, A.K., Crippen, G.M., Ravenkar,G.R., McKernan, P.A., Smee, D.F.
    & Robins, R.K. (1989) *J. Med. Chem. 32*, 746-756.
82. Snyder, J.P., *et al.* (1990) in *Frontiers in Drug Research: Benzon
    Symp. 28* (Jensen, B., *et al.*, eds.), Munksgaard, Copenhagen, 109-120.
83. DeGrado, W.F., Wasserman,Z.R. & Lear, J.D. (1989) *Science 243*, 622-628.
84. Erickson, J., *et al. [15 authors]* (1990) *Science 249*, 527-533.
85. Sheridan, R.P., Rusinko, A., Nilakantan, R. & Venkataraghavan, R.
    (1989) *PNAS 86*, 8165-8168.

#A-3

CYTOKINE SIGNALLING: EXPLORING HOW
INTERLEUKIN 1 ACTIVATES CELLS

[1]L.A.J. O'Neill[†], [1]E. Stylianou, [2]M.R. Edbrooke,
[3]R.W. Farndale and [1]J. Saklatvala

[1]Cytokine Biochemistry Group,
  Strangeways Research Laboratory,
  Worts Causeway, Cambridge CB1 4RN, U.K.

[2]Section of Molecular Rheumatology,
  MRC Clinical Research Centre,
  Harrow, Middlesex HA1 3UJ, U.K.

[3]Department of Biochemistry, University of Cambridge,
  Tennis Court Road, Cambridge CB2 1QW, U.K.

Cytokines such as IL-1* are among the most potent activators of cells so far described. Their molecular mechanism of action is still mainly a mystery; but certain clues are starting to appear. Two new families of receptors have been defined: (1) for cytokines, including IL-2, IL-4 and GMCSF, whose main effects appear to be on leucocytes; (2) receptors for TNF and NGF. The IL-1 receptor has yielded no clues as to a signal transduction mechanism. Early events involving protein kinase activation and stimulation of a G-protein have been described. Second messengers such as cAMP and DAG have been implicated in IL-1 action, but their importance remains uncertain. This article describes what is currently known about cytokine signalling, emphasizing work on IL-1, and discusses how such events may lead to the ultimate cellular response to a cytokine such as IL-1-enhanced gene expression.

Cytokines are proteins which mediate the activation, growth and differentiation of cells involved in the immune and inflammatory responses. Their pleiotropic effects, and the range of cell types which respond to them, have allowed elaborate regulatory networks to be established between haematopoietic cells, lymphoid cells, connective tissue cells and endothelial cells, increasing our understanding of the complex and important events of the inflammatory process. Their importance in cell biology is clear, as they are among the most potent activators of cells so far described. IL-1*, which may be considered the first cytokine [1], is perhaps the most pleiotropic and potent of all cytokines. Its possible

---

† correspondent; now at Biochem. Dept., Trinity Coll., Dublin, Ireland.

*Abbreviations.- DAG, diacylglycerol; IL (as in IL-1), interleukin; PG, prostaglandin; PKC, protein kinase C; PMA, 4β-phorbol myristate acetate. VARIOUS (GROWTH) FACTORS.- EGF, epidermal; NGF, nerve; TNF, tumour necrosis; GMCSF, see list overleaf.

role in inflammatory diseases such as rheumatoid arthritis has led to its activities being intensively studied. Yet very little is known about the signal transduction process for IL-1, and indeed how most cytokines activate cells remains obscure.

BIOLOGICAL EFFECTS [†]

The main cytokines and their biological effects (review: [2]) are as follows.

- IL-1: Endogenous pyrogen; T-lymphocyte activation; induction of PG's, metalloproteinases, acute-phase proteins and other cytokines; increased expression of adhesion molecules on endothelial cells.
- IL-2: T-cell growth factor.
- IL-3: Haematopoietic cell differentiation.
- IL-4: B-cell stimulation (proliferation, IgE production).
- IL-5: Eosinophil activation and maturation
        B-cell stimulation.
- IL-6: Haematopoietic cell differentiation
        B-cell stimulation
        Induction of acute-phase response.
- IL-7: B- and T-cell growth factor.
- IL-8: Chemotactic for neutrophils.
- TNF: Similar to IL-1; also -
        Cytotoxic for certain cells
- CSF's: Granulocyte maturation colony-stimulating factors.

These effects may be divided broadly into two groups: those that involve leucocytes such as lymphocytes of T and B types, and those involving other cell types such as fibroblasts, endothelial cells and hepatocytes. IL-2, IL-3, IL-4, IL-5 and IL-7 have important roles in either the growth or the maturation of T- and/or B-lymphocytes. IL-5 also stimulates eosinophils, and IL-8 is a potent chemotactic agent for neutrophils. The colony-stimulating factors stimulate the formation of colonies of neutrophils, monocytes and eosinophils in bone marrow.

While IL-1, IL-6 and TNF also have effects on T- and B-lymphocytes, their ability to stimulate non-lymphoid cells points to their importance in the inflammatory response. IL-6 is the key stimulator of acute-phase protein release in hepatocytes. IL-1 and TNF share a wide range of activities, including induction of fever, stimulation of release of PGE_2 and collagenase from fibroblasts, and enhanced expression of adhesion molecules on endothelial cells. *In vivo*, IL-1 causes proteoglycan loss and neutrophil accumulation, while TNF is considered the primary mediator of cachexia and causes the haemorrhagic necrosis of certain tumours.

[†] Assay approaches for their study are surveyed by K.D. Rainsford in Vol.18 (1988; ed. E. Reid *et al.*; Plenum); mainly IL bioassay.-*Ed.*

Cytokines in essence may therefore be considered the coordinators of immune and inflammatory responses.

SIGNAL TRANSDUCTION

How cytokines activate cells is unknown. Biological activities depend on target cells; thus IL-1 stimulates IL-2 production from lymphocytes, but from fibroblasts it stimulates PGE_2 production. Target cell responses are determined by the expression of receptors and/or the link between the receptor and the signal transduction machinery in a given cell.

Receptors

A new family of receptors has recently been described for some cytokines. The β chain of the IL-2 receptor, and receptors for IL-3, IL-4, IL-6, IL-7 and GMCSF all show important homologies: they are Ig-like, and have 4 conserved cysteines and a region of striking homology close to the transmembrane domain [3]. The significance of these homologies for signal transduction is unclear, but may involve an interaction with a second protein important for generating a signal (as shown for the IL-6 receptor, which associates with a protein termed gp130 [4]).

Two types of TNF receptor have been described. They also appear to belong to a family, as they show significant homologies with the NGF receptor [5, 6].

The IL-1 receptor is unique. It belongs to the Ig supergene family [7]. As with other cytokine receptors its structure has provided no clues as to a likely signal transduction mechanism. A second receptor, which is present on B-cells, has also been detected [8].

Post-receptor events: IL-1

As mentioned previously, very little is known about the second messenger system for any cytokine. Much work has focused on IL-1 and will now be discussed in detail.

G-protein activation

We and others have reported that IL-1 activates a G-protein [9, 10]. When IL-1α or IL-1β was added to membranes prepared from a T-cell line (the murine thymoma line EL4.NOB-1), an increaase in GTPγS binding was detectable from 1 min (Fig. 1). The effect was due to an increase in affinity for nucleotide, as the enhancement was greater at low concentrations of GTPγS. When the effect was correlated to receptor occupancy, low receptor occupancies (20-30%) were sufficient to stimulate maximal increases in nucleotide binding. This provides an

Fig. 1. IL-1 increases GTPγS binding in EL4 membranes. The membranes ([9]; 20 μg protein) were incubated at 37°, with (o) or without (●) IL-1, in 20 mM Tris-HCl pH 8.0/30 mM MgCl$_2$/1 mM EDTA/1 mM dithiothreitol /200 mM NaCl/0.1% Lubrol 2.5 nM [^{35}S]GTPγS. Reactions were started by membrane addition and terminated with ice-cold buffer containing 25 μM GTPγS. Unbound was separated from bound by filtration through 0.45 μm nitrocellulose filters, which were washed and counted for ^{35}S. A, IL-1α, 10 ng/ml; B, IL-1β, 100 ng/ml. Error bars represent S.E.M. (n = 3).

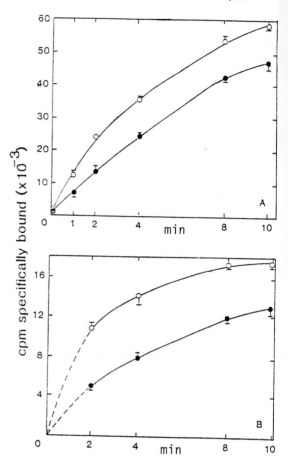

explanation as to why IL-1 is biologically active at concentrations considerably lower than the dissociation constant of the receptor. For example, the Kd for the IL-1 receptor on EL4 cells is 10^{-10} M, whilst the EC$_{50}$ for IL-1-induced IL-2 production from these cells is 10^{-13} M [9]. Signal amplification *via* G-proteins may provide an explanation for this apparent discrepancy.

G_i, cAMP and DAG

Other workers have reported direct activation of a G-protein by IL-1 in a pre-B cell line, 70Z/3 [10]. This cell bears a different receptor (67 kDa) to that on EL4 cells (80 kDa). As pertussis toxin inhibited this activation, and was also inhibitory for a range of IL-1-induced biological responses, the authors implicated a G_i-like protein in IL-1 action. However, evidence was also provided for increases in adenylate cyclase activity in membranes prepared from IL-1-responsive cells. This suggested that cAMP was the second messenger for IL-1; but this is unlikely since agents such as PGE₁ do not mimic IL-1 action. In fact we and others [11] have failed to show changes in either cAMP levels or adenylate cyclase activity in response to IL-1, although the same cells proved remarkably responsive to agents such as PGE$_1$ (10 µM) and forskolin (10 µM) which served as positive controls. With membranes (45 µg membrane protein) prepared [9] from EL4.NOB-1 cells, incubated for 20 min at 30° in 100 µl containing 100 µM [^{32}P]ATP, 10 mM MgCl$_2$ and 500 µM cAMP, radiolabel assays for cAMP generation ([12]; adenylate cyclase activity) gave the following values for pmol cAMP/mg per min:-

Basal:	9.2 ±0.4 (S.E.M.)	
PGE$_1$:	70.6 ±1.5;	+ GTP, 10 µM: 210.9 ±2.5
Forskolin:	470.0 ±15;	+ GTP, 10 µM: 510.0 ±6.4
IL-1α (100 ng/ml):	9.9 ±0.4;	+ GTP, 10 µM: 11.4 ±0.5

We also suggested a G$_i$-like G-protein in the response to IL-1, as pertussis toxin inhibited IL-1-induced IL-2 production by EL4 cells (Fig. 2). Dobson and co-workers [13] also found pertussis sensitivity in an IL-1 signal, namely DAG generation in EL4 cells, suggesting a coupling to phospholipase C. Other workers have reported DAG generation in response to IL-1 [14, 15], but as in the case of cAMP, others have failed to show such changes [16].

The nature of the G-protein and the effector enzyme to which it may be coupled therefore awaits clarification. It is unlikely that the IL-1 receptor is directly coupled to a 'classical' G-protein such as G$_i$. Receptors coupled to such G-proteins have 7 membrane-spanning domains, whereas the IL-1 receptor spans the membrane once. The coupling may therefore be indirect. Interestingly, the cytokines IL-2, TNF and M CSF have all been shown to activate a pertussin-sensitive G-protein [17-19]. None of these receptors has 7 membrane-spanning domains, suggesting that cytokine-G-protein coupling is novel.

Activation of protein kinase C

PMA mimics many of the biological effects if IL-1, suggesting a PKC involvement in IL-1 signalling. There are, however, no direct reports of PKC activation in response to IL-1. Yet IL-1 clearly activates a protein kinase as an early post-receptor event [20]. Two main substrates have been identified: a heat-shock protein hsp27 [20], and the EGF receptor [21]. The effect on the EGF receptor is to decrease its affinity ('transmodulation'). The significance of these phosphorylations is unclear. It is unlikely, however, that they are mediated by PKC, since in the case of the EGF receptor staurosporine fails to inhibit IL-1, as does down-regulation

Fig. 2. Pertussis toxin inhibits IL-1-induced IL-2 production in EL4 cells. Cells (1×10^6/ml) were incubated for 4 h at 37° without toxin (o) or with it (100 ng/ml; ■), washed, then incubated for 24 h with increasing concentrations of IL-1α. Finally supernatants were removed for IL-2 assay on CTLL-2 cells [9] (& K.D. Rainsford in Vol. 18, this series - *Ed*.). Values represent [^{3}H]thymidine incorporated.

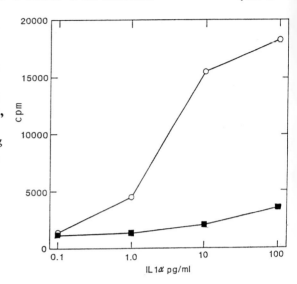

of the enzyme by prolonged exposure to PMA [21]. The nature of the kinase therefore remains unknown. It is noteworthy that TNF, which shares many of the biological activities of IL-1, shows a similar pattern of response.

THIRD MESSENGERS: TRANSCRIPTION FACTORS

Cells respond ultimately to IL-1 with an enhancement of gene expression. Gene expression is controlled mainly by the binding of transcription factors to discrete areas in the 5' flanking regions of genes. PMA and forskolin, activators of protein kinases C and A respectively, activate transcription factors such as AP1 and NFκB [22, 23]. There is also evidence for IL-1 activation of these factors [24, 25]. The early activation of a protein kinase by IL-1 is therefore likely to be important in the regulation of transcription factors such as NFκB. In EL4 cells, we have recently shown that activation of NFκB is an early signal (Fig. 3), probably not involving PKA or PKC (*see* 'Added in Proof').

NFκB has been implicated in the activation of several IL-1-responsive genes, such as those for IL-2 [26], IL-2 receptor [27] and serum amyloid A [28]. This transcription factor is thought to be activated by the phosphorylation of an inhibitory protein, IκB, which upon phosphorylation dissociates from NFκB allowing it to bind to the NFκB sequence on the relevant gene [29]. We are currently attempting to characterize NFκB activation in EL4 cells. Regulation of third messengers such as NFκB and AP1 *via* protein kinase activation is likely to be central to the ultimate response of a cell to IL-1.

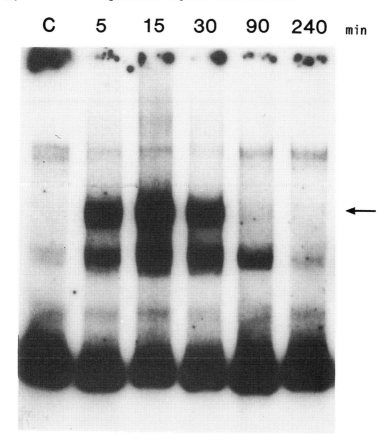

Fig. 3. IL-1 induces NKκB binding in EL4 cells. After incubation (1 × 10^7 cells; 1 ml) for the indicated times, 10 ml ice-cold PBS
was added. Nuclear extracts [24] (4 μg protein) were then incubated with a probe containing the NFκB sequence, and gel retardation
studies [28] performed. Retarded band indicated by *arrow*.

CONCLUSION

Although cytokines are potent activators of cells, the
events that occur once they bind to their receptors are
proving difficult to unravel. As can be seen with IL-1,
understanding cytokine signal transduction is a controversial
area. Different cell types show different responses, and
there may be multiple signalling pathways. Relating early
events such as protein kinase activation to the modulation
of gene expression will be very important for our understanding
of how cytokines activate cells.

References *(JBC = J. Biol. Chem.)*

1. di Giovane, F.S. & Duff, G.W. (1990) *Immunol. Today 11*, 13-20.
2. Arai, K., Lee, F., Miyajima, A., Shoichiro, M., Arai, N. &
   Yokota, T. (1990) *Annu. Rev. Biochem. 59*, 783-836.

3. Gearing, D.P., King, J.A., Gough, N.M. & Nicola, N.A. (1989) *EMBO J. 8*, 3667-3676.

4. Taga, T., Hibi, M., Hirata, Y., Yamasaki, K., Yasukawa, K. & Kishimoto, F. (1989) *Cell 58*, 573-581.

5. Loetscher, H., Pan, Y.E., Lahm, H.W., Gentz, R., Brockhaus, M., Tabuchi, H. & Lesslauer, W. (1990) *Cell 61*, 351-359.

6. Schall, T.J., Lewis, M., Koller, K.J., Lee, A., Rice, G.C., Wong, G.H.W., Gatanaga, T., Granger, G.A., Lentz, R., Raab, H. Kohr, W.J. & Goeddel, D.V. (1990) *Cell 61*, 361-370.

7. Sims, J.E., March, C.J., Cosman, D., Widmer, M.B., Macdonald, H.R., McMahon, C.J., Grubin, C.E., Wignall, J.M., Jackson, J.L., Call, S.M., Friend, D., Alpert, A.R., Gillis, S., Urdal, D.L. & Dower, S.K. (1988) *Science 241*, 585-589.

8. Bomstyk, K., Akahoshi, T., Yamada, M., Furutani & Oppenheim, J. (1986) *J. Immunol. 136*, 4496-4502,   [*JBC 265*, 3146-3152.

9. O'Neill, L.A.J., Bird, T.A., Gearing, A.J.H. & Saklatvala, J. (1990)

10. Chedid, M., Shirakawa, F., Naylor, O. & Mizel, S. (1989) *J. Immunol. 142*, 4301-4306.

11. Carroll, G.J. (1986) *Br. J. Rheum. 25*, 358-365.

12. Saloman, Y., Landos, C. & Rodbell, M. (1974) *Anal. Biochem. 58*, 541-548.

13. Dobson, P.R.M., Plested, C.P., Jones, D.R., Banks, T. & Brown, B.L. (1989) *J. Mol. Endocr. 2*, R5-R8.

14. Rosoff, P.M., Savage, N. & Dinarello, C.A. (1988) *Cell 54*, 73-81.

15. Kester, M., Simonson, M.S., Mene, P. & Sedor, J.R. (1989) *J. Clin. Invest. 83*, 718-723.

16. Abraham, R.T., Ho, S.N., Barna, T.J. & McKean, D.J. (1987) *JBC 262*, 2719-2728.

17. Evans, S.W., Beckner, S.K. & Farrar, W.L. (1987) *Nature 325*, 166-168.

18. Imamura, K. & Kufe, D. (1988) *JBC 263*, 14093-14098.

19. Imamura, K., Sherman, M.L., Spriggs, D. & Kufe, D. (1988) *JBC 263*, 10247-10253.

20. Kaur, P. & Saklatvala, J. (1988) *FEBS Lett. 241*, 6-10.

21. Bird, T.A. & Saklatvala, J. (1990) *JBC 265*, 235-240.

22. Sen, R. & Baltimore, D. (1986) *Cell 47*, 921-928.

23. Shirakawa, F., Chedid, M., Suttles, J., Pollok, B.A. & Mizel, S.M. (1989) *Proc. Nat. Acad. Sci. 85*, 8201-8205.

24. Kovacs, E.J., Oppenheim, J.J. & Young, H.A. (1986) *J. Immunol. 137*, 3649-3651.

25. Osborn, L., Kunkel, S. & Nabel, G.J. (1989) *Proc. Nat. Acad. Sci. 86*, 2336-2340.

26. Hoyos, B., Ballard, D.W., Bohnlein, E., Siekevitz, M. & Greene, W.C. (1989) *Science 244*, 457-460.

27. Crabtree, G.R. (1989) *Science 243*, 355-361.

28. Edbrooke, M., Burt, D.W., Cheshire, J.K. & Woo, P. (1989) *Mol. Cell Biol. 9*, 1908-1916.

29. Baeurle, P.A. & Baltimore, D. (1988) *Science 242*, 540-546.

Added in proof.- PMA and forskolin did not mimic IL-1 in activation of NFκB, which was staurosporine-insensitive [30]; cf. [20] for PK type.

30. O'Neill, L.A.J., Edbrooke, M.R., Stylianou, E., Farndale, R., Woo, P. & Saklatvala, J. (1991) *Br. J. Pharmacol.*, in press.

#A-4

STRATEGIES TO ELUCIDATE THE SPECIFICITY OF RECEPTOR INTERACTIONS WITH GUANINE NUCLEOTIDE BINDING PROTEINS

Graeme Milligan, Fiona M. Mitchell, Steven J. McClue,
I. Craig Carr, Ian Mullaney and Fergus R. McKenzie

Molecular Pharmacology Group,
Departments of Biochemistry and Pharmacology,
University of Glasgow, Glasgow G12 8QQ, Scotland, U.K.

A family of heterotrimeric G-proteins* function to allow communication between agonist-occupied receptors and effector systems which are either ion channels or enzymes which generate intracellular second messengers. The high degree of similarity in primary sequence amongst the various G-proteins makes it pertinent to assess (1) how specific are contacts between receptors and G-proteins, and (2) which effector systems are regulated by each G-protein.

Approaches which involve the use of selective site-specific anti-G-protein antisera to prevent functional contacts between particular G-proteins and receptors are described. Thereby we have demonstrated in neuroblastoma × glioma hybrid, NG108-15 cells, that the inhibitory G-protein of the adenylate cyclase cascade is the product of the $G_i2\alpha$ gene and that G_0 functions to regulate voltage-sensitive Ca^{2+} channels.

The molecular identity of PT-sensitive G-proteins, which interact directly with particular receptors, can also be assessed by the ability of CT to catalyze ADP-ribosylation of such 'inappropriate' G-proteins when they lack a nucleotide at the guanine nucleotide binding site. The rationale for this effect and examples of its use are described.

G-proteins are found widely throughout evolution. Highly conserved G-proteins have been identified *via* either cDNA cloning or immunological means in various mammals, birds, amphibia, invertebrates, yeast, slime moulds and green plants.

Each of the classical heterotrimeric G-proteins, of which some 16 have now been identified, consists of distinct α, β and γ polypeptides. The α subunits appear to define the identity of the G-protein and vary in size from 39 to 46 kDa as calculated from primary sequence information derived from corresponding cDNA clones.

**Abbreviations.-* G-, guanine nucleotide-binding-; CT, cholera toxin; PT, pertussis toxin. DADLE - *see text.*

Whilst the α subunits are distinct, they are highly homologous, with sequence identity in mammals varying from 94% between $G_i1\alpha$ and $G_i3\alpha$ to 41% between $G_s\alpha$ and $G_z\alpha$. Further, there is remarkable sequence conservation of the individual G-protein α subunits between species: thus in $G_i1\alpha$ from human and bovine tissues there is complete conservation of sequence (354 amino acids) and there is but a single amino acid substitution in $G_s\alpha$ (394 amino acids) between rat and human [1]. Even in *Drosophila melanogaster* a form of $G_s\alpha$ is 71% identical with that from mammalian species whilst a $G_o\alpha$ homologue is ~80% identical with mammalian $G_o\alpha$ [2]. This article will discuss approaches which we have shown to be useful in attempts to assess the specific roles of each G-protein and to gain information on the specificity (or otherwise) of receptor-G-protein-effector interactions.

With the exceptions of receptor-control of phospholipase A_2 activity and, to a degree, inhibition of adenylate cyclase activity, it is believed that the α subunits of G-proteins define the nature of the necessary interactions between G-protein and both receptor and effector moieties, although the interaction of α-subunit with receptors appears to require the presence of β/γ subunits. As such, the individual G-protein α subunits must possess domains able to interact selectively with the other components of the signal transduction cascade if individual G-proteins are to function in a specific or selective manner to control cellular responses to exogenous information. As well as these functions, the α subunits of a G-protein must also interact with the β/γ subunit complex. Further, the α subunit contains the site of GTP binding and hydrolysis, and it is this GTPase activity that defines the state of activation/deactivation of the G-protein and hence that of the effector system.

Whilst none of the classical signal-transducing G-proteins have yet been crystallized, and accordingly 3-D structure has not been defined, the crystallizations of the single polypeptide GTP-binding proteins $P21^{N-ras}$ and the bacterial elongation factor EfTu has provided information to allow a tentative model for the structure of the α subunit of an average G-protein to be constructed; this model can be used to test hypotheses related to the location of functional domains of these proteins [3]. Such information is obviously of importance for the rational design of approaches to assess the interactions of receptors, G-proteins and effectors. The extreme C-terminal region of all G-protein α subunits appears to represent the site of contact with receptors, and evidence from the construction of chimaeric G-protein α subunits suggests that the site of interaction with effector systems is also likely to be towards the C-terminus of the α subunit [4].

Information which defines the extreme C-terminal region of G-protein α subunits as a key site for interaction with receptors has been based on two pieces of evidence. (1) The realization that a cysteine residue located 4 amino acids from the C-terminus of a number of G-proteins acts as the acceptor site for addition of ADP-ribose catalyzed by the toxin PT [5] and that this modification prevents receptor activation of G-proteins which are modified by PT. (2) The *unc* mutant of the S49 lymphoma cell line, in which a β-adrenergic receptor and $G_s\alpha$ cannot interact productively, has been demonstrated to be due to a single base substitution. This results in the replacement of an Arg residue which is located 6 amino acids from the C-terminus of $G_s\alpha$ in the wild type by a Pro residue in the mutant [6,7].

Many PT-sensitive G-proteins are found to be co-expressed in particular cells [8], and a major challenge is to assess whether receptors can select specifically between such a population of closely similar G-proteins. Improved resolution by SDS-PAGE gels in the 40 kDa region, which can be achieved by approaches that include (a) lowering the cross-linking of the gel matrix [9], (b) prior alkylation of the membranes [10] and (c) incorporation of deionized urea into the gel [11], in combination with specific immunological probes can be used to define the molecular nature of the various PT-sensitive G-proteins present in a single cell type or tissue. The rank order of apparent M_r for the α subunits of the PT-sensitive G-proteins under such conditions is $G_i1 > G_i3 > G_i2 > G_o$.

USE OF ANTISERA TO THE C-TERMINAL REGIONS OF G-PROTEINS TO DEFINE THE SPECIFICITY OF RECEPTOR SIGNALLING CASCADES

We have made considerable use of anti-peptide antisera which selectively identify the C-terminal decapeptides of the α subunits of the individual PT-sensitive G-proteins to define which of them functions to mediate inhibition of adenylate cyclase and receptor-driven inhibition of voltage-operated Ca^{2+} channels in neuroblastoma x glioma hybrid cells (NG 108-15). Our studies were performed with plasma membranes.

Incubation of the membranes with IgG fractions from each of normal rabbit serum and anti-G_i2, G_i3 and G_o C-terminal antisera, followed by challenge of the membranes with receptor-saturating levels of the synthetic enkephalin, D-Ala^{2}Leu^5-enkephalin (DADLE), indicated that δ opioid receptor-mediated stimulation of high-affinity GTPase activity [12,13] (Fig. 1) and inhibition of forskolin-amplified adenylate cyclase [13] were attenuated specifically by pre-incubation with the IgG fraction from the anti-G_i2 antiserum. None of the other IgG fractions were able to mimic this effect. Accordingly

Fig. 1. Activation of the δ opioid receptor of NG 108-15 cells stimulates the GTPase activity of G_i2.

Membranes from the cells were incubated for 60 min in the absence (**1**) or presence (**2-6**) of IgG fractions derived from normal rabbit serum (**2**), or (**3-6**) antisera which identify the C-terminal decapeptides of - (**3**) $G_i2\alpha$, (**4**) $G_o\alpha$, (**5**) $G_i3\alpha$ or (**6**) $G_s\alpha$. The stimulation of high-affinity GTPase activity by a receptor-saturating concentration (1 μM) of DADLE was then assessed. *Data adapted from ref. [13].*

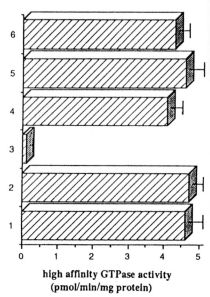

high affinity GTPase activity
(pmol/min/mg protein)

we were able to conclude that G_i2 was the true G_i of the adenylate cyclase cascade, at least in this system. A similar conclusion has been reached by Simonds & co-authors, in studies of the α2A receptor of human platelets [14]. To assess the nature of the G-protein which mediates α2 adrenergic inhibition of voltage-operated Ca^{2+} channels, McFadzean & co-workers [15] injected IgG fractions of the anti-G-protein antisera described above into prostaglandin E1-differentiated NG 108-15 cells and subsequently measured the ability of nor-adrenaline to inhibit Ca^{2+} currents. This ability was blocked by the IgG fraction of the anti-G_o antiserum [15] but not by the other antisera, implying a specific role for G_o in this process.

ASSESSMENT OF INTERACTIONS OF RECEPTORS AND G-PROTEINS USING CHOLERA TOXIN-CATALYZED ADP-RIBOSYLATION OF 'INAPPROPRIATE' G-PROTEINS

Whilst both rod and cone transducins can be substrates, under appropriate conditions, both for PT and for CT, it is generally considered that the other PT-sensitive G-proteins are not substrates for ADP-ribosylation catalyzed by CT. However, if CT-catalyzed ADP-ribosylation is performed on well washed membranes derived from a range of tissues, in the absence of guanine nucleotides, then one can note CT-dependent radiolabel incorporation into a 40 kDa polypeptide as well as into the polypeptides corresponding to forms of $G_s\alpha$. In membranes of the hybrid (NG 108-15) cells, the addition of receptor-saturating concentrations of the enkephalin DADLE produced a marked increase in CT-catalyzed ADP-ribosylation

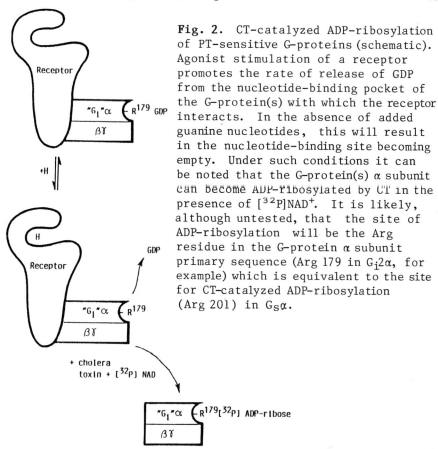

Fig. 2. CT-catalyzed ADP-ribosylation of PT-sensitive G-proteins (schematic). Agonist stimulation of a receptor promotes the rate of release of GDP from the nucleotide-binding pocket of the G-protein(s) with which the receptor interacts. In the absence of added guanine nucleotides, this will result in the nucleotide-binding site becoming empty. Under such conditions it can be noted that the G-protein(s) α subunit can become ADP-ribosylated by CT in the presence of [^{32}P]NAD$^+$. It is likely, although untested, that the site of ADP-ribosylation will be the Arg residue in the G-protein α subunit primary sequence (Arg 179 in $G_i2\alpha$, for example) which is equivalent to the site for CT-catalyzed ADP-ribosylation (Arg 201) in $G_s\alpha$.

of this 40 kDa protein but had no effect on the CT-catalyzed ADP-ribosylation of G_s [16]. Further, DADLE had no effect on PT-catalyzed ADP-ribosylation of G-proteins in these membranes.

The most likely explanation of these results is that the 40 kDa polypeptide represents the G-protein with which the opioid receptor on these cells interacts. A possible scenario to account for these results is that agonist-activation of the receptor promotes the release of GDP from the nucleotide-binding site. Since under the conditions of these experiments GTP was not available to replace the GDP, then a conformational change occurred such that the G-protein became a (weak) substrate for CT (Fig. 2). It is pertinent to this argument to note that all of the G-proteins identified to date have an invariant Arg residue in a position in the primary sequence equivalent to that which is the site of CT-catalyzed ADP-ribosylation in $G_s\alpha$ (Table 1). Further, this amino acid is close to a section of the primary sequence which forms part of the nucleotide-binding site, and the maintenance of this Arg residue is cardinal for G-protein GTPase activity [17, 18].

Table 1. The α subunits of all PT-sensitive G-proteins have a conserved arginine (**R**) residue at the position equivalent to that which is the site for CT-catalyzed ADP-ribosylation in G_s.

The site of CT-catalyzed ADP-ribosylation in $G_s\alpha$ is Arg 201. All of the G-protein α subunits which are substrates for PT-catalyzed ADP-ribosylation have this Arg residue conserved in the equivalent position as it plays a key role in the GTPase activity of the G-protein [17, 18]. It might thus be anticipated that all of these G-proteins would also be potential substrates for CT-catalyzed ADP-ribosylation.

G-protein	Relevant sequence
	CT
G_s	195 ↓ 207 Q D L L R C R V L T S G I
$G_i 1$	172 184 Q D V L R T R V K T T G I
$G_i 2$	173 185 Q D V L R T R V K T T G I
$G_i 3$	172 184 Q D V L R T R V K T T G I
G_o	173 185 Q D I L R T R V K T T G I

As noted above, the opioid receptor in NG 108-15 is of the δ subtype and is known to interact with a PT-sensitive G-protein to cause inhibition of adenylate cyclase activity. As PT prevents effective coupling between receptors and relevant G-proteins, then we argued that PT treatment of NG 108-15 cells, prior to cell harvest and preparation of the membranes, would prevent DADLE-stimulation of CT-catalyzed ADP-ribosylation of the 40 kDa protein. This indeed was the case. Further, pre-treatment of the cells with CT did not prevent opioid-peptide stimulation of CT-catalyzed ADP-ribosylation of the 40 kDa polypeptide [16]. As CT treatment had presumably produced ADP-ribosylation of $G_s\alpha$ using endogenous non-radioactive NAD^+ as substrate, then this peptide was essentially the only protein in membranes from these cells now able to incorporate radioactivity from [^{32}P]NAD^+ in a CT-dependent manner. This reflects the fact that CT does not catalyze ADP-ribosylation of the 'G_i-like' proteins in whole cells.

Similar observations have also been noted in rat glioma C6 cells, where a small, heat-stable factor in foetal calf serum promotes the CT-dependent ADP-ribosylation of a 40 kDa polypeptide which appears to be the PT-sensitive G-protein, $G_i 2$ [19]. Recent experiments on the human leukaemia cell line HL60 has provided convincing evidence that a single agonist (the bacterial chemotactic factor, N-formyl-Met-Leu-Phe) can promote CT-catalyzed ADP-ribosylation of two distinct

PT-sensitive G-proteins in a single cell [20]. Based on the immunological profile of PT-sensitive G-proteins in these cells [21], these are likely to be G_i2 and G_i3. Whilst agonist-promoted, CT-catalyzed ADP-ribosylation of these two G-proteins demonstrated very similar dose-response curves [20], it is not yet possible to conclude that this implies that a single receptor for the agonist interacts with more than a single G-protein as nothing is known about potential hetero-geneity of receptors for this peptide.

In an effort to eliminate potential problems derived from the expression of multiple receptors with very similar pharmacology in a single cell we have expressed a construct containing the human $\alpha2$-C10 adrenergic receptor and the mammalian expression vector pDOL in Rat 1 fibroblasts. Following selec-tion of positive clones on the basis of resistance to geneticin, a positive clone (1C) was then examined for the ability of the $\alpha2$ adrenergic receptor agonist, UK14304, to stimulate both CT-catalyzed ADP-ribosylation and high-affinity GTPase activity in membranes from these cells [22] (Fig. 3, a & b). The dose-response curves for UK14304 regulation of each of these activities were very similar. Furthermore, treatment of 1C cells with PT prior to preparation of the membranes obliterated the effect of UK14304 in both assays. Membranes prepared from non-transfected parental Rat 1 fibroblasts or a clone of the Rat 1 cells (4D) which had been transfected with the receptor construct and selected on the basis of resistance to geneticin, but which did not express measurable numbers of receptors as assessed by the specific binding of [3H]yohimbine, did not display agonist-stimulation of high-affinity GTPase activity or enhanced CT-catalyzed ADP-ribosyla-tion.

These results demonstrated that both of these responses were absolutely dependent upon the presence and activation of the receptor. CT-catalyzed ADP-ribosylation of 'G_i' could not be observed when the nucleotide-binding pocket of the polypeptide contained either GDP or GDP + AlF_4^-. However, it has generally been noted that agonist effects on CT-catalyzed ADP-ribosylation of PT-sensitive G-proteins can be observed in the presence of the guanine nucleotide analogue Gpp[NH]p but not GTPγS. As such, the details of this process require further study.

Recent experiments by Katada & co-workers [23] have revealed that reconstitution of the α subunits of each of G_i1, G_i2 and G_o from brain into membranes of HL60 cells pre-treated with PT allowed the ADP-ribosylation of each of these polypeptides by CT when the assay was performed

Fig. 3.
UK 14304 regulates
both CT-catalyzed
ADP-ribosylation of
'G$_i$' and stimulation
of high-affinity
GTPase activity in
Rat 1 fibroblasts
expressing the α2-C10
adrenergic receptor.
a: UK14304 stimulation
of high-affinity
GTPase activity.
b: UK14304 stimulation
of CT-catalyzed
ADP-ribosylation of a
40 kDa polypeptide
('G$_i$').

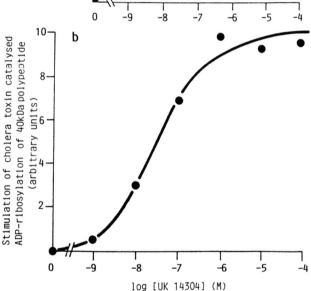

in the presence of FMLP (the chemotactic factor mentioned above), Gpp[NH]p and Mg^{2+}. G$_i$1 and G$_i$2 were ADP-ribosylated more efficiently than G$_o$. This may indicate a rank order of interaction of these G-proteins with the receptor(s) for FMLP [23].

     As an approach whose use has not yet been described, it should be possible to immunoprecipitate G-proteins which have been ADP-ribosylated by CT in an agonist-dependent manner in membranes of cells expressing a defined receptor population. Such studies would allow direct assessment of the specificity of contact of receptors and PT-sensitive G-proteins in native membranes.

Acknowledgements

This work is supported by the Medical Research Council and the Science and Engineering Council.

References

1. Kaziro, Y. (1990) In *Guanine Nucleotide Binding Proteins as Mediators of Cellular Signalling Processes* (Houslay, M.D. & Milligan, G., eds.), John Wiley, Chichester, pp. 47-66.
2. Hurley, J.B. (1990) *Biochem. Soc. Symp. 56*, 81-84.
3. Masters, S.B., Stroud, R.M. & Bourne, H.R. (1986) *Protein Engineering 1*, 47-54.
4. Masters, S.B., Sullivan, K.A., Miller, R.T., Beiderman, B., Lopez, N.G., Ramachandran, J. & Bourne, H.R. (1988) *Science 241*, 448-451.
5. West, R.E. Jr., Moss, J., Vaughan, M., Liu, T. & Liu, T-Y. (1985) *J. Biol. Chem. 260*, 14428-14430.
6. Sullivan, K.A., Miller, R.T., Masters, S.B., Beiderman, B., Heideman, W. & Bourne, H.R. (1987) *Nature 330*, 758-759.
7. Rall, T. & Harris, B.A. (1987) *FEBS Lett. 224*, 365-370.
8. Milligan, G. (1989) *Cell Signalling 1*, 411-419.
9. Mitchell, F.M., Griffiths, S.L., Saggerson, E.D., Houslay, M.D., Knowler, J.T. & Milligan, G. (1989) *Biochem. J. 262*, 403-408.
10. Sternweis, P.C. & Robishaw, J.D. (1984) *J. Biol. Chem. 259*, 13806-13813.
11. Schnefel, S., Banfic, H., Eckhardt, L., Schultz, G. & Schulz, I. (1988) *FEBS Lett. 230*, 125-130.
12. McKenzie, F.R., Kelly, E.C.H., Unson, C.G., Speigel, A.M. & Milligan, G. (1988) *Biochem. J. 249*, 653-659.
13. McKenzie, F.R. & Milligan, G. (1990) *Biochem. J. 267*, 391-398.
14. Simonds, W.F., Goldsmith, P.K., Codina, J., Unson, C.G. & Spiegel, A.M. (1989) *Proc. Nat. Acad. Sci. 86*, 7809-7813.
15. McFadzean, I., Mullaney, I., Brown, D.A. & Milligan, G. (1989) *Neuron 3*, 177-182.
16. Milligan, G. & McKenzie, F.R. (1988) *Biochem. J. 252*, 369-373.
17. Landis, C.A., Masters, S.B., Spada, A., Pace, A.M., Bourne, H.R. & Vallar, L. (1989) *Nature 340*, 692-696.
18. Freissmuth, M. & Gilman, A.G. (1989) *J. Biol. Chem. 264*, 21907-21914.
19. Milligan, G. (1989) *Cell Signalling 1*, 65-74.
20. Gierschik, P., Sidiropoulos, D. & Jakobs, K-H. (1989) *J. Biol. Chem. 264*, 21470-21473.

21. Murphy, P.M., Eide, B., Goldsmith, P. Brann, M.,
    Gierschik, P., Spiegel, A. & Malech, H.L. (1987) *FEBS
    Lett. 221*, 81-86.
22. Milligan. G., Carr, C., Gould, G.W., Mullaney, I. &
    Lavan, B.E. (1991) *J. Biol. Chem. 266*, in press.
23. Iiri, T., Tohkin, M., Morishima, N., Ohoka, Y., Ui, M.
    & Katada, T. (1989) *J. Biol. Chem. 264*, 21394-21400.

#A-5

REQUIREMENT FOR PHOSPHOLIPASES C AND A$_2$ ACTIVATION FOR fMetLeuPhe-STIMULATED SECRETION IN STREPTOLYSIN-O-PERMEABILIZED NEUTROPHILS AND HL60 CELLS

Shamshad Cockcroft

Department of Physiology, University College London, University Street, London WC1E 6JJ, U.K.

The stimulus-secretion pathway when initiated in neutrophils by a ligand such as fMLP* or ATP requires the activation of two phospholipases, PL-C and PL-A$_2$. PL-C as well as PL-A$_2$ activation is coupled to the receptor by the G-proteins designated G$_p$ and G$_a$ respectively. Studies where the pharmacological specificity for the ATP receptor was analyzed for the activation of PL-C, PL-A$_2$ and secretion indicated that secretion occurred only if the agonist were capable of stimulating both PL's. Any ATP-related molecule which could stimulate PL-C but not PL-A$_2$ did not elicit secretion.

The relationship between secretion and PL-A$_2$ activation was also studied in streptolysin O-permeabilized neutrophils. The GTP analogues GTPγS, GppNHp and GppCH$_2$p all stimulated both reactions. From the analysis of the Mg-ATP requirement for secretion and PL-A$_2$ activation, only fMLP-stimulated secretion correlated with PL-A$_2$ stimulation. GTPγS-dependent exocytosis could be observed under conditions where the activation of PL-A$_2$ was minimal. The G-protein that is involved in GTPγS-stimulated secretion is therefore not G$_a$ but another G-protein (G$_e$).

The stimulus-secretion coupling process provides a model for cell activation which is simplified by virtue of its immediacy. For regulated secretion, it commences with the binding of the ligand to its receptor and culminates in the exocytotic release of granule contents, occurring within a fraction of a minute to minutes for most secretory cells. It nonetheless comprises a sequence of steps which can be separated into individual parts. This sequence can be viewed as a minimum of two stages: (1) the generation of intracellular signals that switch the exocytotic machinery, and (2) the reactions that constitute the exocytotic event itself, namely the fusion of the granule with the p.m.*

Abbreviations.- DAG, diacylglycerol; $[Ca^{2+}]_i$, intracellular (cytosolic) free Ca^{2+} concentration (pCa^{2+} being analogous to pH); fMLP, fMet-Leu-Phe; PKC, protein kinase C; PL (as in PL-D), phospholipase; pde, phosphodiesterase; PT, pertussin toxin; p.m., plasma membrane. *See overleaf for (e.g.)* InsP, Ptd.

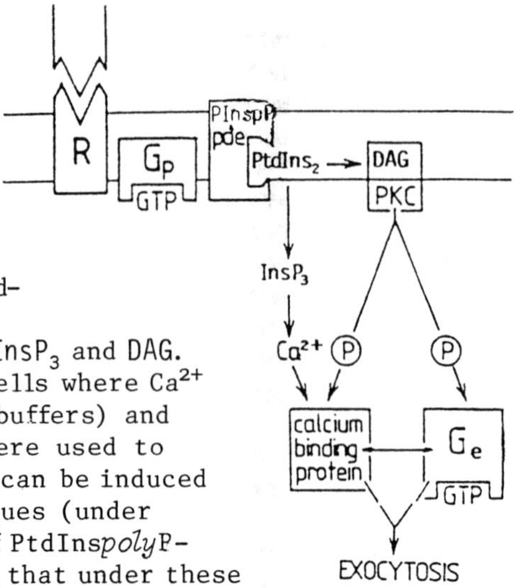

Fig. 1. Interrelationships
(shown schematically) between
the early events in stimulus-
secretion coupling and
subsequent events mediated
by the G-proteins G_p and G_e
respectively. G_p transduces
receptor- (R-)mediated
signals and activates PL-C* (Ptd-
Ins $poly$P-specific phosphodi-
esterase, PInspP pde) to generate $InsP_3$ and DAG.
As studied in permeabilized cells where Ca^{2+}
(controlled by use of Ca-EGTA buffers) and
guanine nucleotide analogues were used to
trigger exocytosis, secretion can be induced
by use of Ca^{2+} $plus$ GTP analogues (under
conditions where activation of PtdIns$poly$P-
pde is suppressed), indicating that under these
conditions the GTP analogues interact with G_e.
If Mg-ATP is provided so that PKC-mediated
reactions can occur, the resulting phosphorylations (P)
can enhance the effective affinity for both Ca^{2+} and
the guanine nucleotide.

The use of permeabilized cell preparations for studying
stimulus-secretion coupling has provided some new insights
into the process of exocytosis, as summarized in Fig. 1.
Two GTP-binding (G) proteins act in series in the regulation
of the reactions which culminate in exocytosis. Early events
involve the binding of an agonist to a receptor, leading
to the G_p-mediated production of DAG and $InsP_3$* by PL-C.
In the intact cell, $InsP_3$, and possibly its subsequent metabolites such as $Ins(1,3,4,5)P_4$, mediate the rise in cytosol Ca^{2+};
but in permeabilized cell preparations $InsP_3$ leaks out of
the cells and is therefore compensated by provision of Ca-EGTA
buffers. The other product of PL-C, DAG, can stimulate PKC-
dependent phosphorylation only when Mg-ATP is provided. There
is no obligatory requirement for Mg-ATP when Ca^{2+} $plus$ GTPγS
is used to trigger exocytosis from permeabilized cells (see
Fig. 4D below). Further downstream is a second G-protein,
G_e, whose target is currently not known. There is some
evidence to suggest that it could be either a protein phosphatase
[1] or PL-A$_2$ [2, 3].

―――――――――――

*Abbreviations are as listed on previous p., and additionally:-
PA, phosphatidic acid; Ptd, phosphatidyl (but PC = Ptd-choline,
and in Fig. 1 PInspP = Ptd-Ins$poly$P). Inositol (poly)phosphates
are termed Ins$(poly)$P's; $InsP_3$ (a trisphosphate) here signifies
$Ins(1,4,5)P_3$. PL (phospholipase) and other abbreviations conform
with those in other arts., some of which are pertinent — notably
#A-6 (Bonser et al., on PL-D).- Ed.

The experimental evidence to support this model comes from studies where it can be demonstrated that with permeabilized cell preparations secretion can be induced when Ca^{2+} and guanine nucleotides are provided (as recently reviewed [4]). The effect of the guanine nucleotide is not dependent on its interaction with G_p but is due to an effect on a different G-protein. Inhibition of PL-C activity by neomycin or due to lack of the substrate, Ptd-InsP$_2$, does not substantially decrease secretion induced by Ca^{2+} *plus* guanine nucleotide. Mg-ATP is thus not essential for the secretory reaction, although it can enhance the effective affinity for both Ca^{2+} and guanine nucleotide in the cnooytotio reaotion.

This model (Fig. 1) is based on studies where secretion from permeabilized cells was triggered by non-hydrolyzable analogues of GTP. It remains to be shown that this model also applies to receptor-mediated exocytosis. We have used human neutrophils and a related cell-line, the HL60, to investigate the intracellular requirements for exocytosis. Studies on intact-cell preparations using the agonist fMLP or ATP demonstrate clearly that receptor-mediated stimulation of secretion requires the activation not only of PL-C but also of PL-A$_2$ [3]. A further investigation of PL-A$_2$ activation in streptolysin O-permeabilized cells has revealed that this PL is regulated by a PT-sensitive G-protein and is required for fMLP-stimulated secretion but *not* GTPᵧS-induced exocytosis.

RECEPTOR-STIMULATED EVENTS IN INTACT CELLS

fMLP stimulates at least 3 PL's within seconds when added to neutrophils (or HL60 cells). These are PL-C (specific for Ptd-InsP$_2$), PL-A$_2$ which releases arachidonic acid from PC and Ptd-inositol, and PL-D which hydrolyzes PC to form PA and choline. Secretion from these cells occurs over a period of 10-20 sec [5] as does the activation of PL-A$_2$ [6, 7] and PL-C [5, 8]. For phosphatidate formation (the initial product of PL-D activation) the time-course is similar (Fig. 2). These observations pose the question whether all three PL's are essential for exocytotic secretion.

Since an increase in $[Ca^{2+}]_i$ is required for exocytosis, it is clear that receptor-stimulated PL-C for producing InsP$_3$ is a necessary part of the sequence of events leading to exocytosis. There appears, moreover, to be a good correlation between PL-A$_2$ activation and secretion (Fig. 3). Neutrophils and HL60 cells possess a receptor for ATP which is coupled to both PL-C and PL-A *via* a PT-sensitive G-protein(s). When the pharmacological specificity of the ATP receptor in respect of inducing secretion was compared with the specificity for activating PL-A$_2$ and PL-C, it emerged that only those analogues

Fig. 2. PL-D activation in human
neutrophils stimulated with fMLP
(100 nM, in the presence of 5 µg/ml
cytochalasin B): time courses for
formation of PA (the initial
product of PL-D activation) and of
DAG, derived from subsequent
metabolism of PA. The fatty acid
composition of the stimulated PA
and DAG resembled that of PC.
Data re-drawn from [9] & [10].

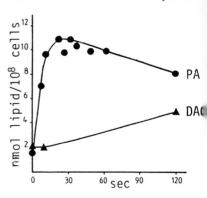

that stimulated PL-C and PL-A_2 stimulated secretion. A study
of PL-D activation, not yet been carried out, would provide
some insight into the relationship between secretion and
PL-D activation.

STUDY OF EXOCYTOSIS IN PERMEABILIZED NEUTROPHILS

Many strategies for p.m. permeabilization have been utilized,
and it is important to appreciate that these generate lesions
differing widely in dimensions and lifetimes. The permeabili-
zing technique used in the experiments described below entails
use of streptolysin O. This induces an efflux of soluble
proteins (e.g. lactate dehydrogenase) which is maximal at
3 min without release of secretory material. Secretion occurs
only when Ca^{2+} and GTPγS are also provided. In our experiments
the cells are treated with metabolic inhibitors (deoxyglucose
plus antimycin A) for 5 min prior to permeabilization. Under
these conditions the intact cells are totally refractory
to receptor-mediated stimulating agents such as fMLP or ATP.

Metabolically inhibited neutrophils can be stimulated
to secrete with fMLP provided that they are permeabilized
in the presence of 10 µM Ca^{2+} (Fig. 4D; for A & B, see later).
There is no requirement for Mg-ATP. Fig. 4D also shows that
with GTPγS, as with fMLP, induced secretion depends on the
presence of 10 µM Ca^{2+}. In the absence of Mg-ATP, however,
PL-C activation cannot occur because the substrate, Ptd-InsP_2,
is depleted in metabolically inhibited cells. Moreover,
protein phosphorylation is unlikely to occur because of
the lack of any phosphorylating agent. However, secretion
can still occur with GTPγS and fMLP, and that due to fMLP
is inhibited by PT pre-treatment.

Adding Mg-ATP (to 1 mM) has several effects on secretion
(Fig. 4C).- (1) Ca^{2+} at 10 µM without agonist becomes stimulatory
by itself. (2) Near-optimal secretion with fMLP as well
as GTPγS is now observed at 1 µM, a Ca^{2+} concentration which
by itself is non-stimulatory. Moreover, (3) GTPγS can promote
secretion at low Ca^{2+}, 10 nM.

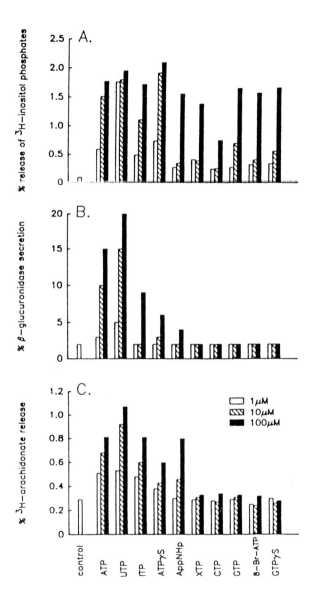

Fig. 3. Specificity of the ATP receptor: effect of various nucleotides on (**A**) PL-C activation, (**B**) secretion, and (**C**) PL-A$_2$ activation. Intact dibutyryl cAMP-differentiated HL60 cells were incubated for 10 min with the indicated nucleotides. PL-C and PL-A$_2$ activations were determined by measuring the production of InsP's and arachidonate respectively. *From [3], courtesy of the Biochemical Journal; likewise for Fig. 5.*

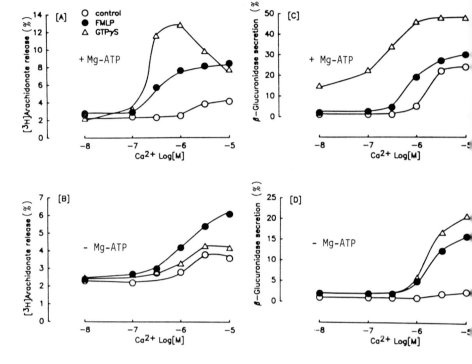

Fig. 4. Relationships between secretion *(right)* and PL-A$_2$ activation *(left)* stimulated by GTPγS or fMLP in permeabilized human neutrophils. Cells were labelled with [^{3}H]arachidonate for 1 h, washed into a glucose-free buffer and, after metabolic inhibition for 5 min with antimycin A and deoxyglucose, treated with streptolysin O (0.4 i.u./ml) in the presence (**A, C**) or absence (**B, D**) of Mg-ATP (1 mM), Ca-EGTA buffers (3 mM EGTA) and fMLP (1 µM; ●) or GTPγS (10 µM; Δ). In the absence of Mg-ATP the stimulation of PL-C and PKC is suppressed. Secretion and arachidonate release were measured after 10 min on the same cells.

        In summary, secretion can be observed under the following conditions:
1) pCa^{2+} 5 plus GTPγS;
2) pCa^{2+} 5 plus fMLP;
3) pCa^{2+} 5 alone provided that Mg-ATP is present;
4) pCa^{2+} 6 plus GTPγS provided that Mg-ATP is present;
5) pCa^{2+} 6 plus fMLP provided that Mg-ATP is present;
6) pCa^{2+} 8 plus GTPγS provided that Mg-ATP is present.

Secretion needs 1 mM Mg-ATP, suggesting that the Mg-ATP is utilized to fuel the inositol lipid kinases to provide the substrate for PL-C. The resulting DAG generated from Ptd-InsP$_2$ hydrolysis would activate PKC and be responsible for reducing the Ca^{2+} requirement for secretion. Thus PKC-mediated phosphorylation is likely to play an important modulatory role in supporting exocytosis.

STUDY OF PL-A_2 ACTIVATION IN PERMEABILIZED CELLS

Arachidonate release was used as a measure of PL-A_2 activation. Many studies on PL-A_2 activation indicate that this enzyme is regulated by a G-protein, G_a, in a similar manner to PL-C. The evidence for this is as follows.-
1) Guanine nucleotides stimulate PL-A_2 activity in permeabilized cell systems and membrane preparations. Examples include neutrophils [11] (& see Fig. 5), rat basophilic leukaemia cells (RBL) [12], mast cells [13, 14] and platelets [15].
2) Fluoride (an activator of G-proteins) has been shown to stimulate PL-A_2 [11, 16].
3) PT pre-treatment inhibits receptor-mediated PL-A_2 activation [16-18].

Activation of PL-A_2 and PL-C appear to be independently regulated processes. Both phorbol ester and PT treatment have been used to dissociate the two PL's [3, 19-24]. Thus it could be surmised that a separate G-protein is responsible for regulating PL-C (G_p) and PL-A_2 (G_a).

Fig. 5 illustrates that streptolysin O-permeabilized neutrophils respond to the GTP analogues (GTPγS, GppNHp and GppCH$_2$p) by stimulating secretion as well as arachidonate release. The rank order of the GTP analogues in stimulating PL-A_2 is identical to that for other G-protein-regulated processes such as secretion or activation of PL-C, adenylate cyclase or cAMP-pde.

Fig. 4, A & B, illustrates the effect of Mg-ATP on arachidonate release. In the absence of Mg-ATP, release is still stimulated by fMLP, albeit at a reduced level. This reduction correlates with the concomitant reduction in secretion. It is noteworthy that the Ca^{2+} requirement shifts to higher concentrations for both processes. In the presence of Mg-ATP, GTPγS stimulates maximal arachidonate release at 1 μM Ca^{2+}. A further increase in Ca^{2+} to 10 μM causes inhibition. This biphasic effect of GTPγS is also noticeable with all the GTP analogues (Fig. 5). Secretion, on the other hand, does not show a similar effect. In contrast to fMLP, GTPγS-stimulated arachidonate release appears to be dependent on the presence of Mg-ATP (Fig. 4B) - which is surprising. Seemingly the mechanism for stimulating PL-A_2 is different for fMLP and GTPγS. Despite the inhibition of arachidonate release with GTPγS in the absence of Mg-ATP, secretion is still evident. From this one can conclude that PL-A activation may be essential for fMLP-induced but not for GTPγS-induced secretion. PL-A_2 activity is thus unlikely to be the target for G_e.

Fig. 5. Activation of PL-A$_2$
and exocytosis by GTP analogues
in streptolysin O-permeabilized
human neutrophils, treated
generally as for Fig. 4.
The permeabilization, after
labelling and metabolic
inhibition, was done in
Ca-EGTA buffers along with
1 mM Mg-ATP and 100 μM
GTPYS, GppNHp or GppCH$_2$p.
Release of labelled arachidonate after 10 min at 37°
served as a measure of PL-A$_2$
activation as well as
secretion from the same cell
preparation.

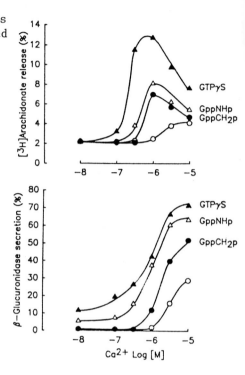

CONCLUSIONS

A major factor to be taken into account when discussing
secretion stimulated by GTPYS and fMLP from permeabilized
cells is that whilst fMLP can interact only with the receptor,
GTPYS can activate any G-protein at any location.   Thus
GTPYS can bypass many of the transmembrane signalling systems
and help point the way to the reactions that constitute
the exocytotic event itself.   In this context, PL-A$_2$ activation
can be clearly understood as a transmembrane signalling system.
The Ca^{2+} requirement for PL-A$_2$ activation when stimulated
with fMLP (Fig. 4A) is 1 μM, 10-fold higher than the resting
level of Ca^{2+} in cells.   This would suggest that although
PL-A$_2$ activation is stimulated in parallel to that of PL-C,
there must be an obligatory requirement for Ca^{2+} to increase
before PL-A$_2$ is switched on.

The schematic model in Fig. 1 illustrates the working
hypothesis that at least two G-proteins are involved in the
steps leading to the exocyotic event.   For receptor-mediated
secretion, evidently another event – the activation of PL-A$_2$
– has to be taken into account.   PL-A$_2$ activation is important
only for providing intracellular messengers and is not a
prerequisite for the exocytotic process itself (GTPYS can trigger
secretion in the absence of PL-A$_2$ activation).   Whether G$_e$
is involved in receptor-mediated exocytosis is still not

clear, but nothing in the results discussed here excludes
that possibility.

As regards the identity of the G-protein that regulates
PL-A$_2$, it is most likely a G-protein that belongs to the
heterotrimeric family, probably G$_{i2}$ or G$_{i3}$, the two PT substrates
found in neutrophils [25]. The fMLP receptor is known to
interact with G$_{i2}$ (26-28] and is the probable candidate for
G$_p$ in these cells. It cannot be totally excluded that
the βγ subunits are responsible for PL-A$_2$ activation, as
has been demonstrated for bovine rod outer segments [18,
29]. There are many candidates available for G$_e$ including
some of the small G-proteins that have been identified recently
in neutrophils, e.g. *rap*1 [30] or G$_{22K}$ [31]. The different
kinds of low mol. wt. G-proteins are increasing apace, and
doubtless there are many new G-proteins in search of a function.
The involvement of G-proteins in regulated secretion has
many parallels with their involvement in constitutive secretion
[32-35]. The identification and role of the G-proteins thus
involved may provide a pointer for future studies in regulated
exocytosis. The availability of yeast defective in secretion
is a powerful tool which is not yet available for mammalian
cells.

Acknowledgements

Work in the author's laboratory has been supported by
the Wellcome Trust, the Medical Research Council and the
Lister Institute. S.C. is a Lister Institute Fellow.

References

1. Tatham, P.E.R. & Gomperts, B.D. (1989) *Biosc. Repts. 9*,
    99-109.
2. Cockcroft, S. & Stutchfield, J. (1988) *Phil. Trans. R. Soc.
    Lond. Ser. B 320*, 247-265.
3. Cockcroft, S. & Stutchfield, J. (1989) *Biochem. J. 263*,
    715-723.
4. Gomperts, B.D. (1990) *Annu. Rev. Physiol. 52*, 591-606.
5. Cockcroft, S., Bennett, J.P. & Gomperts, B.D. (1980)
    Nature 288, 275-277.
6. Wynoop, E.M., Broekman, M.J., Korchak, H.M., Marcus, A.J.
    & Weissman, G. (1986) *Biochem. J. 236*, 829-837.
7. Tao, W., Molski, T.F.P. & Sha'Afi, R.I. (1988) *Biochem. J.
    257*, 633-637.
8. Burgess, C.M., McKinney, J.S., Irvine, R.F. & Putney, J.J.W.
    (1985) *Biochem J. 232*, 237-243.
9. Cockcroft, S. & Allan, D. (1984) *Biochem J. 222*, 557-559.
10. Cockcroft, S. (1984) *Biochim. Biophys. Acta 795*, 37-46.
11. Nakashima, S., Nagata, K-I., Ueda, K. & Nozawa, Y. (1988)
    Arch. Biochem. Biophys. 261, 375-383.

12. Narasimhan, V., Holowka, D. & Baird, B. (1990) *J. Biol. Chem.* *265*, 1459-1464.
13. Okano, Y., Yamada, K., Yano, K. & Nozawa, Y. (1987) *Biochem Biophys. Res. Comm.* *145*, 1267-1275.
14. Churcher, Y., Allan, D. & Gomperts, B.D. (1990) *Biochem. J.* *266*, 157-163.
15. Silk, S.T., Clejan, S. & Witcom, K. (1989) *J. Biol. Chem.* *264*, 21466-21469.
16. Bokoch, G.M. & Gilman, A.G. (1984) *Cell* *39*, 301-308.
17. Corda, D. & Kohn, L.D. (1986) *Biochem. Biophys. Res. Comm.* *141*, 1000-1006.
18. Jeselma, C.L. & Axelrod, A. (1987) *J. Biol. Chem.* *262*, 3623-3627.
19. Burch, R.M., Luini, A. & Axelrod, J. (1986) *Proc. Nat. Acad. Sci.* *83*, 7201-7205.
20. Slivka, S.R. & Insel, P.A. (1987) *J. Biol. Chem.* *262*, 4200-4207.
21. Slivka, S.R. & Insel, P.A. (1988) *J. Biol. Chem.* *263*, 14640-14647.
22. Troyer, D.A., Gonzalez, O.F., Douglas, J.G. & Kreisberg, J.I. (1988) *Biochem. J.* *251*, 907-912.
23. Conklin, B.R., Brann, M.R., Buckley, N.J., Ma, A. & Bonner, T.I. (1988) *Proc. Nat. Acad. Sci.* *85*, 8698-8702.
24. Reynolds, E.E., Mok, L.L.S. & Kurokawa, S. (1989) *Biochem. Biophys. Res. Comm.* *160*, 868-873.
25. Murphy, P.M., Eide, B., Goldsmith, P., Brann, M., Gierschik, P., Spiegel, A. & Malech, H.L. (1987) *FEBS Lett.* *221*, 81-86.
26. Polakis, P.G., Uhing, R.J. & Snyderman, R. (1988) *J. Biol. Chem.* *263*, 4969-4976.
27. Gierschik, P. & Jacobs, K.H. (1987) *FEBS Lett.* *224*, 219-223.
28. Offermans, S., Schafer, R., Hoffmann, B., Bombien, E., Spicher, K., Hinsch, K-D., Schultz, G. & Rosenthal, W. (1990) *FEBS Lett.* *260*, 14-18.
29. Axelrod, J., Burch, R.M. & Jelsema, C.L. (1988) *Trends Neurosc.* *11*, 117-123.
30. Bokoch, G.M. & Quilliam, L.A. (1990) *Biochem. J.* *267*, 407-411.
31. Bokoch, G.M., Parkos, C.A. & Mumby, S.M. (1988) *J. Biol. Chem.* *263*, 16744-16749.
32. Goud, B., Salminen, A., Walworth, N.C. & Novick, P.J. (1988) *Cell* *53*, 753-768.
33. Bourne, H.R. (1988) *Cell* *53*, 669-671.
34. Salminen, A. & Novick, P.J. (1987) *Cell* *49*, 527-538.
35. Segev, N., Mulholland, J. & Botstein, D. (1988) *Cell* *52*, 915-924.

#A-6

ROLE OF PHOSPHOLIPASE D IN SIGNAL TRANSDUCTION

Robert W. Bonser, N.T. Thompson and **L.G. Garland**

Wellcome Research Laboratories,
Beckenham, Kent BR3 3BS, U.K.

PL-D* appears to be important in intracellular signal transduction. PL-D activity is readily quantitated in intact cells by measuring the formation of Ptd-alcohols as produced by a unique PL-D-dependent transphosphatidylation reaction. Thereby several groups have demonstrated PL-D activation by various stimulatory agonists in different cell types. The enzyme's major substrate appears to be PC which is hydrolyzed to PA. By phosphatase action PA yields DAG, which activates PKC. Like DAG, PA is a second messenger: it is mitogenic, it increases intracellular Ca^{2+} and cGMP concentrations, it activates PL-A_2, and it inhibits adenylate cyclase.

It is unclear precisely how PL-D is activated, but its activity in intact cells and membrane fractions is increased by Ca^{2+} ionophores, phorbol esters, unsaturated fatty acids and guanine nucleotides. Bearing on the obscure physiological significance of PL-D activation, we have shown, in human neutrophils, absolute PL-D-dependence for cytochalasin B-primed superoxide production stimulated by fMLP; the dependence possibly hinges on PL-D-derived PA rather than DAG. Evidence is slowly emerging on the role, notably in mitogenesis, of PL-D activation; thus, it is potently blocked by wortmannin (a fungal metabolite), which may help define the importance of PL-D in cell signalling.

DAG AS A SIGNALLING AGENT

The importance of the receptor-coupled PtdInsP$_2^*$-specific PL-C in intracellular signalling is now clearly established (as reviewed [1]). The products of this phospholipase have well recognized second-messenger roles. InsP$_3$ mobilizes intracellular Ca^{2+}, and DAG activates PKC. The stimulated hydrolysis of PtdInsP$_2$ by PL-C has been regarded by many as an important source of DAG in activated cells. More recently the source of the DAG that activates PKC has been questioned.

----
*Abbreviations.- Inositol phosphates are termed InsP's (not IP's), e.g. InsP$_3$ = trisphosphate, Ins(1,4,5)P$_3$ in this text; InsP$_2$ = bisphosphate, Ins(4,5)P$_2$ herein; Ptd (prefix) = phosphatidyl; PA, phosphatidic acid; PC, phosphatidylcholine. AA, arachidonic acid; DAG, (1,2)diacylglycerol; DRG, diradylglycerol; fMLP, fMet-Leu-Phe; cGMP, cyclic GMP; PKC, protein kinase C; PL, phospholipase, e.g. PL-D. 'DRG' is amplified in legend to Fig. 3.

The development by Preiss & co-authors [2] of a radioenzymatic assay for DAG provided the first convenient means of measuring the amount of DAG produced in activated cells. The assay uses membranes from *E. coli* which have been induced to over-express DAG-kinase. The conversion of DAG to PA by DAG-kinase, in the presence of [^{32}P]ATP, has formed the basis of a simple and accurate measure of DAG levels in cells.   It was soon apparent to those investigators who measured the amount of DAG produced and also its fatty acid composition that the PL-C-dependent degradation of PtdInsP$_2$ could not be the sole source of DAG in stimulated cells [3-9[*]].   This point of view has been reinforced by a number of reports showing that PKC is activated and DAG levels are elevated in the absence of measurable InsP$_3$ formation [10-18].

PHOSPHATIDYLCHOLINE (PC) AS A DAG SOURCE

Much of the evidence now points to PC as the major source of DAG in activated cells. The fatty acid composition of the released DAG closely resembles that of PC, and moreover a stimulated turnover of PC has been observed in many cell types in response to a variety of stimuli (as reviewed: [19-21]). What is unclear at present is the contribution made by PL-C and/or PL-D to the stimulated turnover of PC.   PL-C releases DAG and phosphorylcholine, whereas PL-D generates PA and choline.

The stimulated formation of these products has been studied by a number of investigators, including ourselves, in a variety of cell types [22-30]. Most of the kinetic evidence indicates that choline is produced before phosphorylcholine and that PA levels increase before DAG, suggesting that it is mainly PL-D that is responsible for the stimulated degradation of PC.   Our own data, from studies with human neutrophils stimulated by the chemotactic peptide fMLP, are in total agreement with this concept and clearly show that (i) choline release is rapid, (ii) it occurs over the same time course as PA production, and (iii) formation of both products precedes DRG generation [30] (Fig. 1). Since neutrophils do not release phosphorylcholine, it must be assumed that PL-C does not contribute to the stimulated hydrolysis of PC in human neutrophils [31].

These and kindred studies have provided strong evidence that DAG release from PC in activated cells is *via* a PL-D-phosphatidate phosphatase pathway [26-30, 32].   Indeed, the slow sustained phase of DAG production, observed in some cell types, may arise *via* this mechanism [11, 14, 30-34]. It is

[*]Here and later, the refs. list has been curtailed, still leaving each group 'represented'.- *Ed.*

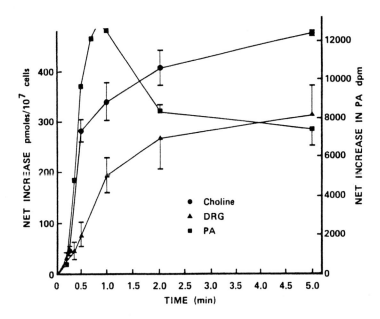

Fig. 1. Human neutrophils were pre-incubated with $1\text{-}O\text{-}[^3H]\text{-}$ alkyl-2-lyso-PC, and the formation of $1\text{-}O\text{-}[^3H]$alkyl-PA ('PA' *symbols*) was measured as in [32]. Choline release and DAG production were measured radioenzymatically as in [30]. The 'PA' results are from a single representative experiment. Those for choline release and DAG production are means ±S.E.M. for 3-5 experiments using neutrophils from different donors. fMLP was used at 100 nM throughout.

noteworthy that a rapid agonist-stimulated production of PA, which preceded DAG release, was observed in neutrophils some 7 years ago [35]. The fatty acid composition of the released DAG suggested that it was not derived from the inositol phospholipid pool, and one of the conclusions drawn from these experiments was that PL-D probably contributes to PA formation in the neutrophil.

REINFORCING EVIDENCE FOR A PL-D-DEPENDENT PATHWAY

Although compelling, the foregoing evidence for a PL-D-dependent pathway is mostly circumstantial. More definitive evidence has come from the elegant studies of Billah & co-workers [36-38] on PC metabolism in myeloid cells. Using $1\text{-}O\text{-}$alkyl-2-acetyl-*sn*-glycerol, DAG-kinase from *E. coli* and $[^{32}P]$ATP in combination with some synthetic chemical steps, they generated $1\text{-}O\text{-}$alkyl-2-lyso-*sn*-glycero-3-$[^{32}P]$phosphoryl-choline. This lipid was acylated and incorporated into

the PC pool of HL60 cells and human neutrophils. When
these cells were stimulated with fMLP, they released [^{32}P]PA.
Since there was no radiolabel in the cellular ATP pools,
the [^{32}P]PA generated could not have been formed through
a PL-C-DAG-kinase pathway and must have been produced by
a PL-D-dependent mechanism.

This work provided the first unambiguous evidence that
a receptor-linked PL-D was present in mammalian cells, and
supported earlier conclusions from transphosphatidylation
reactions in synaptosomes and hepatocytes [39, 40]. Transphosphatidylation is a unique PL-D-dependent reaction in which
the phosphatidyl moiety of a phospholipid is transferred
to aliphatic alcohols to produce Ptd-alcohols. This property
has now been exploited by many groups to measure PL-D activation
in a number of cell types in response to a variety of
stimulatory agonists [24, 29, 30, 32, 36-50]. Thus there
is an increasing body of evidence directly implicating a
receptor-linked PL-D in the stimulated turnover of PC in
mammalian cells. In contrast, the evidence in favour of
a receptor-linked PL-C-dependent hydrolysis of PC remains
circumstantial.

ROLES OF PHOSPHATIDIC ACID (PA)

The importance of the PL-D-phosphatidate phosphatase
pathway as a source of DAG, as a second messenger, in activated
cells has already been discussed. However, it is conceivable
that PA itself may have a second-messenger function since
it is known (i) to be mitogenic, (ii) to increase intracellular
Ca^{2+} and cGMP concentrations, (iii) to activate PL-A_2, and
(iv) to inhibit cellular adenylate cyclase [51-55]. Moreover,
PL-D activation may be the first step of one or more
putative pathways leading to AA. PC is the major source
of AA in activated cells, and AA is very likely to be
present in the PA released from PC by PL-D activation.
Lapetina and co-authors [56] postulated that AA release in
thrombin-stimulated horse platelets was by a PA-specific PL-A_2
mechanism. Furthermore, the same group showed that in horse
neutrophils a pool of PA that contained AA esterified in
the sn-2 position was rapidly acylated and deacylated [57].
AA could conceivably be released from the PL-D-derived DRG
pool by DAG-lipase in a manner analogous to that proposed
for the PtdInsP$_2$-specific PL-C-derived DAG pool [58].

At present there is no definitive evidence to support
any of the different conjectured pathways for AA release.
Nevertheless, they do provide a different viewpoint to the
current dogma that AA release in activated cells occurs
by a receptor-linked PL-A_2 dependent mechanism. The possibility
that PL-D activation is important for eicosanoid production
warrants follow-up.

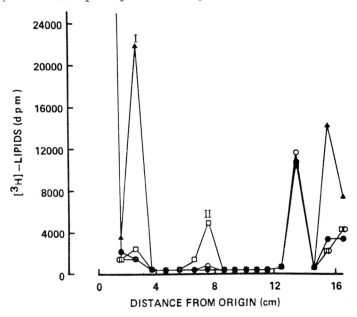

Fig. 2. Formation of 1-*O*-[³H]alkyl-PA and -Ptd-butanol by fMLP-stimulated human neutrophils. After pre-labelling, with the ³H–'lyso' compound as in Fig.1 (details in [32]), cell suspensions were pre-incubated for 5 min without (●,▲) or with (o,□) butanol before adding vehicle (●,o) or fMLP to 100 nM (▲,□). Lipids were separated by TLC and bands 1 cm wide were scraped off the chromatogram and the ³H radioactivity measured by scintillation spectrometry. I, 1-*O*-alkyl-PA; II, 1-*O*-alkyl-Ptd-butanol. *Courtesy of Biochemical Journal (Fig. 3 also).*

MECHANISM AND SIGNIFICANCE OF PL-D ACTIVATION

How PL-D becomes activated is unclear, but it is known that Ca^{2+} ionophores, phorbol esters, DAG, unsaturated fatty acids, NaF and guanine nucleotides will increase PL-D activity in intact cells and membrane preparations [38-40, 42, 45-49, 59-62]. Phorbol ester activation strongly suggests that a PKC-dependent step might be involved, and the action of fluoroaluminates and guanine nucleotides indicates that a GTP-binding protein may also be required. At present there is no direct evidence to suggest that PL-D is coupled to receptors through a GTP-binding protein in a manner analogous to the PtdInsP₂-specific PL-C [1].

The physiological significance of PL-D activation in the human neutrophil has been investigated in our laboratory. As shown in Fig. 2, neutrophils stimulated with fMLP in the presence of ethanol or butanol generate, by a PL-D-dependent transphosphatidylation reaction, Ptd-alcohols at the expense

Fig. 3. Effects of butanol and ethanol (5-min preincubation, 37°) in human neutrophils after stimulant addition. **A** (assays as in [32]): DRG, determined after 5 min (□,o), and Ins(1,4,5)P$_3$, determined after 20 sec (■,●), following addition of fMLP (to 100 nM). Butanol: ■,□; ethanol: ●,o.

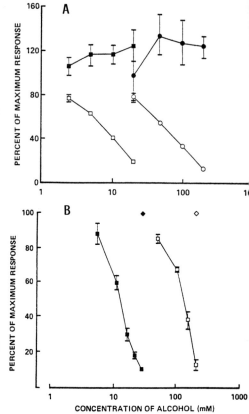

B (assay as in [32]): superoxide production, following addition (to 100 nM) of fMLP (■,□) or phorbol myristate acetate (◆,◇). Butanol: ■,◆; ethanol: □,◇.

Results are means ±S.E.M. for 3 or 4 (**A**) or 3-5 (**B**) separate experiments using neutrophils from different donors; symbols in **A** lacking error bars encompass the S.E.M.

Note on 'DRG': the term refers to 1-O-alkyl, 1-O-alkenyl and diacyl substituted glycerols.

of PA. Under these conditions the production of DRG was almost totally abolished (Fig. 3A). Furthermore, production of superoxide by these cells was also inhibited in a concentration-dependent manner by the alcohols (Fig. 3B). Butanol and ethanol had no effect on superoxide production stimulated by phorbol esters, indicating that the alcohols were not non-specifically blocking the NADH oxidase and disrupting membrane-dependent events (Fig. 3B). Moreover, the alcohols had no effect on InsP$_3$ production, indicating that they did not block PtdInsP$_2$-specific PL-C (Fig. 3A).

These observations strongly suggest that PL-D activation is functionally linked to superoxide production in the human neutrophil. Furthermore, they indicate that most of the DRG released by activated neutrophils arises by the PL-D-phosphatidate phosphatase pathway and that the stimulated degradation of PtdInsP$_2$ by PL-C contributes very little to the total DRG pool. A recent report [63] has suggested that PL-D-derived PA and not DAG is important for superoxide generation by neutrophils. This conclusion was based on the use of propanolol, which inhibits phosphatidate phosphatase; it blocked DRG but not PA production nor O$_2$ consumption.

The importance of PL-D activation in other cells or tissues is currently a matter of speculation. Mitogens are known to activate PL-D in cultured cells, and since PA can stimulate the growth of cells *in vitro* it is possible that PL-D plays an important role in mitogenic signalling [23, 46, 51, 52, 55]. Synaptosomal membranes are enriched in PL-D, and it is also present in some cell lines derived from neuronal tissue; furthermore, PL-D can be activated by the neurotransmitter acetylcholine [28, 45, 48, 59, 60, 64]. Qian & Drewes [28] have suggested that PL-D is part of a novel signal transduction process coupling neuronal muscarinic receptors to cellular responses. It seems a plausible supposition that PL-D plays a role in synaptic transmission events.

Such conclusions about the role of PL-D are currently based upon studies of enzyme distribution and the known action of stimulatory agonists. The generation of more definitive data will ultimately depend upon finding a potent and selective inhibitor for PL-D. A fungal metabolite, wortmannin, has recently been shown to be a potent blocker of PL-D activation in human neutrophils [65]. This compound may prove to be a useful pharmacological tool for examining the physiological significance of PL-D activation in a variety of cells and tissues.

CONCLUDING COMMENTS

The discovery that Ca^{2+}-mobilizing receptors are coupled to a $PtdInsP_2$-specific PL-C was a major advance in our understanding of signal transduction and phospholipid-derived second-messenger generation. Research throughout the 1980's has produced a wealth of information about the InsP's and calcium homeostasis and has been the decade of the inositol phospholipids.[@] The emerging importance of PC metabolism and PL-D in second-messenger production suggests that the 1990's will see our understanding of intracellular signalling advance even further, and the next few years are likely to be the era of the choline phospholipids.

References

1. Berridge, M.J. & Irvine, R. (1989) *Nature 341*, 197-205.
2. Preiss, J., Loomis, C.R., Bishop., W.R., Stein, R., Neidel, J.E. & Bell, R.M. (1986) *J. Biol. Chem. 261*, 8597 8600.
3. Banschbach, M.W., Geison, R.K. & Hokin-Neaverson, M. (1981) *Biochim. Biophys. Acta 663*, 34-45.

[@]*Note by Ed.*- Arts. in which PL-C and $InsP_3$ feature, with diagrams, include #A-1, #B-6 & #B-7. For PC and AA see #B-3.

4. Grove, R.I. & Schimmel, S.D. (1982) *Biochim. Biophys. Acta 711*, 272-280.

5. Ragab-Thomas, J.M-F., Hullin, F., Chap, H. & Douste-Blazy, L. (1987) *Biochim. Biophys. Acta 917*, 388-397.

6. Polverino, A.J. & Barritt, G.J. (1988) *Biochim. Biophys. Acta 970*, 75-82.

7. Uhing, R.J., Prpic, V., Hollenbach, P.W. & Adams, D.O. (1989) *J. Biol. Chem. 264*, 9224-9230.

8. Augert, G., Bocckino, S.B., Blackmore, P.F. & Exton, J.H. (1989) *J. Biol. Chem. 264*, 2574-2580.

9. Horwitz, J. (1990) *J. Neurochem. 54*, 983-991.

10. Farese, R.V., Davis, J.S., Barnes, D.E., Standaert, M.L., Babischkin, J.S., Hock, R., Rosic, N.K. & Pollet, R.J. (1985) *Biochem. J. 231*, 269-278.

11. Truett, A.P., Verghese, M.W., Dillon, S.B.& Snyderman, R. (1988) *Proc. Nat. Acad. Sci. 85*, 1549-1553.

12. Whetton, A.D., Monk, P.N., Consalvey, S.D., Huang, S.J., Dexter, T.M. & Downes, C.P. (1988) *Proc. Nat. Acad. Sci. 85*, 3284-3288.

13. Rosoff, P.M., Savage, N. & Dinarello, C.A. (1988) *Cell 54*, 73-81.

14. Wright, T.M., Ranagan, L.A., Shin, H.S. & Raben, D.M. (1988) *J. Biol. Chem. 263*, 9374-9380.

15. Martins, T.J., Sugimoto, Y. & Erikson, R.L. (1989) *J. Cell Biol. 108*, 683-691.

16. Kester, M., Simonson, Y., Mene, P. & Sedor, J.R. (1989) *J. Clin. Invest. 83*, 718-723.

17. Brenner-Gati, L., Trowbridge, J.M., Moucha, C.S. & Gershengorn, M.C. (1989) *Endocrinology 126*, 1623-1629.

18. Morris, J.D.H., Price, B., Lloyd, A.C., Self, A.J., Marshall, C.J. & Hall, A. (1989) *Oncogene 4*, 27-31.

19. Pelech, S.L. & Vance, D.E. (1989) *Trends Biochem. Sci. 14*, 28-30.

20. Loffenholz, K. (1989) *Biochem. Pharmacol. 38*, 1543-1549.

21. Exton, J.H. (1990) *J. Biol. Chem. 265*, 1-4.

22. Liscovitch, M., Blusztajn, J.K., Freese, A. & Wurtman, R.J. (1987) *Biochem. J. 241*, 81-86.

23. Cook, S.J. & Wakelam, M.J.O. (1989) *Biochem. J. 263*, 581-587.

24. Martin, T.W. & Michaelis, K. (1989) *J. Biol. Chem. 264*, 8847-8856.

25. Wright, T.M., Shin, H.S. & Raben, D.M. (1990) *Biochem. J. 267*, 501-507.

26. Cabot, M.C., Welsh, C.J., Cao, H. & Chabbott, H. (1988) *FEBS Lett. 233*, 153-157.

27. Dunlop, M. & Metz, S.A. (1989) *Biochem. Biophys. Res. Comm. 163*, 922-928.

28. Qian, Z. & Drewes, L.R. (1990) *J. Biol. Chem., 265*, 3605-3610.

29. Halenda, S.P. & Rehm, A.G. (1990) *Biochem. J. 267*, 479-483.

30. Thompson, N.T., Tateson, J.E., Randall, R.W., Spacey, G.D., Bonser, R.W. & Garland, L.G. (1990) *Biochem. J. 271*, 209-213.

31. Truett, A.P., Snyderman, R. & Murray, J.J. (1989) *Biochem. J.* *260*, 909-913.
32. Bonser, R.W., Thompson, N.T., Randall, R.W. & Garland, L.G. (1989) *Biochem. J. 264*, 617-620.
33. Griendling, K.K., Delafontaine, P., Rittenhouse, S.E., Gimbrone, M.A. & Alexander, R.W. (1987) *J. Biol. Chem. 262*, 14555-14562.
34. Sunako, M., Kawahara, Y., Kariya, K., Araki, S., Fukuzaki, H. & Takai, Y. (1989) *Biochem. Biophys. Res. Comm. 160*, 744-750.
35. Cockcroft, S. & Allan, D. (1984) *Biochem. J. 222*, 557-559.
36. Pai, J-K., Siegel, M.I., Egan, R.W. & Billah, M.M. (1988) *J. Biol. Chem. 263*, 12472-12477.
37. as for 36., *Biochem. Biophys. Res. Comm. 150*, 355-364.
38. Billah, M.M., Pai, J-K., Mullmann, T.J., Egan, R.W. & Siegel, M.I. (1989) *J. Biol. Chem. 264*, 9069-9077 (& see 17069-17077).
39. Kobayashi, M. & Kanfer, J.N. (1987) *J. Neurochem. 48*, 1597-1603.
40. Bocckino, S.B., Wilson, P.B. & Exton, J.H. (1987) *FEBS Lett. 225*, 201-204.
41. Mullmann, T.J., Siegel, M.I., Egan, R.W. & Billah, M.M. (1990) *J. Immunol. 144*, 1901-1908.
42. Kinsky, S.C., Loader, J.E. & Benedict, S.H. (1989) *Biochem. Biophys. Res. Comm. 162*, 788-793.
43. Gruchalla, R.S., Dinh, T.T. & Kennerly, D.A. (1990) *J. Immunol. 144*, 2334-2342.
44. Rubin, R. (1988) *Biochem. Biophys. Res. Comm. 156*, 1090-1096.
45. Liscovitch, M. (1989) *J. Biol. Chem. 264*, 1450-1456 (also 11762-11767).
46. Ben-Av, P. & Liscovitch, M. (1989) *FEBS Lett. 259*, 64-66.
47. Tettenborn, C.S. & Mueller, G.C. (1988) *Biochem. Biophys. Res. Comm. 155*, 249-255.
48. Gustavsson, L. & Hansson, E. (1990) *J. Neurochem. 54*, 737-742.
49. Hii, C.S.T., Kokke, Y.S., Pruimboom, W. & Murray, A.W. (1989) *FEBS Lett. 257*, 35-37.
50. Domino, S.E., Bocckino, S.B. & Garbers, D.L. (1989) *J. Biol. Chem. 264*, 9412-9419.
51. Yu, C-L., Tsai, M-H. & Stacey, D.W. (1988) *Cell 52*, 63-71.
52. Imagawa, W., Bandyopadhyay, G.K., Wallace, D. & Nandi, S. (1989) *Proc. Nat. Acad. Sci. 86*, 4122-4126.
53. Ohsaka, S. & Deguchi, T. (1983) *FEBS Lett. 152*, 62-65.
54. Harris, R.A., Schmidt, J., Hitzemann, B.A. & Hitzemann, R.J. (1981) *Science 212*, 1290-1291.
55. Van Corven, E.J., Groenink, A., Jalink, K., Eichholtz, T. & Moolenaar, W.H. (1989) *Cell 59*, 45-54.
56. Lapetina, E.G., Billah, M.M. & Cuatrecasas, P. (1981) *Nature 292*, 367-369.
57. as for 56. (1980) *J. Biol. Chem. 255*, 10966-10970.
58. Bell, R.L., Kennerly, D.A., Stanford, N. & Majerus, P.W. (1979) *Proc. Nat. Acad. Sci. 76*, 3238-3241.
59. Chalifour, R. & Kanfer, J.N. (1982) *J. Neurochem. 39*, 299-305.
60. Hattori, H. & Kanfer, J.N. (1985) *J. Neurochem. 45*, 1578-1584.
61. Anthes, J.C., Eckel, S., Siegel, M.I., Egan, R.W. & Billah, M.M. (1989) *Biochem. Biophys. Res. Comm. 163*, 657-664.

62. Qian, Z. & Drewes, L.R. (1989) *J. Biol. Chem. 264,* 21720–21724.

63. Rossi, F., Grzeskowiak, M., Della Bianca, V., Calzetti, F. & Gandini, G. (1990) *Biochem. Biophys. Res. Comm. 168,* 320–327.

64. Martinson, E.A., Goldstein, D. & Heller Brown, J. (1989) *J. Biol. Chem. 264,* 14748–14754.

65. Reinhold, S.L., Prescott, S.M., Zimmerman, G.A. & McIntyre, T.M. (1990) *FASEB J. 4,* 208–214.

#A-7

ASSESSMENT OF RECEPTOR-LINKED PHOSPHOLIPASE D
ACTIVITY USING A NOVEL ASSAY

Neil T. Thompson, Robert W. Bonser
and Lawrence G. Garland

Wellcome Research Laboratories,
Beckenham, Kent BR3 3BS, U.K.

Hydrolysis of PC* by a receptor-linked PL-D has been observed in various types of cell. PL-D catalyzes a unique transphosphatidylation reaction which in the presence of aliphatic alcohols produces Ptd-alcohols. This property has been exploited by others to measure agonist-dependent activation of PL-D [1]. With this technique, cellular phospholipids must be pre-labelled before cells are stimulated in the presence of a high concentration of an alcohol (e.g. 200 mM ethanol). We have improved the assay by using [³H]butan-1-ol of high specific activity and measuring its conversion to Ptd-But in human neutrophils. Time- and concentration-dependent conversion occurred rapidly in the presence of fMLP and opsonized zymosan; the reaction preceded the release of diradylglycerols. Advantages of the use of high specific activity [³H]butanol are discussed. Besides neutrophils, human platelets and mouse fibroblasts in culture have been assayed by this technique.

PL-D* has been identified in a variety of mammalian cells (e.g. rat brain [2] and human eosinophils [3]), but enzyme activity is generally revealed only in the presence of a detergent or free fatty acid [1]. Recent evidence that a PL-D activity specific for PC may be involved in signal transduction events has generated much interest in this poorly characterized enzyme [4]. PL-D normally hydrolyzes phospholipids to PA and the free base. Besides, the enzyme also catalyzes transphosphatidylation reactions [5] whereby the phosphatidyl portion of the phospholipid is transferred to a nucleophilic acceptor other than a water molecule (Scheme 1).

Short-chain aliphatic alcohols are very good acceptors and are converted to the corresponding Ptd-alcohol. Bases such as ethanolamine are very poor acceptors, thus distinguishing the transphosphatidylation reaction from base exchange.

**Abbreviations.-* PL (as in PL-D), phospholipase; PA, phosphatidic acid; Ptd *(as prefix)* = phosphatidyl, or merely P as in PC (choline) & PE (ethanolamine); But, butanol. DMSO, dimethylsulphoxide; fMLP, fMet-Leu-Phe; LT (as in LTB₄), leukotriene; PAF, platelet activating factor; PMA, phorbol 12-myristate-13-acetate. *Media: see text;* HB(B) = Hepes-buffered (Hank's balanced salt solution, 'B'); DMEM = Dulbecco's modified Eagle medium; PBS = phosphate-buffered saline. BSA, bovine serum albumin.

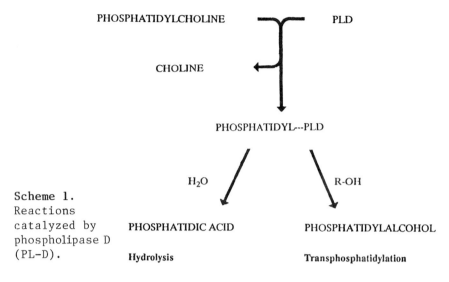

Scheme 1.
Reactions
catalyzed by
phospholipase D
(PL–D).

Transphosphatidylation is unique to PL–D [5]; the physiological significance of this property is, however, unknown because Ptd-alcohols are not normally found in mammalian tissues. Billah & co-workers have exploited transphosphatidylation as a means to identify agonist-dependent PL–D activity and have demonstrated that HL60 granulocytes [6] and human neutrophils [7] stimulated in the presence of 85 mM ethanol generate Ptd-ethanol. In cells in which the 1-*O*-alkyl-2-acylphosphatidylcholine (1-*O*-alkyl-PC) pool had been radiolabelled both with ^{32}P and with ^3H, they demonstrated the formation of dual-labelled 1-*O*-[^3H]alkyl-[^{32}P]phosphatidylethanol, indicating that the molecule was derived from the PC pool by PL–D action [1, 7].

Measurement of transphosphatidylation using pre-labelled cells of various types has identified an agonist-dependent PL–D, suggesting that this enzyme has a significant role in signal transduction (see accompanying art. by Bonser *et al.*). A detailed evaluation of this role requires an assay which can be used reliably in a variety of tissues and cells, for which the pre-labelling method has been used successfully whilst having some disadvantages. These arise from the difficulties associated with labelling phospholipid pools in certain sample types and the likelihood that the high concentrations of alcohol required to produce measurable levels of Ptd-alcohols may be toxic to some cells. Other methods of measuring PC hydrolysis by PL–D rely on following the formation of either PA or choline. Measuring the formation of these products is not a satisfactory assay for PL–D activity because both may be derived by alternative pathways. We have therefore attempted to improve upon existing transphosphatidylation assays by using [^3H]butan-1-ol of high specific activity and following the agonist-dependent incorporation of this label into Ptd-[^{3}H]But. It is anticipated that the modified assay will serve for a wider variety of sample types as a measure of agonist-dependent PL-D. This article describes the characterization of the assay in human neutrophils and its use in a number of different sample types including vascular smooth-muscle rings and monolayers of cultured fibroblasts.

MATERIALS AND METHODS

Materials were as follows.- Radiochemicals, from Amersham International: [^{3}H]butan-1-ol (12 Ci/mmol; 100 mCi/ml solution in PBS kept at 4°), 1-stearoyl-2-[^{14}C]arachidonoylphosphatidyl-choline (56 mCi/mmol) and 1-O-[^{3}H]octadecyl-2-lysophosphatid-ylcholine (80 mCi/mmol). Organic solvents, of AnalaR grade: BDH Chemicals (Poole, U.K.). Other reagents: Sigma (Poole).

Sample preparation and incubation

Human peripheral blood **neutrophils** were purified [8] and suspended in 'H' buffered with 30 mM Hepes, pH 7.2 (HBH). Where indicated, the 1-O-alkyl-PC pool was labelled by incuba-tion with 1-O-[^{3}H]octadecyl-2-lysophosphatidylcholine as described by Pai *et al*. [6]. As already described [9], cells (10^7) were incubated with 5 μM cytochalasin B and, where indicated, 100 μCi [^{3}H]butan-1-ol for 5 min at 37° before adding the agonist. fMLP, PMA and LTB were dissolved in DMSO such that the final DMSO concentration did not exceed 0.2% (v/v). Opsonized zymosan and PAF were dissolved in HBH, which for the latter contained 2 mg/ml BSA.

Mononuclear cells (human peripheral) were prepared by differential centrifugation on Ficoll-Paque and treated essenti-ally as for neutrophils (above). **Leucocytes** (rat peritoneal) were isolated by peritoneal lavage as described [10] except that the rats were killed by asphyxiation with CO_2. Incubations were carried out as for neutrophils. **Washed platelets** (human [11]) were suspended in Tyrode solution (no Ca^{2+}) containing 5 mM Hepes pH 7.4 (HBT). Platelets (0.5×10^8) were incubated as for neutrophils except that 200 μCi [^{3}H]butan-1-ol was used and $CaCl_2$ was adjusted to 1 mM.

Arterial rings (1 mm; 3 per assay) were prepared from rabbit femoral arteries washed in Krebs' solution, the endo-thelium being removed by rubbing the internal surface with a fine-gauge wire. Pre-incubation was in 0.5 ml Krebs' solution containing 200 μCi [^{3}H]butan-1-ol for 5 min at 37° before stimulation with PMA. **Fibroblasts** (Swiss 3T3) were grown to confluence in 10 cm petri dishes, then washed for 20 min in DMEM buffered with 20 mM Hepes pH 7.4 containing 1% BSA (HBD). HBD (5 ml) containing 500 μCi [^{3}H]butan-1-ol was then added to each dish; after incubation for 5 min, the stimulus was introduced.

Extraction, TLC and choline measurement

Reactions with fibroblasts were terminated by aspiration of the HBD containing the unincorporated [^{3}H]butan-1-ol followed by the addition of 1.5 ml cold (-80°) methanol. Cells were scraped from the dish and washed with a further 1.5 ml methanol into a polypropylene tube to which 3 ml of chloroform was added. For the other sample types the reaction was stopped by adding 2 ml ice-cold buffer (HBH, HBT or Krebs' as appropriate) followed by a brief centrifugation at 700 g. The supernatant containing the bulk of the [^{3}H]butan-1-ol was removed, and the pellet dispersed in 1.5 ml of chloroform/methanol (1:2 by vol.) and 0.5 ml of the appropriate buffer. For the arterial rings extraction was effected by homogenization in a glass homogenizer with a teflon plunger.

Lipids from all types of sample were then extracted [12], and the chloroform layers evaporated to dryness under a N_2 stream. Residues were dissolved in 1 ml chloroform and washed 3 times with 2 ml of the upper phase from the solvent system chloroform/methanol/1 M NaCl/H_2O (2:2:1:1 by vol.); radiolabelled products were separated by TLC on silica gel plates (Whatman) in solvent system 1 which comprised the organic phase of 2,2,4-trimethylpentane/ethyl acetate/acetic acid/H_2O (5:11:2:10). For the analysis of 1-O-[^{3}H]alkylPA from 1-O-[^{3}H]alkylPC-labelled neutrophils, the chloroform layers from the solvent-extracted cells were analyzed directly by TLC on silica gel plates using solvent system 2, which comprised chloroform/methanol/propan-1-ol/ethyl acetate/33 mM (0.25%) KCl (25:13:25:9).

Choline was measured by our modification [13] of the method of Wang & Haubrich [14].

RESULTS AND DISCUSSION

Studies with neutrophils

Neutrophils stimulated in the presence of as little as 16 µM [^{3}H]butan-1-ol produced several labelled products. At this concentration <0.01% of the radiolabelled substrate was incorporated into these products, thus necessitating the use of a very high specific activity [^{3}H]butan-1-ol and an efficient procedure for removing non-volatile contaminants. This was achieved by removing the cells from the bulk of the [^{3}H]butan-1-ol (e.g. by centrifugation) before the solvent-extraction step, followed by extensive washing of the chloroform phase and thorough drying under a N_2 stream. This resulted in the TLC profile shown in Fig. 1, wherein the background radioactivity was <100 dpm [^{3}H]/cm.

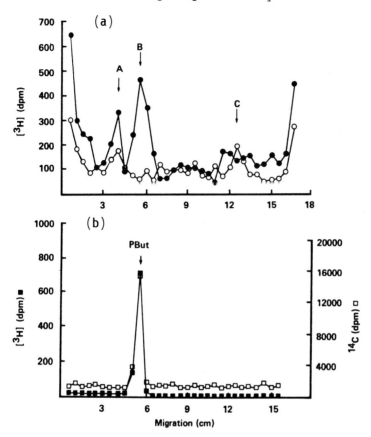

Fig. 1. TLC profile of lipids from resting and fMLP-stimulated human neutrophils (**a**), pre-incubated with [^{3}H]butan-1-ol and cytochalasin B for 5 min prior to addition of fMLP (to 0.3 μM; ●) or vehicle control (o). After a further 5 min the reaction was stopped, then the lipids were extracted and washed (see text) before TLC in solvent system 1. Scraped-off 1 cm bands were counted for [^{3}H]. In (**b**), 1-stearoyl-2-[^{14}C]arachidonoyl-sn-phosphatidyl[^{3}H]butanol was similarly analyzed (see text).

When extracts from stimulated cells were analyzed by TLC using solvent system 1, we identified two sharp peaks of radioactivity with R_F's of 0.24 (A) and 0.33 (B), one broad peak (C) of R_F between 0.66 and 0.84, and (number unknown) other radiolabelled products which either remained at the origin or ran at the solvent front (Fig. 1a). Peak B was identified as Ptd-But from its co-migration with a dual-labelled 1-stearoyl-2-[^{14}C]arachidonoylphosphatidyl-[^{3}H]-butanol standard prepared by incubating cabbage PL-D with 1-stearoyl-2-[^{14}C]arachidonoyl-PC in the presence of [^{3}H]butan-1-ol (Fig. 1b). The formation of this novel product in

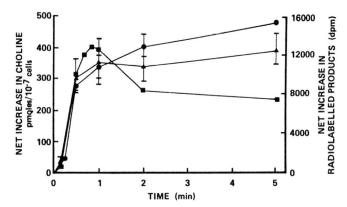

Fig. 2. Time course of product formation due to PL-D in fMLP-stimulated neutrophils, pre-incubated (before adding fMLP) as for Fig. 1 (▲; formation of Ptd-[³H]But) or with the ³H precursor (see text) to label the 1-O-alkyl-PC pool (■; formation of 1-O-[³H]alkyl-PA). Products including choline (●), were analyzed at the indicated times. Values are mean ±S.E.M. (n = 3) or, for 1-O-[³H]alkyl-PA, averaged duplicates from 1 typical expt. (out of 3).

the human neutrophil was absolutely dependent upon stimulation. Peak A, on the other hand, was present in both resting and stimulated cells although in some cases the levels were slightly higher following stimulation. The identity of this peak is unknown. The less polar material (peak C) could also be found in extracts prepared from cell-free incubations and was identified as a contaminant of the [³H]butan-1-ol stock which was not completely removed by the washing procedures (Fig. 1a).

Ptd-[³H]But accumulated rapidly in fMLP-stimulated neutrophils, reaching maximum levels after 1 min (Fig. 2). The amount of choline also increased rapidly under these conditions, becoming maximal after 1-2 min (Fig. 2). Production of PA, monitored by measuring formation of 1-O-[³H]alkyl-PA pre-labelled in the PC pool with 1-O-[³H]alkyl-2-lysoPC (lysoPAF), followed a similar time course, peaking 50 sec after stimulation. At later times, 1-O-[³H]alkyl-PA levels decreased, in contrast to Ptd-[³H]But and choline levels which remained elevated (Fig. 2). In all sample types we have studied to date, the level of Ptd-[³H]But is sustained, consistent with the suggestion that this novel metabolite is resistant to further breakdown [15].

Agents besides fMLP stimulated [³H]butan-1-ol incorporation into Ptd-[³H]But by neutrophils. Stimulation occurred after treatment with PAF, LTB₄, opsonized zymosan, NaF and PMA, supporting the proposal that PL-D may be widely involved in signal trans-

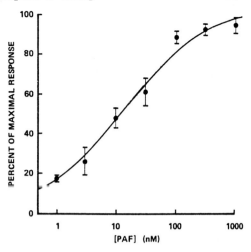

Fig. 3. PAF-stimulated [^{3}H]Ptd-But production in human neutrophils, pre-incubated as for Fig. 1. After 5 min with PAF at the indicated concentrations, extracted lipids were analyzed for Ptd-[^{3}H] But (see text). Data are mean ±S.E.M. from at least 4 different experiments.

duction. The actions of fMLP, LTB$_4$, C$_{5a}$ and PMA have been seen previously by measuring production of Ptd-[^{3}H]But in neutrophils labelled with 1-O-[^{3}H]alkyl-2-1ysoPC or [^{3}H]oleate [7, 16-18]. The [^{3}H]butan-1-ol technique was, however, particularly useful for measuring the response induced by PAF. Labelling of the 1-O-alkylPC pool requires pre-incubation with [^{3}H]1ysoPAF, a process which could generate significant amounts of PAF itself and so lead to down-regulation of PAF receptors. Making use of [^{3}H]butan-1-ol we could demonstrate that PAF induced a concentration-dependent activation of PL-D, with half-maximal activation occurring at 13 nM (Fig. 3). Furthermore, the extent of PL-D activation by PAF was similar to that caused by fMLP.

    Sample types besides neutrophils were investigated for stimulus-dependent Ptd-[^{3}H]But formation, to explore the useful-ness of the [^{3}H]butan-1-ol technique as a measure of cellular PL-D activity. Human platelets, human peripheral mononuclear cells, Swiss 3T3 fibroblasts, rat peritoneal leucocytes and smooth-muscle rings prepared from denuded rabbit femoral artery all respond to PMA by incorporating [^{3}H]butan-1-ol into Ptd-[^{3}H]-But (Table 1). Although the significance of PMA-stimulated PL-D activation is not known, these results demonstrate the capacity of the technique to detect PL-D activity in different sample types including cell monolayers and an intact tissue (the muscle rings). Agonist-dependent formation of Ptd-[^{3}H] But has also been detected in human platelets stimulated by thrombin or collagen [9], in rat peritoneal leucocytes stimulated by PAF and in bombesin-stimulated fibroblasts (S. Cook and M.J.O. Wakelam, pers. comm.). The amount of [^{3}H]butan-1-ol in the platelet and vascular smooth-muscle experiments was increased to 400 μCi/ml to increase the levels of Ptd-[^{3}H]But formed, hardly affecting the TLC profile.

Table 1. Ptd-[^{3}H]But (dpm) in cells of different types, after the usual pre-incubation with [^{3}H]butan-1-ol and exposure for 30 min, prior to solvent extraction, to PMA or DMSO vehicle. Each value is an averaged duplicate from 1 experiment but is representative of at least 3 experiments in all.

Cellular material	PMA, nM	Product, ^{3}H: + Vehicle	+ PMA
Human mononuclear cells	100	97	500
Human platelets	1000	1183	5994
Swiss 3T3 fibroblasts	1000	1688	20844
Rabbit femoral artery	1000	123	503
Rat peritoneal leucocytes	1000	348	3725

CONCLUDING COMMENTS

Production of novel Ptd-alcohols by transphosphatidylation provides a very useful and specific assay for cellular PL-D. We have demonstrated here how high specific activity [^{3}H]butan-1-ol may be used to exploit this property and measure PL-D activity in a range of cell types. In support of the proposal that PL-D mediates Ptd-But formation we found stimulus-dependent formation of choline and of Ptd-[^{3}H]But from [^{3}H]butan-1-ol have similar time courses (Fig. 2). The rate of formation of PA, as judged from 1-O-[^{3}H]alkyl-PA accumulation in 1-O-[^{3}H]alkyl-PC labelled cells, was also consistent with this.

The advantages of the use of [^{3}H]butan-1-ol as a measure of transphosphatidylation can be summarized thus.-
- 1. The method avoids the need for a labelling step and is not limited by existing labelling techniques.
- 2. The responsiveness of the test material is not compromized by prolonged incubation in the presence of potential bioactive mediators.
- 3. The use of [^{3}H]butan-1-ol does not distinguish between the phospholipid substrates and therefore allows total PL-D activity to be measured.
- 4. The high specific activity of the [^{3}H]butan-1-ol allows its concentration in the medium to be as low as 16 μM, reducing toxicity problems.
- 5. With a low alcohol concentration, only a small percentage of the phospholipid substrate is converted to Ptd-But, and hence total phospholipid metabolism through the PL-D pathway is not substantially altered.

The use of [^{3}H]butan-1-ol should be a useful addition to existing techniques for evaluating the physiological significance of PL-D activation in different cell types.

References

1. Pai, J-K, Siegel, M.I., Egan, R.W. & Billah, M.M. (1988) *J. Biol. Chem. 263*, 12472-12477.
2. Kobayashi, M. & Kanfer, J.N. (1987) *J. Neurochem. 48*, 1597-1603.
3. Kater, L.A., Goetzel, E.J. & Austen, K.F. (1976) *J. Clin. Invest. 57*, 1173-1180.
4. Exton, J.H. (1990) *J. Biol. Chem. 265*, 1-4.
5. Dawson, R.M. (1967) *Biochem. J. 102*, 205-210.
6. Pai, J-K., Siegel, M.I., Egan, R.W. & Billah, M.M. (1988) *Biochem. Biophys. Res. Comm. 150*, 355-365.
7. Billah, M.M., Eckel, S., Mullmann, T.J., Egan, R.W. & Siegel, M.I. (1989) *J. Biol. Chem. 264*, 17069-17077.
8. Tateson, J.E., Randall, R.W., Reynolds, C.H., Jackson, W.P., Bhattacherjee, P., Salmon, J.A. & Garland, L.G. (1988) *Br. J. Pharmacol. 94*, 528-539.
9. Randall, R.W., Bonser, R.W., Thompson, N.T. & Garland, L.G. (1990) *FEBS Lett. 264*, 87-90.
10. Blackwell, G.J., Carnuccio, R., Di Rosa, M., Flower, R.J., Parente, L. & Persico, P. (1980) *Nature 287*, 147-149.
11. Thompson, N.T., Scrutton, M.C. & Wallis, R.B. (1986) *Eur. J. Biochem. 161*, 399-408.
12. Bligh, E.G. & Dyer, W.J. (1959) *Can. J. Biochem. Physiol. 37*, 911-917.
13. Thompson, N.T., Tateson, J.E., Randall, R.W., Spacey, G.D., Bonser, R.W. & Garland, L.G. (1990) *Biochem. J. 271*, 209-213.
14. Wang, F.L. & Haubrich, D.R. (1975) *Anal. Biochem. 63*, 195-201.
15. Billah, M.M., Pai, J-K., Mullman, T.J., Egan, R.W. & Siegel, M.I. (1989) *J. Biol. Chem. 264*, 9069-9076.
16. Mullman, T.J., Siegel, M.I., Egan, R.W. & Billah, M.M. (1990) *J. Immunol. 144*, 1901-1908.
17. Gelas, P., Ribbes, G., Record, M., Terce, F. & Chap, H. (1989) *FEBS Lett. 251*, 213-218.
18. Reinhold, S.L., Prescott, S.M., Zimmerman, G.A. & McIntyre, T.M. (1990) *FASEB J. 4*, 208-214.

#A-8

TRANSPORT OF HYDROPHILIC PROTEINS ACROSS BIOLOGICAL MEMBRANES: USE OF DIPHTHERIA TOXIN AS A MODEL PROTEIN

I.H. Madshus & J.Ø. Moskaug

Department of Biochemistry, Institute for Cancer Research,
The Norwegian Radium Hospital,
Montebello, 0310 Oslo 3, Norway

Bacterial and plant toxins are the only examples of proteins known to enter the cytosol of eukaryotic cells when added in the medium. DT* has for several years been employed as a model protein in studies of protein translocation across biological membranes. DT is synthesized as a single polypeptide chain by pathogenic strains of Cornybacterium diphtheriae, *but the toxin is activated by proteolytic cleavage, yielding disulphide-linked fragments A and B. Entry can easily be assayed, because fragment A has ADP-ribosyl transferase activity, which inactivates elongation factor 2. Fragment B is responsible for receptor binding and membrane interaction.*

Both native DT and fragment B alone form cation-selective channels, measured as an influx of $^{22}Na^+$, *when cells with surface-bound protein are exposed to low pH. The capacity for channel formation serves as a measure of whether the protein can be inserted properly into the membrane, and can be used in the repertoire of functional tests for DT mutants and DT-derived fusion proteins.*

Translocation of DT to the cytosol normally occurs from endosomes, which have low pH due to proton-translocating ATPases in the limiting membrane [1]. Entry of surface-bound DT can be induced directly across the p.m.* by incubating cells in medium of low pH, thereby mimicking conditions inside endosomes [2, 3]. Low pH is required for a change of conformation of the toxin that causes hydrophobic regions to be exposed and thus allows its insertion into lipid membranes [4-7].

Protein translocation can be studied in greater detail by using radiolabelled material and investigating resistance to proteolytic enzymes added externally to intact cells. When radiolabelled DT is bound to specific cell-surface receptors and pulsed through the surface membrane at low pH,

---
Abbreviations: DT, diphtheria toxin, usually 'nicked' (nDT); SDS, sodium dodecyl sulphate; PAGE, polyacrylamide gel electrophoresis; p.m., plasma membrane.

the 25 kDa C-terminal portion the 37 kDa fragment B becomes inserted into the membrane and thereby protected against proteolytic enzymes in the medium, while fragment A (21 kDa) is released to the cytosol [8, 9]. Both native DT and fragment B alone form cation-selective channels that allow influx of $^{22}Na^+$, when cells with surface-bound protein are exposed to low pH [10, 11].

EXPERIMENTAL SYSTEMS FOR STUDYING MEMBRANE TRANSLOCATION OF DIPHTHERIA TOXIN

Cytotoxicity

When DT is added to sensitive cells with specific receptors, protein synthesis declines in a dose-dependent manner after a lag time because of the inactivation of elongation factor 2 [12]. This decline is measurable by comparing [^3H]leucine incorporation into intoxicated cells with that into control cells.

Entry of endocytosed DT into the cytosol can be inhibited by lysosomotropic drugs, such as NH_4Cl, because low pH is required for the toxin to change conformation and insert into the membrane [4-7]. When NH_4Cl is added to the medium at neutral or slightly alkaline pH, NH_3 will rapidly diffuse across the membranes, and within acidic vesicles NH_3 associates with H^+, thereby causing an increase in intravesicular pH (Fig. 1).

Since entry of this toxin can be induced directly across the surface membrane at low pH, it is possible to study in more detail what conditions are required for translocation [2, 3]. A pre-requisite for actually studying entry from the surface is that endocytosis should not have occurred simultaneously with the binding. To avoid endocytic uptake, the binding step can be performed at 4°, since endocytosis is blocked at this temperature.

In some experiments that have appeared in the literature, the 'binding' of DT was at 37° in the presence of NH_4Cl, and the cells were subsequently acid-pulsed in the presence of NH_4Cl, the idea being that the presence of NH_4Cl would keep intravesicular pH elevated. However, since the net effect of decreasing extracellular pH in the presence of NH_4Cl is the same as that of removing NH_4Cl from the medium (see Fig. 1), the phenomenon studied in these cases was entry of DT both from intracellular organelles and directly from the cell surface.

When pH is lowered in the presence of NH_4^+, the intracellular pH decreases, because the equilibrium between NH_3 and

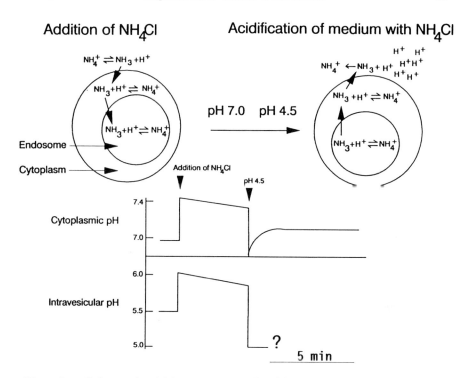

Fig. 1. Schematic illustration of effects on pH, cytosolic and intravesicular, of NH_4^+ added to the medium - preserving its neutrality *(left)* or rendering it acidic *(right)*.

NH_4^+ is shifted towards NH_4^+ resulting in very small amounts of NH_3 outside the cell. The immediate effect of this shift is a steep gradient for NH_3 out of the cell. As NH_3 penetrates membranes, it will rapidly diffuse out of the cell, thereby driving the dissociation of NH_4^+ into NH_3 and H^+. H^+ will be left behind, and the result is that intravesicular pH will be even lower than prior to addition of NH_4^+ [13]. *Cytosolic pH* will be regulated back to its normal value due to pH-regulatory ion-transport mechanisms in the membrane, while such mechanisms have not been demonstrated in the regulation of *intravesicular pH*. It is likely that due to leakage phenomena and a reduced H^+-ATPase activity endosomal pH will return in time to the original value.

Membrane insertion/translocation assays

When radiolabelled DT is added to sensitive cells at 4°, radiolabelled full-length DT can be recovered after cell lysis and visualized using SDS-PAGE and autoradiography. If the gel is run under non-reducing conditions, it can be seen that the toxin has not been reduced (Fig. 2, left).

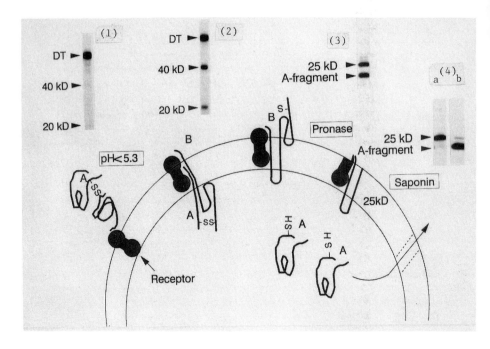

Fig. 2. Illustration of the translocation assay (A and B being the 2 portions of nDT): schematic outline of the translocation process *(below)*, and *(above)* lane patterns seen when the cells are lysed and the acid-precipitable material is analyzed by SDS-PAGE and autoradiography after each of the 4 steps, viz.: (1) binding of ^{125}I-nDT (at 4°); (2) exposure to low pH (at 37°); (3) pronase E treatment; then (4) saponin fractionation of the cells.
#The 'kD' values (really kDa) 40 and 20 correspond to the 37 (B) and 21 (A) kDa values mentioned early in the text.
#nDT, nicked diphtheria toxin (trypsin treatment; cf. [14]).

When cells with surface-bound toxin are exposed to medium of pH 4.5 at an elevated temperature (37°), a fraction of the toxin becomes exposed to cell-mediated reduction and will appear, after SDS-PAGE, as two fragments even under non-reducing conditions (Fig. 2, lane 2).

Pronase E is a non-specific proteolytic enzyme which can be used to remove, by degradation, proteins outside the cell's limiting membrane. When cells with surface-bound radiolabelled nDT are incubated at pH 4.5 for 2 min and then treated with pronase, two protected fragments can be observed when the cell lysate is subjected to SDS-PAGE and autoradiography (Fig. 2, lane 3). In order to investigate whether

a polypeptide is associated with the membrane fraction or
with the supernatant (cytosolic) fraction, cells can be treated
with concentrations of saponin high enough to release marker
enzymes from the cytosol. One of the bands (25 kDa) was
retained in the membrane fraction of permeabilized cells,
while the other band (21 kDa) was associated with the supernatant fraction (Fig. 2, lane 4b). The 21 kDa band was identified as fragment A, because it has ADP-ribosylating activity
[8]. This demonstrates that fragment B of nDT gives rise
to a fragment inserted into the membrane at low pH, while
fragment A is translocated to the cytosol when the toxin
is exposed to low pH and the inter-chain disulphide bridge
is reduced.

It should be noted that resistance to proteolysis does
not necessarily mean that proteins are *in* the membrane or
have passed through it. There is always the possibility
that some part of the protein is exposed at the surface
even if it cannot be degraded by non-specific proteolytic
enzymes. The protein can be tightly folded up or closely
associated with its cellular receptor. In our hands the
efficiency of proteolytic digestion also shows day-to-day
variations. It should also be noted that, in the case of
DT, the sensitivity to proteases changes with low pH, due
probably to the conformational changes of the molecule. It
is conceivable that removal of non-translocated toxin by
pronase E to some extent occurs by degradation of the cellular
receptor. However, since the receptor has not yet been
identified, this point cannot be addressed.

Channel formation assays

DT was shown to form cation-selective channels in the
membrane of sensitive cells at low pH [10, 11]. The channels
released intracellular K^+, resulting in reduced intracellular
concentrations of K^+. Moreover, the channels were open
to Na^+, shown by influx of $^{22}Na^+$. When indicators of intracellular pH were used, it could be observed that at low
pH intracellular pH declined faster in the presence than
in the absence of DT, demonstrating that also protons could
pass the channel. Anions did not flow through it, nor
did sucrose or other uncharged solutes [10, 11].

Apparently, specific receptors are indispensable for toxin-
induced formation of cation-selective channels in cells
[15]; yet DT has previously been shown to form channels
in lipid bilayer membranes not possessing specific receptors
[5, 16]. This apparent contradiction may be due to different
sensitivities of the assay systems used. In a planar lipid
bilayer system, formation of *one single* channel by DT insertion
creates a detectable current, while the same sensitivity

cannot be achieved when $^{22}Na^+$ flux is assayed. Perhaps DT bound non-specifically to cells, as at low pH, occasionally becomes inserted into the p.m., and maybe such insertion occurs much less efficiently than that of specifically bound toxin (toxin associated with lipid only). Electrophysiological studies on whole cells exposed to DT and low pH would be useful to clarify this issue.

Recently it was shown that fragment B alone was capable of channel formation, and that this channel was even more efficient than that of whole DT [17]. The capacity for channel formation is a valuable measure of proper insertion into membranes and can be included in the repertoire of functional tests for DT mutants and fusion proteins derived from DT.

References

1. Morris, R.E., Gerstein, A.S., Bonventre, P.F. & Saellinger, C.B. (1985) *Infect. Immun. 50*, 721-727.
2. Sandvig, K. & Olsnes, S. (1980) *J. Cell Biol. 87*, 828-832.
3. Draper, R.K. & Simon, M.I. (1980) *J. Cell Biol. 87*, 849-854.
4. Sandvig, K. & Olsnes, S. (1981) *J. Biol. Chem. 256*, 9068-9076.
5. Kagan, B.L., Finkelstein, A. & Colombini, M. (1981) *Proc. Nat. Acad. Sci. 78*, 4950-4954.
6. Blewitt, M.G., Chung, L.A. & London, E. (1985) *Biochemistry 24*, 5458-5464.
7. Montecucco, C., Schiavo, G. & Tomasi, M. (1985) *Biochem. J. 231*, 123-128.
8. Moskaug, J.Ø., Sandvig, K. & Olsnes, S. (1988) *J. Biol. Chem. 263*, 2518-2525.
9. Moskaug, J.Ø., Stenmark, H. & Olsnes, S. (1991) *J. Biol. Chem. 266*, 2652-2659.
10. Papini, E., Sandona, D., Rappuoli, R. & Montecucco, C. (1988) *EMBO J. 7*, 3353-3359.
11. Sandvig, K. & Olsnes, S. (1988) *J. Biol. Chem. 263*, 12352-12359.
12. Pappenheimer, A.M., Jr. (1977) *Annu. Rev. Biochem. 46*, 69-94.
13. Boron, W.F. & De Weer, P. (1976) *J. Gen. Physiol. 67*, 91-112.
14. Moskaug, J.Ø., Sandvig, K. & Olsnes, S. (1987) *J. Biol. Chem. 262*, 10339-10345.
15. Stenmark,, H., Olsnes, S. & Sandvig, K. (1988) *J. Biol. Chem. 263*, 13449-13455.
16. Donovan, J.J., Simon, I., Draper, R.K. & Montal, M. (1988) *Proc. Nat. Acad. Sci. 78*, 172-176.
17. Stenmark, H., McGill, S., Olsnes, S. & Sandvig, K. (1989) *EMBO J. 8*, 2849-2853.

#ncA

NOTES and COMMENTS relating to

THE SIGNALLING SCENE, AND RESPONSE INITIATION

This subsection opens with supporting ('nc') articles.

From p. 97   there is Forum discussion material.

From p. 100 there is supplementary material provided by
the Editor (see also p. 97).

Consult the start of the main Contents list concerning
the book structure and inescapable compromises in the section
assignment for certain contributions whose wide-ranging
subject-matter does not fall neatly into one section.
In particular, some 'supplementary material' that arguably
falls in topic-area **#A** will be found elsewhere, e.g. in
subsection '**#ncE**'.

#ncA-1

A Note on

THE INTERACTION OF CHOLERA TOXIN WITH VERO CELLS

G. Bastiaens, H.J. Hilderson[†], E. Dams, G. Van Dessel,
A. Lagrou, W. Dierick and M. De Wolf

RUCA-Laboratory for Human Biochemistry and
UIA-Laboratory for Pathological Biochemistry,
University of Antwerp,
Groenenborgerlaan 171, B-2020 Antwerp, Belgium

Subcellular fractionation techniques have been used to assess the localization of ^{125}I-labelled CT* taken up by Vero cell cultures[⊗] in the light of the hypothesis [1-3] that internalization of the toxin is required for the generation of the active A_1 peptide. CT internalization and its endocytic pathway were followed by differential pelleting in the classical mode that yields **N, M, L, P** and **S** fractions, three of which were re-run isopycnically (legend to Fig. 1). At the outset, labelled CT (or CT_B) was allowed to bind to the cells for 30 min at 4°, thus minimizing endocytosis, whereafter the cells were re-cultured (Fig. 1 legend).

Recoveries.- After differential pelleting it was found that as a function of re-culture time increasing amounts of radioI were recovered from **P** up to 60 min re-culture (7.6-15%), whereafter radioI dropped to zero-time levels (9.6% after 10 h). **M** showed a similar pattern except that the maximum (6.4-14.4%) was at 30 min. Recovery in **S** was maximal (5-26%) after 2 h and back to zero-time levels at 10 h. In **L**, however, radioI recovery was 70% at zero time and fell to 30% at 2 h and 24% at 10 h. During re-culture much of the radioI was progressively released into the culture medium: 5% at 30 min and up to 65% at 10 h; 80% of the radioI released was non-TCA-precipitable.

Marker distributions.- As the radiolabel was moving from **L** to **P**, **M** and **S** during re-culture, it was necessary to subject the particulate fractions involved to density-gradient centrifugation in order to dissect the intracellular pathways in more detail. Zero-time distribution profiles for various marker enzymes are shown in Fig. 1 (a-g & j). As shown (in h, i, k & l) for several of them, most showed little change upon re-culture. In **L**, containing most of the cell-surface membranes, their markers distributed as in the homogenate, but

[†]addressee for any correspondence

Abbreviations (by Editor; see also Fig. 1 legend).- CT, cholera toxin (holotoxin unless denoted CT_B); **d**, density (g/ml); radioI or (in Fig. 1) I*, radioactivity; TCA, trichloroacetic acid; AcG, *N* -acetylglucosaminidase; AcP/AlkP, acid/alkaline phosphatase; 5'N, 5'-nucleotidase. [⊗Origin: green African monkey kidney.

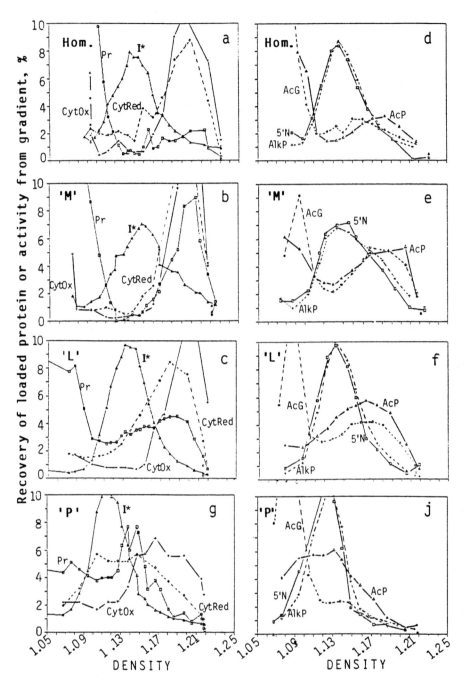

Fig. 1. Distribution profiles after isopycnic sucrose gradient centrifugation of individual fractions at zero time (no re-culture; *this page*) and after 30 min of re-culture (*opposite p.*, *where Legend continues*).

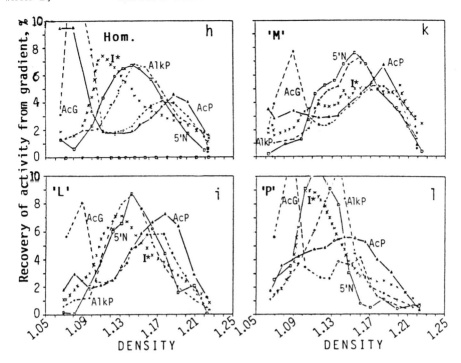

Fig. 1 *continuation:* profiles after 30 min of re-culture.
Pr, protein (Lowry); CytOx, cytochrome oxidase; CytRed,
cytochrome reductase; also *(above, as well as in* a, b, c & g
opposite) I* = ^{125}I radioactivity. Hom. = homogenate; M, L & P
are subcellular fractions. For 5'N, AlkP, AcP and AcG, see
footnote at start of article. Centrifugation was in a Kontron
vertical rotor, 3 h at 40,000 rpm; sucrose gradient 0.5 → 2.0 M.
Re-culture conditions: medium 199, 2 mM glutamine, antibiotics,
5% CO_2, 90% humidity; 37° for different times (30 min illustrated).
The sediments applied as a negative gradient on top of the
sucrose gradient were derived from 40 boxes of confluent cells
(~10^8 cells; 2-3 ml).

were not completely coincident (AlkP equilibrated at a somewhat
higher **d** than 5'N), pointing to the existence of domains
in the plasma membrane. The profile differences between
AcG and AcP were more pronounced: AcG but not AcP penetrated
the sucrose gradient at its top, forming a peak at **d** ~1.09,
suggesting the presence of particles of very low **d** (endosomes?)
containing AcG (K-J. Andersen, pers. comm.). In **P** most
marker enzymes equilibrated at lower **d** values.

Radioactivity, for which time-related between-fraction
distribution is given in Fig. 1 (cf. art. #F-3) showed **H**
and subfraction profiles as in Fig. 2 after re-culture for 0 min-10 h.

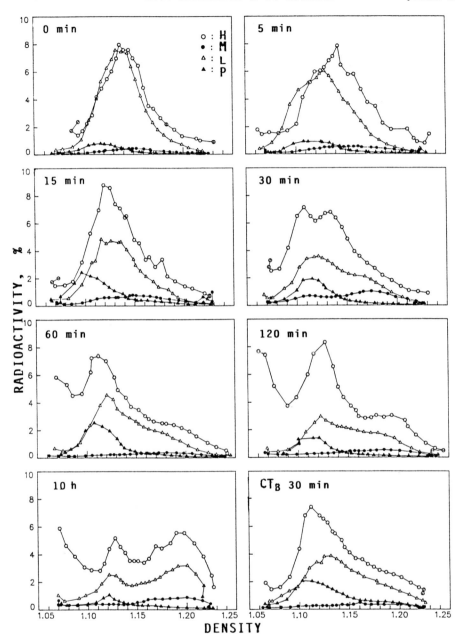

Fig. 2. ^{125}I-CT (or -CT$_B$) distribution profiles in homogenate (H) and subfractions after re-culture for the times shown. The surface under each profile corresponds to the amount of protein recovered from the respective subcellular fractions.

The radioI profile at zero time could be regarded as a
marker for the cell-surface membrane. After 5 min of re-culture
the peak in **L** had shifted towards lower **d**'s (shoulder at
~1.10:, coated vesicles?).   RadioI in **P** began a rise at
the same **d**, and (at 1.16) **M** similarly.   No radioI was
released into the culture medium.   After 15 min there was
an increased proportion of radioI in **P** (**d** now 1.09-1.10)
as well as in **M** (**d** 1.15 and 1.22) at the expense of that
in **L**.   This means that radioI was now being transferred
into low-d particles (early endosomes?) and to lysosomes.

After 30 min and 60 min more radioI in **L** was lost,
that in **P** remaining constant.   In **M** an increase could be
observed at **d** 1.18, and likewise in **L**.   A limited amount
of radioI was recovered from the culture medium.   Probably
a steady state of radioI was attained in the low-d particles,
radioI being transferred to lysosomes where it was converted
into products not precipitable by TCA.   Concomitantly in
the distribution profiles for the homogenates loaded onto
the gradients (Fig. 2) radioI become manifest at the top
of the gradients, corresponding to the radioI recovered in
the soluble (**S**) fractions after differential pelleting.

After 120 min, radioI was reduced in **P** and had shifted
to higher **d**'s (1.11-1.25; high-**d** endosomes?).   In **L**, radioI
remained at **d** 1.19 (as after 30 and 60 min), but radioI at
lower **d**'s was reduced considerably.   After 10 h of re-culture
it was obvious that a relatively high amount of radioI (~35%
of total) still persisted within the cells, especially in lysosomes
(now at **d** 1.20) and in high-**d** endosomes.

Whether or not the Golgi apparatus is bypassed in the
transfer processes during re-culture was not clear from these
studies.   The data do not conflict with the protein translocation
model for diphtheria toxin (Madshus, #A-8), but for CT (De Wolf,
#ncA-2) internalization does not suffice for A_1 polypeptide
generation and subsequent adenylate cyclase activation.

References

1. Houslay, M.D. & Elliot, K.R.F. (1979) *FEBS Lett. 104*, 359-
   363.
2. Janicot, M. & Desbuquois, B. (1987) *Eur. J. Biochem. 163*,
   433-442.
3. De Wolf, M.J.S., Van Dessel, G.A.F., Lagrou, A.R.,
   Hilderson, H.J. & Dierick, W.S.H. (1987) *Biochemistry
   26*, 3799-3806.

#ncA-2

A Note on

CHOLERA TOXIN ACTION AND SIGNAL TRANSDUCTION

M. De Wolf[†], E. Dams, G. Bastiaens.
H.J. Hilderson and W. Dierick

RUCA - Laboratory for Human Biochemistry,
University of Antwerp,
Groenenborgerlaan 171, B-2020 Antwerp,   Belgium

CT^*, an enterotoxin secreted by *Vibrio cholerae*, exerts its pathological effects by increasing the cAMP level in intestinal epithelial cells.   It is an oligomeric protein ($M_r \approx 84,000$) composed of two structural and functional sub-units, CT_A and CT_B ($M_r \approx 29,000$ and 55,000 respectively). CT_B contains 5 identical polypeptide β chains, arranged in a ring-like configuration. CT_A consists of 2 different polypep-tides, A_1 or α chain ($M_r \approx 23,000$) and A_2 or γ chain (M_r 5,500) linked by a single disulphide bridge [1, 2].

CT action is initiated by rapid binding of GM_1 to the CT_B on the cell surface, this interaction being multivalent. After   a lag phase the toxin catalyzes the activation of adenylate cyclase through the NAD-dependent ADP-ribosylation of the stimulatory component of the enzyme, $G_{s\alpha}$ [3]. During this lag period the A subunit is believed to be translocated across the membrane and reduced to form A_I, the catalytic active form [4].

In the present study we examined whether: (i) receptor binding sites are shared between adjacent β chains, and (ii) multi-valent binding is essential for CT_A to traverse the lipid bilayer.

RECEPTOR RECOGNITION DOMAINS

In order to test the hypothesis that receptor recognition domains are located at the interfaces of adjacent β chains of the CT_B pentamer and that binding requires joint participa-tion of amino acid residues from adjoining chains, we construc-ted hybrid CT_B pentamers from inactive chemically modified parental β chains.   One derivative consisted of CT_B specially modified,   by formylation,   in the   single essential [5, 6] Trp 88 residue of each β chain; this treatment preserves the structural integrity of CT_B (fCT_B).   The other inactive derivative consisted of CT_B specially modified in 3 amino groups located in or near the receptor binding site.   Selective modification of these amino groups was achieved by first

[†]addressee for any correspondence
Abbreviations.- cAMP, cyclic AMP; CT, cholera toxin (see text for subunits and for fCT_B & $sssCT_B$); GM_1, monosialoganglioside; *prefix* oligo relates to the carbohydrate moiety.

blocking, through reversible acylation with citraconic anhydride, available amino groups in CT_B complexed with its receptor, followed by the irreversible acylation with succinic anhydride, of amino groups exposed upon removal of the ligand and finally removal of the blocking groups by decitraconylation.

Hybrid CT_B was prepared by incubating equal molar amounts of formylated and site-specific succinylated CT_B (sssCT_B) in 0.1 M glycine/HCl buffer pH 3.2 containing 6 M urea and 0.2 M NaCl for 24 h at 4°, and subsequently removing the denaturant by stepwise dialysis against Tris-acetate buffers pH 7.5 containing urea in decreasing amounts. Application of this reconstitution procedure to the inactive derivatives of CT_B leads to a marked increase in GM_1 binding as evidenced by the ability of hybrid CT_B to compete with GM_1 binding of native CT_B in a solid-phase radiobinding assay. The values that follow represent receptor binding activity, defined as a ratio: IC_{50} for native CT_B as % of IC_{50} for hybrid CT_B, where IC_{50} is the concentration of competitor which produces a 50% reduction of ^{125}I-CT_B bound to GM_1 in the assay (fCT_B and sssCT_B were taken through the same denaturation and renaturation procedure as used for the preparation of hybrids).-

- fCT_B: <0.0001%
- sss CT_B: 0.001%
- hybrids from fCT_B and sssCT_B: 24%
- hybrids from fCT_B and native CT_B: 49%
- hybrids from sssCT_B and native CT_B: 50%
- mixture of equal amounts of sssCT_B and fCT_B: 0.001%

When either fCT_B or sssCT_B is mixed with native CT_B and subjected to a denaturation/renaturation cycle, no increase in receptor binding activity is observed. Therefore it is unlikely that the marked increase in binding of the hybrid CT_B preparation is the result of a conformational correction mechanism. Assuming a binomial distribution, $1/16$th of CT_B pentamers should have a composition similar to that of the parent compounds, $10/16$th should have one active binding site and $5/16$th should have two active binding sites. The ability of oligo-GM_1 to reverse the quenching of the fluorescence of hybrid CT_B by iodide was used to determine the number of active binding sites in the hybrid CT_B preparation. Maximal reversal of iodide quenching of CT_B and hybrid CT_B occurred at oligo-GM_1/protein molar ratios of 5:1 and 1:3 respectively. This agrees very well with the assumption that one out of four reconstituted binding sites of the CT_B pentamer is active.

Taking into account that the receptor binding activity of hybrid CT_B is ~25% that of native CT_B (see the IC_{50} values above), one must assume that the active reconstituted binding sites have an affinity similar to that of the native sites. This successful reconstitution of the binding sites provides strong evidence for the location of receptor recognition domains at the interfaces of adjacent β chains of CT_B.

MULTIVALENT BINDING AND EXPRESSION OF TOXIC ACTIVITY

The CT_B pentamers having one or two functional binding sites were used to address the question of whether multivalent binding of CT is essential for expression of toxic activity. This activity was estimated by measuring the ability of CT or CT-containing hybrid CT_B to elevate the cAMP level in intact Vero cells in culture. When Vero cells were incubated with CT at 37°, intracellular cAMP began to increase after a 20 min delay. For CT as well as the hybrid, the maximal response was obtained at a toxin concentration of 0.1 µg/ml. However, the extent of activation was much lower (4-fold) with the hybrid CT.

Studies are in progress in order to determine whether the decreased ability of hybrid CT to activate adenylate cyclase is a consequence of impaired internalization of the toxin or a defect at the level of translocation of A_1 through the lipid bilayer. Preliminary experiments indicate that the internalization of hybrid CT is normal. Therefore it appears that endocytosis of CT, although required, is not sufficient for the generation of the A_1 polypeptide and subsequent activation of the adenylate cyclase, and that multivalent binding of CT is a prerequisite for translocation of A_1 across the lipid bilayer.

Acknowledgements

This work was supported by Belgian NFWO grant No. 3.0083.87 and 'Geconcerteerde aktie' No. 87/92-119 from the Belgian Government.

References

1. Gill, D.M. (1977) *Cyclic Nucleotide Res.* 8, 85-118.
2. Lai, C-Y. (1980) *CRC Crit. Rev. Biochem.* 9, 171-206.
3. Gilman, A.G. (1984) *J. Cell Biol.* 36, 577-579.
4. Fishman, P.H. (1982) *J. Membr. Biol.* 69, 85-97.
5. De Wolf, M.J.S., Fridkin, M. & Kohn, L.D. (1981) *J. Biol. Chem.* 256, 5489-5496.
6. De Wolf, M.J.S., Fridkin, M., Epstein, M. & Kohn, L.D. (1981) *J. Biol. Chem.* 256, 5481-5488.

SOME LITERATURE PERTINENT TO 'A' THEMES, noted by Senior Editor

Noted in *Biochem. Pharmacol.*- (1) 'The new biology of drug receptors': Lefkowitz, R.J., Kobilka, B.K. & Caron, M.G. (1989) *38*, 2941-2948; (2) Transphosphorylation & G-protein activation; a p.m. NDP kinase reconverts GDP to GTP: Otero, A. de S. (1990) *39*, 1399-1404.
....

Comments on #A-1: J.M. Pfeilschifter- RENAL SIGNALLING PATHWAYS
        #A-2: H. LeVine - RECEPTOR INVESTIGATION
        #A-3: L.A.J. O'Neill - CYTOKINE SIGNAL TRANSDUCTION

Pfeilschifter, replying to (1) R. Bruzzone, (2) G.J.Barritt: (1) all the PKC inhibitors we used - H-7, staurosporine and K252a - acted in the same way and augmented agonist-stimulated InsP$_3$ generation; (2) it is not known how raising extracellular Ca^{2+} increases the frequency of Ca^{2+} transients, such as occurs with agonist action. **Le Vine, answering Barritt:** it is controversial whether some DNA binding receptors reside in the p.m., as some cell fractionation studies suggest. **Remarks by G. Milligan.-** PCR approaches have led to the isolation of many cDNA genes which obviously appear to encode G-protein-linked receptors, but the ligand for these receptors is not known.   Given the range of receptors already identified, is sufficient information available to allow predictions as to potential ligands or class of effectors regulated by these receptors? **Reply.-** Not really. The conservation of a specific asparagine residue in receptors for biogenic amines might be used to predict further receptors for these ligands.

Pfeilschifter, to O'Neill: Does the time-course of IL-1-induced inhibition of adenylate cyclase correspond to the very rapid effect of IL-1 on GTP-binding in EL-4 cells? **Reply.-** At 20 min (the shortest time investigated) there was already a significant inhibition. **S.O. Døskeland.-** You suggested that G-proteins act as amplifiers.   How many GTPγS binding sites are generated by IL-1 compared with the number of occupied IL-1 receptors? **Reply.-** There might be 20-30 G-proteins per receptor; we haven't calculated the stoichiometry. **P.A. Kirkham:** have you tried doing co-precipitation analysis to see whether G-proteins or any other proteins are associated with the IL-1 receptor?   **Reply:** no; but other groups have not found any protein associated with this receptor. **Kirkham:** might there be a very weak protein association that occurs only when the receptor is activated? **Reply:** that is very possible.   **Remark by J.T. O'Flaherty.-** It would be worth checking the effects of GTP on IL-1 binding to membranes, and whether, say, G$_1$ or G$_0$ is present in purified receptor preparations. **Reply.-** There is no effect, nor are G-proteins present. **Reply to J.W.C.M. Jansen,** who asked about self-induced IL-2 synthesis: NFκB effects begin in <5 min, whereas API-sensitive genes need ~4 h to respond.

Comments on #A-4: G. Milligan – G–PROTEIN/RECEPTOR INTERACTIONS
#A-5: S. Cockcroft – PHOSPHOLIPASES C/A_2 & SECRETION

Pfeilschifter, to Milligan.– Have you any ideas on effector enzymes to which G_{i2} and G_{i3} might be coupling? **Reply.–** G_{i2} is evidently the true 'G_i', i.e. the inhibitory G-protein of the adenylate cyclase cascade. This conclusion is based on experiments both by ourselves and by Spiegel & co-workers who have shown that Ab's which identify G_{i2} specifically prevent receptor-mediated inhibition of adenylate cyclase. The situation with G_{i3} is rather less clear. Brown & co-workers have provided some good evidence for a role of G_{i3} in K^+-channel regulation. **Bruzzone.–** You suggest that α_2-adrenergic receptors are coupled to 2 G-proteins in fibroblasts. Don't you think that experiments from Capon's lab. show that muscarinic M1 receptors are also coupled, in CHO cells transfected with M1 cDNA, to 2 distinct G-proteins? **Reply.–** Yes, I do. The essential difference between these experiments and our own is that the M1 receptor has been shown to interact in the transfected CHO cells with both a PT-sensitive and a PT-insensitive G-protein. In our studies we demonstrated interaction of the transfected α_2-CIO receptor with 2 separated PT-sensitive G-proteins, G_{i2} and G_{i3}.

Barritt asked Cockcroft whether she had tested inhibitors of arachidonic acid metabolism to elucidate the role of PL-A_2 (reply: no!), and whether there are any other systems in which GTPγS stimulates a response involving a low-M_r G-protein. **Reply.–** I don't know of another example of such activation. **Remark by O'Flaherty.–** There is evidence recently published (*JBC*) that PL-A_2 (in some cell types, possibly macrophages), is activated by Ca^{2+}: rises in $[Ca^{2+}]_i$ or $[Ca^{2+}]_e$ apparently cause cytosolic PL-A_2 to move to p.m. and become active in deacylation. This system seems to represent a PL-A_2-activating mechanism opposite to yours, and different cell types vary widely in mechanisms. **Cockcroft, answering P-O. Berggren.–** It is not feasible to block secretion, e.g. peptide-stimulated, by PT pre-treatment of the cells. **Bruzzone asked** about the suitability of neutrophils as a model for exocytotic secretion, since they do not exocytose granules *in vivo* unless treated with cytochalasin B.– (Reply) The latter serves to disrupt a dense network that prevents granule-p.m. fusion.

Comment on #A-6/7: R.W. Bonser, N.T. Thompson – PHOSPHOLIPASE D

Reply to R. Heath.– The PL-D mechanism is found in a large and growing range of cell types, especially erythropoietic. **Pfeilschifter asked** whether, insofar as phorbol esters activate PL-D in many cell systems, PKC is the physiological activator of PL-D. What are the effects of PKC inhibitors and PKC down-regulation on agonist-induced PL-D activation? **Reply.–** PKC inhibitors, e.g. staurosporine or K252a, block 50-70% of the PMA response (PL-D activation), but not the remainder. Down-regulation of PKC by prolonged exposure to phorbol esters can also block the response, but likewise incompletely. Interestingly, PKC inhibitors potentiate PL-D activation by the chemotactic peptide in human neutrophils. PKC evidently has both positive and negative effects on PL-D activation, and PL-D regulation is complex and so far ill-defined. **Barritt asked** whether DAG/PA interconversion is tightly controlled.- Is there a futile cycle between them? How is the observation that synthetic DAG added to cells is rapidly converted to PA reconciled with a pathway of DAG formation *via* PL-D and PA? **Reply.-** There will indeed be tight control; also, PA itself may act as an intracellular messenger. **Answer to Bruzzone.-** Our data clearly show that even at high concentrations as used to inhibit PA and DRG production, the alcohols don't inhibit fMLP-stimulated InsP$_3$ production and so don't block PtdInsP$_2$-specific PL-C activation. Also, neither ethanol nor butanol affects PMA-induced superoxide production, suggesting that neither inhibits PKC or NADPH oxidase activation. Ptd-alcohol formation is unaffected by pre-incubation time, suggesting that inhibition of PA and DRG production is not due to conversion of the alcohols into active metabolites or to a time-dependent non-specific action on membrane integrity. Neither alcohol causes LDH release from neutrophils, and even at high levels the cells still exclude trypan blue. Inhibition reversibility awaits study.

Comments on #A-8: I.H. Madshus - DIPHTHERIA TOXIN BEHAVIOUR
     #ncA-2: M.J.S. De Wolf - CHOLERA TOXIN SIGNALLING ROLE

O'Flaherty asked Madshus whether DT cleavage is due to host-cell cytosolic enzymes, and DT entry into cells is reversible before the cleavage step. **Reply.-** DT is usually cleaved (nicked) by serum proteases. It can probably also be cleaved by proteases inside endosomes in some cell types. Usually DT is nicked by proteases before binding to surface receptors. Nicked bound DT is in equilibrium with non-bound DT in solution; the binding is reversible. **Reply to Milligan.-** There is no evidence on the possibility that, insofar as the DT 'receptor' is sensitive to both protease and PL-C, it is an inositol glycan-linked extracellular protein. In fact (answers to Bruzzone) there is no information as to its structure; the evidence that DT binds to a receptor is that only sensitive cells bind DT and concentrate it on the p.m., and binding of radiolabelled DT is displaced by adding cold DT. In transducing the A-part of DT (answer to Hilderson) the receptor can play an active role, but no other proteins.

De Wolf, answering Milligan: within cells the $\frac{1}{2}$-life is >24 h for B and ~6 h for A. **Question by Julie Smith.-** CT in the natural situation is presented to enterocytes at the brush border, yet the basolateral membrane is the locus for adenylate cyclase. How is it accessed by CT in these polarized epithelial cells? **Reply.**- Presumably G-proteins known to be present in brush borders move through the lateral plane of the membrane to influence the basolaterally located cyclase. **J.S. comment:** maybe this is the case, but a pre-requisite of polarized cells is that proteins don't move laterally between the different membrane domains.

===

SOME LITERATURE noted by Senior Editor, ctd. from p. 97

Genty, N., Salesse, R. & Garnier, J. (1987) *Biol. Cell 59*, 129-136.- 'Internalization and recycling of **lutropin** receptors upon stimulation by porcine lutropin and human choriogonadotropin in porcine Leydig cells'.

Garbers, D.L. (1989) *J. Androl. 10*, 99-107.- 'Molecular basis of signalling in the **spermatozoon**'; cell-surface receptors for various egg-associated molecules including guanylate cyclase.

Ashkenazi, A., *et al.* (1991) *Cell 56*, 487-493.- 'Functionally distinct G-proteins [PT-sensitive **G$_p$'s**; CHO cells] selectively couple different receptors to PI hydrolysis in the same cell'.

Lo, W.W.Y. & Hughes, J. (1987) *FEBS Lett. 220*, 327-331.- 'A novel cholera toxin-sensitive G-protein (**G$_c$**) regulating receptor-mediated phosphoinositide signalling....' (pituitary clonal cells).

Loffelholz, K. (1989) *Biochem. Pharmacol. 38*, 1543-1549.- Commentary: 'Receptor regulation of choline phospholipid hydrolysis'. A novel source of DAG and PA, maybe a PC derivative and unconnected with InsP$_3$ generation although this may occur in parallel; PKC activation may inhibit the hydrolysis.

Baumgold, J. & White, T. (1989) *Biochem. Pharmacol. 38*, 1605-1616.- 'Pharmacological differences between **muscarinic receptors** coupled to phosphoinositide turnover and those coupled to adenylate cyclase inhibition'; the two receptors, studied in neuroblastoma cells, were distinguishable by use of carbachol, acetylcholine or bethanechol.

Avissar, S., Murphy, D.L. & Schreiber, G. (1991) *Biochem. Pharmacol. 41*, 171-175.- Mg^{2+} reversal of **Li$^+$ inhibition** of β-adrenergic and muscarinic **receptor binding to G-proteins**.

Backer, J.M. & King, G.L. (1991) *Biochem. Pharmacol. 41*, 1267-1277.- Commentary: 'Regulation of receptor-mediated **endocytosis** by phorbol esters'. Receptors considered include: transferrin, insulin, EGF, asialoglycoprotein; PKC role uncertain.

Reisine, T. (1990) *Biochem. Pharmacol. 39*, 1499-1504.- Commentary: '**Pertussis toxin** in the analysis of receptor mechanisms'; focus on how G-proteins recognize and interact with receptors and effector systems. Whilst Ca^{2+} entry into neuroblastom cells may not need a G-protein, a PT-insensitive G-protein mediates muscarinic regulation of InsP generation.- Lambert, D.G. & Nahorski, S.R. (1990) *Biochem. Pharmacol. 40*, 2291-2295.

#B

CYTOPLASMIC TRANSMISSION SYSTEMS, AND SOME AGONIST EFFECTS

#B-1

CRITERIA USED TO JUDGE THAT A CELLULAR RESPONSE IS MEDIATED BY cAMP

[1]Stein Ove Døskeland[†], [1]Roald Bøe, [1]Torunn Bruland,
[1]Olav Karsten Vintermyr, [2]Bernd Jastorff
and [3]Michel Lanotte

[1]Cell Biology Research Group, Department of Anatomy,
University of Bergen, Årstadveien 19,
N-5009 Bergen, Norway

[2]Fachbereich Biologie-Chemie, Universität Bremen,
2800 Bremen 33, Germany

[3]Institute of Hematology, Centre Hayem,
Hôpital St.-Louis, 75010 Paris, France

Criteria (#1-#10) for involvement of cAMP or cAMP-dependent phosphophorylation events in biological actions are reviewed and updated. Criteria #1-#3 (rise in cAMP; first-messenger action mimicked by cAMP analgoues and augmented by phosphodiesterase inhibitor) date back to E.W. Sutherland. Criterion #8 (E.G. Krebs and H.C. Nimmo & P. Cohen) comprises cAMP-dependent phosphorylation of key proteins in the first messenger's target cell. Examples are given of the use of synergistic actions of cAMP analogues to determine not only involvement of cAK* but also whether cAK I or cAK II is responsible (#2). Examples are given of micro-injection of the catalytic subunit of cAK to mimic cAMP action (#4), the reversal of C subunit effect by mutagenized R subunit (#5), the use of the cAMP antagonist (Rp)-cAMP (#6), and down-regulation of cAK by cAMP (#7). Consideration is given in CONCLUSIONS to the scope and merits of the different criteria. Fulfilment of more than one criterion may be requisite to establish whether an action is mediated by cAK-stimulated phosphorylation.

cAMP was the first 'second messenger' discovered [1]. Its intracellular level is controlled by the balance between synthesis rate by adenylate cyclase, conversion to AMP by a family of phosphodiesterases and egress through the p.m.* [2]. Although it has been proposed that cAMP signalling may result

---

[†]addressee for any correspondence

Abbreviations.*- cAK, cAMP-dependent protein kinase, the two forms being denoted **I and **II** and each having C and R subunits (and A and B sites: see text); CRE, cAMP-responsive element(s) of DNA; EGF, epidermal growth factor; p.m., plasma membrane.

Table 1 lists abbreviations for various cAMP analogues; note that N- (Editor's rendering) implies N^6-. For (Rp)-cAMPS (cAMP agonist) and (Sp)-cAMPS (antagonist) see **Criterion #6.**

from turnover-linked effects on high-energy compounds or protons
[3, 4], most - if not all - effects of cAMP are believed to
be *via* its interaction with target molecules, i.e. cAMP
receptors. cAMP is especially well suited to induce conformational changes in target molecules because of the inherent
high potential energy of its cyclic phosphate bond.

In *E. coli* the DNA-binding CAP protein mediates the
induction of the β-galactosidase gene by cAMP [5]. In
the slime mould *Dictyostelium discoideum* a cAMP-binding surface
receptor is responsible for the aggregating effect of extracellular cAMP [6, 7]. In animal cells Kuo & Greengard
[8] postulated that all effects of cAMP were mediated by
phosphorylations carried out by the ubiquitously present cAMP-
dependent kinase cAK. Although it does not follow from
the ubiquitous presence of one effector system that the
existence of others is excluded, the postulate of Greengard
remained viable until the demonstration in mammalian olfactory
epithelium of an ion channel regulated directly by cAMP binding
[9, 10]. Furthermore, it has repeatedly been postulated
[11-13] that cAMP may act *via* the free regulatory subunits
(R I, R II) rather than the C subunit of cAK.

Early efforts to judge whether a phenomenon was mediated
by cAMP were made by Sutherland [1]. Criteria for mediation
by cAMP-stimulated phosphorylation events were proposed by
Krebs [14], and later re-examined and extended by Nimmo
& Cohen [15]. In the following, 10 updated and/or novel
criteria are listed, and each is considered. They take account
of recent advances concerning (a) the molecular biology of
the cAMP-dependent protein kinases (see [16, 17] for recent
reviews), (b) the use of cAMP analogue combinations to mimic
cAMP effects [18, 19] (Table 1 and Fig. 1), (c) progress
in the microinjection of components of the cAMP effector
system [20, 21] (Figs. 2 & 3), (d) the advent of cAMP analogues
which antagonize cAMP [22, 23] (Fig. 4), (e) the development
of methods to measure cAMP bound in intact cells [24], and
(f) the availability of novel inhibitors of protein phosphatases
[25, 26].

CRITERIA FOR INVOLVEMENT OF cAMP IN RESPONSES

Criterion #1.- The first of Sutherland's criteria is
the demonstration of increased cAMP in response to the first
messenger. This criterion still appears valid and will
be fulfilled whether cAMP acts through cAK or not. A pitfall
is that cAMP may increase in one compartment and the biological
reaction to the primary agent occur in another compartment.
Compartments can be subcellular, or represent one cell type
in a mixed population or one phase of the cell cycle in

a non-synchronized population of one cell type. Where the action is believed to be *via* cAK, we propose determination of the endogenous occupation of the R subunit by cAMP [24] as an improved criterion, since this will measure that fraction of cAMP which actually activates cAK. Alternatively, the fractional activity of cAK can be determined [27]. Since cAK saturation is limited to 100%, the degree of saturation of R will be less sensitive than total cAMP to biologically irrelevant 'overshoot' of cAMP in any small compartment.

Criterion #2.- This involves demonstrating that cAMP analogues mimic the effect of first messenger. For high potency in activating cAK [28] an intact 2'-hydroxyl of the ribose moiety of cAMP is important. However, for activation of the *Dictyostelium* cAMP receptor the 2'-hydroxyl is not essential. In order to activate the bacterial CAP protein, analogues must satisfy stricter criteria: both the 2-hydroxyl and N-amino groups must be intact to allow formation of hydrogen bonds. Furthermore, only the *anti* conformation of the glycosidic bond is tolerated; hence bulky substituents at the C-8 of the adenine ring (which force cAMP into the *syn* conformation) tend to be biologically inactive. The analogue specificity of the cAMP-regulated olfactory channel protein is still not known.

When action through cAK is suspected, demonstration of cAMP **analogue synergism** [19, 29] (Figs. 1 & 2) is a better criterion than use of single analogues. Demonstration of the expected synergism virtually excludes the possibilities that the analogues act by phosphodiesterase inhibition, by stimulation of cGMP-dependent protein kinase, or *via* a metabolite (e.g. butyrate in the case of dibutyryl cAMP or 8-chloroadenosine in the case of 8-chloro-cAMP), or non-specifically. Moreover, one can tell whether the action is *via* isozyme I or II of cAK, since these differ in cyclic nucleotide analogues specificity [30] (Table 1). A procedure for quantitative analysis of analogue synergism is given by Lanotte and co-authors [29], and a semi-quantitative procedure is given by Beebe and co-authors [19]. It should be noted that some cells are very impermeable to cAMP analogues of low hydrophobicity, and that 8-amino- and 8-chloro-cAMP can produce metabolites having toxic effects on cells (data not shown). If analogue degradation is a serious problem, there should be trial of analogues with high resistance towards breakdown [N-Bz-cAMP and especially (Sp)-cAMPS].

Criterion #3.- Inhibitors of cAMP phosphodiesterase should augment a submaximal challenge by first messenger. This approach has several pitfalls [31]. The inhibitors available are not always efficient and may inhibit the degradation of cGMP as well as that of cAMP. Besides they may have other effects, such as adenosine receptor antagonism. Preferably, therefore, it should be demonstrated that a positive correlation exists between the potency of a range of such agents as phosphodiesterase inhibitors and as inducers of the effect believed to be mediated by cAMP, and that there is no correlation with the potency of the agents as, e.g., adenosine receptor antagonists [32].

Criterion #4.- Instead of relying on the activation of the endogenous cAMP effector system, the C subunit of cAK can be over-expressed by either transient or stable transfection

Table 1. Induction of cytolysis in leukaemic cells, and inhibition of hepatocyte DNA replication, by 13 cAMP analogues, listed in order of decreasing site A *vs.* B selectivity for cAK I, as tabulated along with the selectivity values for cAK II. (The values are from ref. [30] or, for 8-chloro-cAMP, [33].)
Comment.- Comparison of cAK I and cAK II selectivities shows that site A in both is selected by N-Bu- and N-Bz-cAMP, which are therefore not expected to complement each other in activating either isozyme (cf. Fig. 1A). 8-AHA-cAMP selects site B of both isozymes and should therefore complement N-Bz-cAMP in activating both cAK I and cAK II (cf. Fig. 1B). 8-Pip-cAMP is site A-selective for cAK I and site B-selective for cAK II. This difference is the basis for its use to discriminately activate cAK II in combination with N-Bz-cAMP (Fig. 1C), and cAK I in combination with the site B-selective 8-MA-cAMP (Fig. 1D).

Derivative of cAMP (& abbreviation) [always cAMP as suffix]	Cytolysis: EC$_{50}$, µM (1)	Replication: EC$_{50}$, µM (2)	Site selectivity A/B	
			cAK I	cAK I
N^6-monobutyryl- (N-Bu-)	305	25	39	18
N^6,O'-2-dibutyryl-	280	not determined	39$^\otimes$	18
8-piperidino- (8-Pip-)	1000	14	35	1/70
N^6-benzoyl- (N-Bz-)	295	10	19	12
8-chlorophenylthio- (8-CPT-)	30	1.6	2.0	1/340
N^6-aminohexylcarbamoyl-	825	not determined	1.5	2.
8-chloro-	105	"	1.4	1/90
8-bromo-	860	"	1.3	1/62
8-S-ethyl-	215	"	1/2.9	1/265
8-S-methyl-	375	"	1/3.5	1/35
8-aminohexylamino- (8-AHA-)	1100	35	1/15	1/14
8-amino-	10	not determined	1/26	1/34
8-methylamino- (8-MA-)	2100	30	1/47	1/62

µM values are for ½-maximal (1) lysis of IPC-81 leukaemic cells, measured as 50% decrease of mitochondrial dehydrogenase activity after 40 h incubation with analogue, **or** (2) decrease in [^{3}H]methyl-thymidine incorporation into DNA (2-h pulse 12 h after adding the analogue to primary hepatocytes cultured in presence of EGF). |$^\otimes$for monobutyryl (the analogue's active form)

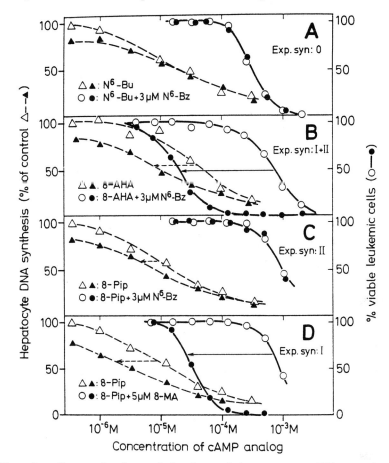

Fig. 1. Synergistic modulation of hepatocyte DNA synthesis and induction of ICT-81 cytolysis by pairs of cAMP analogues. Rat hepatocytes were grown in primary culture, and the promyelocytic cell line ICT81 in suspension [29]. They were exposed for 12 and 40 h respectively to cAMP analogues, alone or in combination, as shown. Their effect on hepatocyte DNA replication was determined by scintillation counting of labelled DNA after pulse labelling with [³H]thymidine, and IPC-81 cell viability by an enzyme assay (see heading to Table 1, also for amplification of the following comments). **A** shows that analogues which do not complement each other for activation of either cAK I or cAK II also fail to synergize to decrease DNA synthesis or viability. **B**: analogues that complement each other in activating both cAK I and cAK II synergize strongly for either parameter tested. **C**: 2 analogues that complement each other for cAK II but not cAK I activation synergize moderately in decreasing hepatocyte DNA synthesis but not in induction of cytolysis. **D**: analogues complementary for cAK I activation act synergistically in induction of cytolysis as well as inhibition of DNA replication. Note the difference in the steepness of the dose–response curves for cAMP analogue action on hepatocytes and leukaemic cells, reflecting a very high cooperativity in the induction of cytolysis.

Fig. 2. Rounding of
fibroblasts microinjected
with C is inhibitable by
coinjection with mutant RI.
A 3T3 cell line (kindly
made available by Dr. G.S.
McKnight, Washington Univ.,
Seattle) was grown in
monolayer culture and
injected with buffer alone
(**A**), homogeneous C subunit
from bovine heart cAK II (**B**)
or a 1:1 mixture of C and
bovine R I with a mutation
in cAMP binding site B
(Arg 333 replaced by Lys)
rendering R I somewhat
less sensitive to cAMP
(**C**). The effect of C was
already evident 15 min
after injection, and was
maximal after ~1 h. The
phase-contrast micro-
graphs (× 315) were taken
1 h after injection. Note
that coinjection of C with
a 10-fold excess of a
peptide corresponding to
the active portion of the
heat-stable inhibitor of C
[34] or synthetically
prepared capped mRNA
coding for RI and efficient
in *in vitro* translation
assay (Døskeland & McKnight
& G. Cadd, G. Hauge: unpublished) did not impair the
C action, presumably due
to short biological ½-lives
of the coinjected materials.

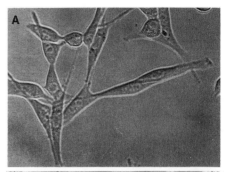
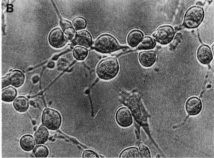
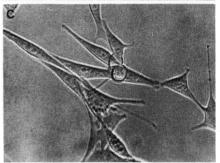

[16, 35]. A problem with that approach can be the counter-regula-
tory production of endogenous R I subunit [16]. Microinjection
of the C subunit has been used to probe for cAK control
of cell shape [20, 21] (Figs. 2 & 3). Fig. 3 shows schematically
the use of microinjection of components of cAK and of combinations of cAMP analogues in modulating cell functions.

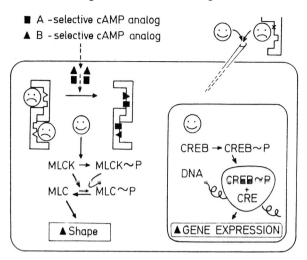

Fig. 3. Presumed steps in the mechanism of action of cAMP analogues and microinjected C:- a simplified scheme, intended to explain the effect of site-selective analogues on activation of endogenous cAK *(upper left)*, the effect of active C (endogenous or microinjected) in promoting rounding of non-muscle cells, and one pathway of C subunit activation of gene expression *(lower right)*. Synergism between cAMP analogues selective for sites A and B is demonstrated by the data of Fig. 1. Rounding of fibroblasts by microinjected C and its inhibition by mutated R I is illustrated in Fig. 2.

Microinjected C has been shown to increase the phosphorylation of myosin light chain kinase (MLCK) and decrease that of the chain (MLC) in non-muscle cells [36], and to activate cAMP-responsive genes [37] presumably by increasing the phosphorylation of proteins (CREB's) interacting with CRE of DNA.

Criterion #5.- An effect mediated by cAK should be suppressible by inhibitors of C. Obvious inhibitors are the regulatory subunits of cAK:- if over-expressed in their native form they will be expected to increase the cAMP concentration required for an action [38, 39]. If cells are made to express mutant forms of R in which one or both cAMP binding sites are deficient, the inhibition of C will hardly be overcome by cAMP [16]. In fact, a number of dominant mutations of R I mapping to the cAMP-binding sites are known in mutant cAMP-resistant S49 cell lines [40].

Another inhibitor with sequence similarity to R is the so-called heat-stable inhibitor protein [41], which specifically inhibits the C subunit of cAK. The entire protein, or peptide fragments therefrom [34], have efficiently counteracted cAMP effects after microinjection [42, 43], extracellular

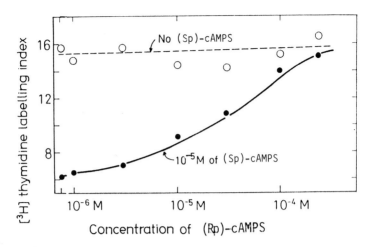

Fig. 4. Reversal of the effect of the cAMP agonist
(Sp)-cAMPS by the antagonist (Rp)-cAMPS. Hepatocytes were
grown in primary culture and their DNA replication scored
as for Fig.1. The inhibition of DNA synthesis brought about
by (Sp)-cAMPS (10 μM; control) can be partially overcome by
(Rp)-cAMPS, tested at increasing concentrations.

presentation [44] or intracellular over-expression [45]. Moreover, a newly developed kinase inhibitor termed H-12 has
shown promising inhibition of cAK [46].

   Criterion #6.- A sulphur-substituted analogue, (Rp)-cAMPS,
provides an alternative way of inhibiting cAK. It binds
to the regulatory moiety of cAK, and counteracts cAMP effects
in short-term studies on hepatocytes [22] and thyrocytes
[23]. Fig. 4 shows that (Rp)-cAMPS counteracts its agonistic
diastereoisomer (Sp)-cAMPS in long-term investigation of hepatocyte
DNA replication. New, more potent derivatives of (Rp)-cAMPS
with ability to discriminate the cAMP sites of cAK isozymes
[47] may be helpful future tools, being potentially usable
to selectively inhibit one isozyme of cAK. It is uncertain
why (Rp)-cAMPS does not decrease basal cAMP effects in hepatocytes [22]. One possibility is that the compound inhibits
cAMP phosphodiesterases [23] so that the endogenous cAMP
level becomes sufficiently high to partly overcome the (Rp)-
cAMPS antagonism. Another possiblity is that (Rp)-cAMPS
is a partial agonist of cAK. Some (Rp)-cAMPS derivates
are able to partially activate cAK II in an *in vitro* assay
(data not shown).

   Criterion #7.- A useful criterion to probe for involvement
of protein kinase C in cell response to a primary agent
has been to test whether the response still occurs after

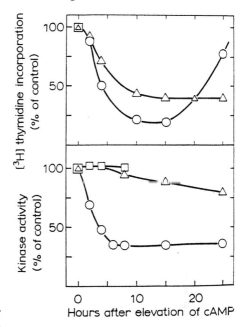

Fig. 5. Temporal effects of elevated cAMP on DNA replication and on the level of catalytic subunit of cAK in hepatocytes and MCF-7 cells (as % of control - *squares* for non-downregulated hepatocytes). Hepatocytes (o) were grown in primary culture as for Fig. 1, and MCF-7 cells (Δ) in standard monolayer culture.   The DNA synthesis (*upper panel*; [³H]thymidine present for 2 h before cell harvesting) was determined in cells harvested at various times after adding forskolin (10 μM)/*iso*butylmethylxanthine (50 μM). Parallel cell plates were removed for determination [39] of activity of the C subunit of cAK (*lower panel*); this activity was also determined in leukaemic cells grown in the presence of 10 nM cholera toxin (I).   *Comment.-* Hepatocyte DNA replication is resumed ~15 h after cAMP elevation (*upper panel*), probably as a result of downregulation of C (*lower panel*). The inhibitory effect of cAMP elevation on MCF-7 DNA replication is more sustained, possibly because of much less pronounced down-regulation in such cells.   In the IPC-81 cells no down-regulation occurs before the onset of cell death.

down-regulation of protein kinase C by prolonged incubation with phorbol esters (#A-1 & #B-2, this vol:- J. Pfeilschifter, and U. Lang & co-authors; & [48]).   This approach can also be used for cAK:- Fig. 5 shows the cAMP effect on DNA replication in hepatocytes correlates temporally with the amount of C subunit present.   Partial down-regulation was accomplished by prolonged exposure to elevated cAMP.   The expression of C can presumably be decreased also by introduction into cells of anti-sense cDNA for C subunit.   It should be noted that cAMP-induced down-regulation of cAK is only partial, and so far has been demonstrated only for a few cell types [39, 49-51].

Criterion #8.- Krebs [14] suggested as criteria for cAMP-mediated phosphorylation that a substrate protein phosphorylated by cAK *in vitro* should be subject to cAMP-dependent phosphorylation in the target cell, and that the resulting change in protein function should be relevant for the action of cAMP on the target cell. Nimmo & Cohen [15] proposed an even stricter criterion: not only should the protein be phosphorylated, but the site phosphorylated in the intact cell should be the same as that phosphorylated by cAK in a cell-free system. These criteria are very useful when the exact locus of the cAMP action can be guessed, but fulfilment is not proof that an action is mediated by cAMP since, e.g., phenylalanine hydroxylase and liver pyruvate kinase are phosphorylated on exactly the same site by cAK and by the multifunctional Ca^{2+}/calmodulin-dependent kinase [52, 53]. As another complication, phosphorylation can increase with no cAK-activity change because an agent can alter the concentration of ligands, e.g. phenylalanine [54], which alter the susceptibility of kinase substrates to phosphorylation [55].

Criterion #9.- An approach used elegantly by, amongst others, Garrison and co-workers [56] is the analysis of phosphorylated proteins on 2-D polyacrylamide gel electrophoresis. By comparing the patterns of phosphopeptides in response to cell stimulation by an agent with unknown second messenger and by agents known to act via cAK, protein kinase C or the Ca^{2+}/calmodulin kinases, an educated guess can be made as to which second messenger(s) is involved.

Criterion #10.- Finally, if increased phosphorylation of a protein is suspected and the phosphatase catalyzing the dephosphorylation is known, inhibition of the relevant phosphatase should mimic the action of increased kinase activity. Although quite specific phosphatase inhibitors are known [25, 26], their usefulness is so far limited by the broad substrate specificity of the major known phosphatases.

CONCLUSIONS

The above criteria should allow cAMP involvement to be determined with a high degree of certainty in most systems. However, fulfilment of more than one criterion is often required before an action mediated by cAK-stimulated phosphorylation can be ascertained. Which approach is optimal depends in large measure on the target system under investigation. Microinjection techniques, use of cAMP analogues and phosphodiesterase inhibitors are suitable when a response can be measured in only a few cells. In that case measurement of bulk biochemical parameters such as cAMP level or phosphoproteins is less feasible. In the case of 'cross-talk' between signalling systems (e.g. Ca^{2+}, cAMP, activation of protein kinase C) the results can be puzzling. If there is strong synergism (mutual interdependence) between cAMP and another pathway, the action of cAMP will be apparent only if the other pathway is active, and inhibition of the cAMP system will inhibit also the effect of the other pathway.

References (JBC = J. Biol. Chem.; PNAS = Proc. Nat. Acad. Sci.)

1. Robison, G.A., Butcher, R.W. & Sutherland, E.W. (1971) *Cyclic AMP*, Academic Press, New York.
2. Barber, R. & Butcher, R.W. (1988) *Meths. Enzymol. 159*, 50–60.
3. Walseth, T.F., Gander, J.E., Eide, S.J., Krick, T.P. & Goldberg, N.D. (1983) *JBC 258*, 1544–1558.
4. Graeff, R.M., Walseth, T.F. & Goldberg, N.D. (1987) *Neurochem. Res. 12*, 551–560.
5. Ullmann, A. & Danchin, A. (1983) *Adv. Cyclic Nucl. Res. 15*, 1–53.
6. Devrotes, P. (1982) in *The Development of Dictyostelium discoideum* (Loomis, W.F., ed.), Academic Press, New York, 117–168.
7. Mann, S.K. & Firtel, R.A. (1987) *Mol. Cell Biol. 7*, 458–469.
8. Kuo, J.F. & Greengard, P. (1969) *PNAS 64*, 1349–1355.
9. Nakamura, T. & Gold, G.H. (1987) *Nature 325*, 442–444.
10. Dhallan, R.S., Yau, K.W., Schrader, K.A. & Reed, R.R. (1990) *Nature 347*, 184–187.
11. Constantinou, A.I., Squinto, S.P. & Jungmann, R.A. (1985) *Cell 42*, 429–437.
12. Shabb, J.B. & Miller, M.R. (1986) *J. Cyclic Nucl. Prot. Phosphoryl. Res. 11*, 253–264.
13. Cho-Chung, Y.S. (1989) *J. Nat. Cancer Inst. 81*, Commentary.
14. Krebs, E.G. (1973) in *Proc. 4th Internat. Congr. Endocrinol.*, Excerpta Medica, Amsterdam, pp. 17–29.   [–266.
15. Nimmo, H.G. & Cohen, P. (1977) *Adv. Cyclic Nucl. Res. 8*, 145–
16. McKnight, G.S., Cadd, G.G., Clegg, C.H., Otten, A.D. & Correll, L.A. (1988) *Cold Spr. Harb. Symp. Quant. Biol. 53 (Molecular Biol. of Signal Transduction)*, 111–119.
17. Taylor, S.S. (1989) *JBC 264*, 8443–8446.
18. Øgreid, D., Ekanger, R., Suva, R.H., Miller, J.P., Sturm, P., Corbin, J.D. & Døskeland, S.O. (1985) *Eur. J. Biochem. 150*, 219–227.
19. Beebe, S.J., Blackmore, P.F., Chrisman, T.D. & Corbin, J.D. (1988) *Meths. Enzymol. 159*, 118–139.
20. Roger, P.P., Rickaert, F., Huez, G., Authelet, M., Hofmann, F. & Dumont, J.E. (1988) *FEBS Lett. 232*, 409–413.
21. Riabowol, K.T., Gilman, M.Z. & Feramisco, J.R. (1988) *Cold Spr. Harb. Symp. Quant. Biol. 53*, 85–90
22. Rothermel, J.D., Jastorff, B. & Parker-Bothello, L.H. (1984) *JBC 259*, 8151–8155.
23. Erneux, C., van Sande, J., Jastorff, B. & Dumont, J.E. (1986) *Biochem. J. 234*, 193–197.
24. Ekanger, R. & Døskeland, S.O. (1988) *Meths. Enzymol. 159*, 97–104.

25. Cohen, P. (1989) *Annu. Rev. Biochem. 58*, 453-508.
26. Cohen, P., Holmes, C.F.B. & Tsukitani, Y. (1990) *Trends Biochem. Sci. 15*, 9
27. Corbin, J.D. (1983) *Meths. Enzymol. 99*, 227-232.   [1
28. Jastorff, B., Abbad, E.G., Petridis, G., Tegge, W., de Witt, R.J.W., Erneux, C., Stec, W.J. & Morr, M. (1981) *Nucl. Acids Res. Symp.Ser. 9*, 2
29. Lanotte, M., Riviere, J.B., Hermouet, S., Houge, G.,   [22 Giersten, B.T. & Døskeland, S.O. (1991) *J. Cell. Physiol.*, in pre
30. Øgreid, D., Ekanger, R., Suva, R.H., Miller, J.P. & [& see (19ℓ Døskeland, S.O. (1989) *Eur. J. Biochem. 181*, 19-31. |147,371-38
31. Wells, J.N. & Kramer, G.L. (1981) *Mol. Cell. Endocrin. 23*, 1-9
32. Dunwiddy, T.V. & Fredholm, B. (1985) *Adv. Cyclic Nucl. Prot. Res 19*, 259-272.
33. Ally, S., Tortora, G., Clair, T., Grieco, D., Merlo, G., Katsaros, D., Øgreid, D., Døskeland, S.O., Jahnsen, T. & Cho-Chung, Y.S. (1988) *PNAS 85*, 6319-6322.   [4383
34. Scott, J.D.,Fischer, E.H., Demaille, J.G. & Krebs, E.G. (1985) *PNAS 82*, 4379–
35. Mellon, P.I., Clegg, C.H., Correll, L.A. & McKnight, S. (1989) *PNAS 86*, 4887–48
36. Lamb, N.J.C., Fenandez, A., Conti, M.A., Adelstein, R., Glass, D.B., Welch, W.J. & Feramisco, J.R. (1988) *J. Cell Biol. 106*, 1955-1971.
37. Riabowol, K.T., Fink, J.S., Gilman, M.Z., Walsh, D.A., Goodman, R.H & Feramisco, J.R. (1988) *Nature 336*, 83-86.
38. Beavo, J.A., Bechtel, P.J. & Krebs, E.G. (1974) *PNAS 71*, 3580–3583.
39. Houge, G., Vintmeyer, O.K. & Døskeland, S.O. (1990) *Mol. Endocrin. 4*, 4ℓ
40. Steinberg, R.A. (1984) *Biochem. Actions of Hormones 11*, 25-65. [4
41. Walsh, D.A., Ashby, C.D., Gonzales, C., Demaille, J.C., Calkins, D. Fischer, E.H. & Krebs, E.G. (1971) *JBC 246*, 1977-1985.
42. Maller, J.L. & Krebs, E.G. (1977) *JBC 252*, 1712-1718.
43. Saez, J.C., Spray, D.C., Nairn, A.C., Hertzberg, E., Greengard, P. & Bennett, V.L. (1986) *PNAS 83*, 2473-2477.
44. Büchler, W., Walter, U.B.J., Jastorff, B. & Lohmann, S.M. (1988) *FEBS Lett.*
45. Grove, J.R., Price, D.J., Goodman, H.M. & Avruch, J.   [2 (1987) *J. Cell. Physiol. 112*, 530-533.
46. Chijiwa, T., Mishima, A., Hagiwara, M., Sano, M., Hayashi, K., Inoue, T Naito, K., Toshioka, T. & Hidaka, H. (1990) *JBC 265*, 5267-5272.
47. Dostmann, W.R.G., Taylor, S., Genieser, H.G., Jastorff, B., Døskeland, S.O. & Øgreid, D. (1990) *JBC 265*, 10484-10491.
48. Collins, M.K.L. & Rozengurt, E. (1982) *J. Cell. Physiol. 112*, 42-50.
49. Hemmings, B.A. (1986) *FEBS Lett. 196*, 126-130.
50. Schoch, G. (1987) *Biochem. J. 248*, 243-250.
51. Richardson, J.M., Howard, P., Massa, J.S. & Maurer, R.A. (1990) *JBC 265*, 13635-13640.
52. Døskeland, A.P., Schworer, C.M., Døskeland, S.O., Christman, T. Soderling, T.R., Corbin, J.D. & Flatmark, T. (1984) *Eur. J. Biochem. 1*
53. Schworer, C.M., El-Magrabi, M.R., Pilkis, S.J. &   [31- Soderling, T.R. (1985) *JBC 260*, 13018-13022.
54. Døskeland, A., Vintmyr, O.K., Flatmark, T., Cotton, G.H., & Døskeland, S.O. (1990) in *Amino Acids. Chemistry, Biology and Medic* (Lubec, G. & Rosenthal, G.A., eds.), Escom Science Publ., Leiden, 867-8
55. Engstrøm, L. (1978) *Curr. Top. Cell. Regul. 13*, 29-51.
56. Garrison, J.C., Johnsen, D.E. & Campanile, C.P. (1984) *JBC 259*, 3283-329

#B-2

MODULATORY EFFECTS OF PROTEIN KINASE C ACTIVATION IN AORTIC SMOOTH MUSCLE AND IN ADRENAL GLOMERULOSA CELLS DURING ANGIOTENSIN II STIMULATION

U. Lang[†], C. Daniel, D. Chardonnens, A.M. Capponi and M.B. Vallotton

Division of Endocrinology, University Hospital, CH-1211 Geneva 4, Switzerland

Responses of aortic smooth muscle cells and adrenal glomerulosa cells to ANG-II[*] include production of prostacyclin and aldosterone respectively. In both cell types, studied in vitro, ANG-II significantly increased membranous PKC activity which was likewise increased, transiently, by PMA.

In smooth muscle cells prostacyclin production was induced by PMA, with concentration-dependence (10 nM-1 µM), and PMA augmented by 60% the maximal values of ANG-II-stimulated production. Prolonged exposure (48 h) to PMA almost completely suppressed membranous and cytosolic PKC activity. In these PKC-depleted cells ANG-II-stimulated prostacyclin production was diminished by 70%.

With adrenal glomerulosa cells, acute exposure to PMA did not affect basal steroidogenesis but greatly inhibited ANG-II-stimulated aldosterone production. Prolonged treatment (24 h) of cells with PMA decreased membranous and cytosolic PKC activity by 90%. However, basal and ANG-II-induced aldosterone production was increased by 30% compared to untreated cells, and the PMA-induced inhibition of ANG-II-stimulated aldosterone production was no longer observed in PKC-down-regulated cells. These results suggest that activation of PKC plays a positive regulatory role in prostacyclin production by aortic smooth muscle cells but exerts a negative-feedback modulatory effect on aldosterone production in adrenal glomerulosa cells.

The binding of ANG-II to its plasma-membrane receptor in smooth muscle and adrenal glomerulosa cells results in phospholipid turnover that yields $InsP_3$ and DAG [1-5]. Whereas $InsP_3$ induces Ca^{2+} mobilization, DAG activates the phospholipid-dependent and Ca^{2+}-sensitive PKC [6, 7]. ANG-II was found

[†]addressee for any correspondence

[*]*Abbreviations.-* ANG-II, angiotensin II; BSA, bovine serum albumin; DAG, diacylglycerol; DMEM, Dulbecco's modified Eagle's medium; DTT, dithiothreitol; FCS, foetal calf serum; $InsP_3$, myoinositol-1,4,5-trisphosphate; PG, prostaglandin; PKC, protein kinase C; PMA, phorbol-12-myristate-13-acetate; PDBu, phorbol-12,13-dibutyrate; TCA, trichloroacetic acid; RIA, radioimmunoassay.

to induce Ca^{2+} mobilization and DAG formation in adrenal glomerulosa cells ([3-5, 8, 9], & A. Capponi *et al.,* in Vol. 19, this series) and in vascular smooth muscle cells [1, 2, 10-13]. Moreover, both cell types respond to ANG-II receptor stimulation by an increase in membranous PKC activity [14-17]. However, PKC activation appears to play a different role in ANG-II-stimulated adrenal glomerulosa cells than in aortic smooth muscle cells. Whereas in the latter the phorbol ester PMA mimics the effect of peptide hormones and augments ANG-II-induced prostacyclin production [14], this is not the case for aldosterone production in adrenal glomerulosa cells [5].

The aim of this study was to investigate the role of PKC during ANG-II stimulation in both types of cell. ANG-II- and PMA-induced changes in PKC distribution and activity in both cell types were measured, and the effect of these changes on production of aldosterone (adrenal) and prostacyclin (smooth muscle) was studied.

MATERIALS AND METHODS

Procurement was as follows.- Ile^5-angiotensin II: Bachem (Basel). PMA, PDBu, ATP, histone III-S, histone II-A, 1,2-diolein, phosphatidylserine, DTT, γ-globulin, collagenase type I, elastase, Triton X-100 and leupeptin: Sigma. DEAE 52-cellulose: Whatman (Maidstone, U.K.). DMEM, FCS and trypsin: Gibco. [$α^{32}$P]ɣATP and [^{3}H]PDBu: Amersham International. 6-Keto-$PGF_{1α}$ antiserum: courtesy of Dr. M. Dunn (Divn. of Nephrology, Case Western Reserve Univ., Cleveland, OH). 'Medium 199' (Hank's salts with glutamine; no bicarbonate, NaCl or KCl): Seromed.

Preparation and culture of smooth muscle cells.- Rat aortic smooth muscle cells were prepared by an enzymatic dispersion method [10, 14], then plated in 90 mm petri dishes and cultured in DMEM containing 10% (v/v) FCS [10, 14]. Confluent monolayers were usually obtained after 7 days, with a cell density of $\sim 10^7$ cells/dish.

At this stage cells were passaged by trypsinization [10, 14] and subcultured in 90 mm petri dishes. Subsequent passages were performed at 48- to 72-h intervals upon confluence, and cells of the 2nd to 5th passages were used in the present study. For studies of PKC activity cells at confluence in 90 mm petri dishes were used, while prostacyclin production was studied in confluent cultures in multiwell plates.

Preparation of glomerulosa cells.- Isolated adrenal glomerulosa cells were prepared by collagenase digestion ([18], with slight modifications [8]) from bovine adrenal glands

obtained at the slaughterhouse.   After removal of adherent
fat, 0.5-mm thick slices (capsular and glomerulosa tissue)
were cut with a Staddie-Riggs microtome.   The slices were
minced, and incubated twice for 30 min at 37° under constant
gentle agitation in medium 199, pH 7.4, containing 3 mM K^+,
0.2% collagenase type I and 0.2% BSA.   After each incubation
period cells were dispersed by pipetting them ~30 times through
70 µm nylon gauze and washed (×3) with medium 199 containing
3 mM K^+ and 0.2% BSA.

Purification of PKC.- After exposure to the agents tested,
cells were homogenized, and cytosolic and soluble fractions
were prepared [14, 15].   Cytosolic PKC and membranous PKC
were partially purified by DEAE-cellulose chromatography using
a linear NaCl gradient (0-0.3 M) in 20 mM Tris-HCl, pH 7.5,
containing 2 mM EDTA, 2 mM EGTA and 10 mM DTT.   PKC activity
was eluted with NaCl between 0.05 and 0.12 M.   PKC activity
was almost undetectable in crude cell extracts, but was
expressed after DEAE-cellulose chromatography, suggesting that
crude cell fractions contain an inhibitor of PKC and/or
a Ca^{2+}-dependent proteinase.   Inhibitors and/or phosphatases
are removed by this procedure [19].

PKC assay.- An aliquot (100 µl) of each fraction from
the column was assayed for PKC activity by measuring phosphorylation of the substrate histone III-S in the presence and
absence of phosphatidylserine and diolein [14].

Assay of phorbol ester binding activity.- This activity
was assayed in 250 µl of a reaction mixture containing 25 mM
Tris-HCl, pH 7.5, 2.5 mM $CaCl_2$, 4 mg/ml BSA, 400 µg/ml phosphatidylserine and 15 nM [³H]PDBu.   To determine non-specific
binding, incubations were made in the presence of 3 µM unlabelled
PDBu.   The incubation was carried out for 1 h at 4° and
terminated by adding 1.5 ml of ice-cold buffer containing
10 mM $Mg(NO_3)_2$ and 1 mM $CaCl_2$.   Bound [³H]PDBu was determined
by rapid filtration on Whatman GF/C filters pre-soaked in
the reaction-terminating Tris buffer [20].   After rapid washing
with 8 ml Tris-buffer, filters were subjected to liquid scintillation counting after addition to each vial of 10 ml scintillation fluid (Pico-Fluor 15; Packard).

Functional tests.- Cultured rat aortic smooth muscle
cells at the 2nd to 5th passages were washed and equilibrated
at 37° for 20 min in Krebs-Ringer buffer containing 0.2%
BSA and 0.2% glucose.   The medium was then replaced with
identical fresh medium containing the test agents.   After
further incubation at 37° in 95% air/5% CO_2, prostacyclin
production was determined in the medium by a specific RIA
of its stable metabolite 6-keto-$PGF_{1\alpha}$ [21].

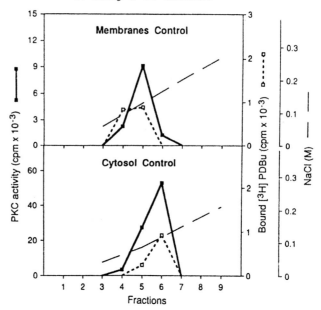

Fig. 1. DEAE-cellulose chromatographic analysis of PKC-activity distribution and [³H]PDBu binding in rat aortic smooth-muscle cell cytosol and detergent-solubilized membranes, prepared from subcultured cells at confluence in a 90-mm petri dish (5.8 ±0.6 mg protein/dish). For chromatography, see text. PKC activity is expressed as cpm ^{32}P incorporated into histones/10 min per 100 µl of each fraction. [³H]PDBu binding represents total binding *less* non-specific binding, as cpm bound to 100 µl of each fraction.

Adrenal glomerulosa cells were resuspended to a final concentration of $3-6 \times 10^5$ cells/ml and incubated under constant agitation at 37° with 95% O_2/5% CO_2 in the presence of various stimulators of steroidogenesis. Finally the aldosterone content of the media was measured by direct RIA [22].

Statistical analysis was performed with analysis of variance, followed by Student's t-test where appropriate. Results are expressed as the mean ±S.E.M.

RESULTS

Effects of ANG–II and PMA on subcellular PKC–activity distribution

PKC activity and [³H]PDBu binding in cytosolic and membranous fractions from aortic smooth muscle cells after DEAE-cellulose chromatography is shown in Fig. 1. A phospholipid-dependent peak of activity for both the membranous and the cytosolic forms of PKC was eluted between 0.05 and 0.12 M NaCl. By subtracting the phospholipid-independent activity from this peak, the specific PKC activity was calculated and expressed

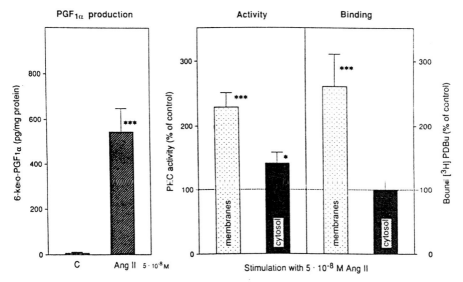

Fig. 2. PKC activity and [³H]PDBu binding activity in the
cytosol and membranes of cultured aortic smooth muscle cells as
affected by incubation with 50 nM ANG-II for 20 min. Portions
of the incubation media were removed for prostacyclin determination, and the cytosolic and membranous fractions of the cells
were assayed for PKC activity and [³H]PDBu binding activity (see
text). Values are means ±S.E.M. as % of controls (n = 7).
[*P <0.05; ***P <0.001.

as cpm ³⁴P incorporated into histone per 10 min by 100 µl
of each fraction. [³H]PDBu binding activity co-eluted with
the PKC peak (Fig. 1).

The membranous and soluble PKC activities represent ~15%
and ~85% respectively. This was also the case in adrenal
glomerulosa cells as we showed previously [15].

Figs. 2 and 3 show that stimulation of smooth muscle
cells with 50 nM ANG-II for 20 min resulted in an increase
in both the membrane-associated and the cytosolic activities
by 150% and 40% respectively. However, [³H]PDBu binding
was increased only in the membranes. In adrenal glomerulosa
cells ANG-II (50 nM) decreased cytosolic PKC activity by
65% within 20 min and correspondingly increased that in membranes (Fig. 3). In both cell types, incubation with 50 nM
PMA for 15 min decreased cytosolic activity by ~75% and
correspondingly increased that in membranes (Fig. 3).

After 30 min of exposure to PMA, membranous PKC activity
fell to near-control levels in adrenal glomerulosa cells
but in aortic smooth muscle cells it was further increased,
to ~340% of controls, although after exposure for 3 h the
activity likewise fell, to 130% of controls (Fig. 3).

Fig. 3. Time-dependent effect of PMA on PKC activity in aortic smooth muscle and adrenal glomerulosa cells, incubated with 50 nM ANG-II or 100 nM PMA for the indicated times. For PKC assay of membranous and cytosolic fractions, see text. Values are means ±S.E.M. as % of controls where n= 4-6; individual values shown where n = 2.

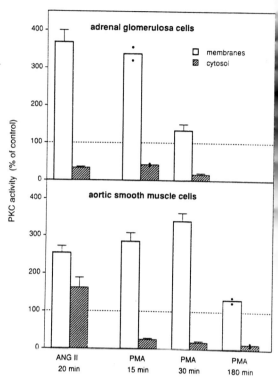

PMA effects on prostacyclin and aldosterone production

Fig. 4 shows that in aortic smooth muscle cells PMA induced a concentration-dependent (10 nM-1 μM) prostacyclin production, whereas 5 μM PMA was needed to induce even a small steroidogenic response in adrenal glomerulosa cells. Fig. 5 shows the influence of PMA on the concentration-dependence of ANG-II-stimulated prostacyclin production. Exposure of smooth muscle cells to 10 nM PMA for 1 h increased basal production by 15% (Fig. 5, on left). Incubation with 0.1 nM ANG-II alone did not change basal production, but stepping-up to 10 nM increased production, by 295% over control and by 113% over PMA-treated cells. Moreover, with 10 nM PMA maximal values of ANG-II-stimulated prostacyclin production were increased by 60% (Fig. 5).

In contrast, exposure of adrenal glomerulosa cells to 100 nM PMA for 1 h did not affect basal aldosterone production (Fig. 6, on left); but incubation with PMA markedly inhibited ANG-II-induced aldosterone production, by 75% if half-maximal and 58% if maximal (Fig. 6).

Cellular responses in long-term incubations with PMA

Down-regulation of PKC activity occurred in both cell types with prolonged PMA treatment. With 50 nM PMA for

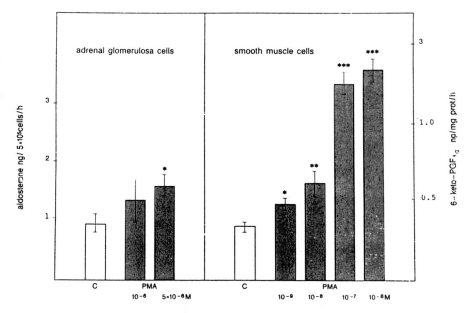

Fig. 4. PMA-induced prostacyclin and aldosterone production
in smooth muscle and adrenal glomerulosa cells, incubated for
1 h without stimulating agent (**C**, *open bars*) or with the
indicated concentrations of PMA (*striped bars*). Units are
ng/h, per 5×10^5 cells for aldosterone (*left panel*) and per
mg cell-homogenate protein for prostacyclin (*right*). Values
are means ±S.E.M. (3-4 expts.). [* $P < 0.05$, **$P < 0.01$, ***$P < 0.001$.

24 h, PKC activity fell by 87 ±2% in the membranes and by
89 ±2% in the cytosol from adrenal glomerulosa cells and
by ~80% in both fractions from aortic smooth muscle cells.
With PMA for 48 h, membranous PKC activity was completely
suppressed and cytosolic activity was decreased by 98%.

In these PKC-depleted smooth muscle cells, basal prostacyclin production was decreased by 26%. PMA (100 nM) did
not induce prostacyclin production, and ANG-II-stimulated production was decreased by 73% compared with untreated cells
(Table 1). In contrast, in PKC-depleted adrenal glomerulosa
cells basal aldosterone production was enhanced by 39 ±4%
and ANG-II-stimulated production by 28 ±6% compared to untreated
cells. Moreover, ANG-II-induced aldosterone production was
not inhibited in the presence of 100 nM PMA, as was the
case in untreated cells (Table 1).

DISCUSSION

This study demonstrates that PKC is present in aortic
smooth muscle and in adrenal glomerulosa cells and that
its activity can be modulated by ANG-II and PMA. In both

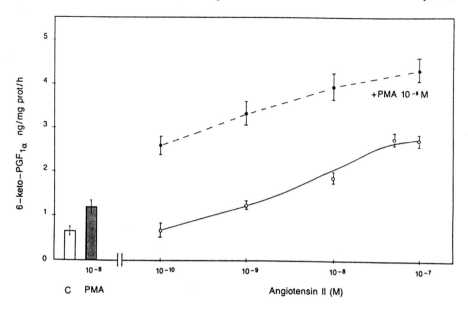

Fig. 5. Effect of PMA on the concentration dependence of
ANG-II-stimulated prostacyclin production in smooth muscle
cells, incubated for 1 h with the indicated concentrations of
ANG-II in the absence (o) and presence (●) of 10 nM PMA. Prostacyclin production by control cells (**C**, *open bars*) and cells
treated with 10 nM PMA alone (*striped bars*) is shown *on left.*
Values are means ±S.E.M. (n =3) as ng/h per mg protein. *From
ref. [14], by permission.*

cell types the enzyme was found to be mostly (85-95%) in
a soluble rather than particulate form. While the activities
of both forms were increased by stimulation of aortic smooth
muscle cells with ANG-II, only in the membranes was specific
[^{3}H]PDBu-binding increased by ANG-II. Thus, ANG-II appears
to enhance cytosolic PKC activity without increasing the
amount of cytosolic enzyme.

Activation of adrenal glomerulosa cells with ANG-II rapidly
decreased cytosolic PKC activity, to a minimum of ~30-40%,
and correspondingly increased the membrane-bound form of the
enzyme. Membrane-association of PKC has been observed in
various cells after hormonal stimulation (e.g. vasopressin
in hepatocytes [23]), in agreement with the possibility that
translocation of PKC from cytosol to membranes, or more
probably conversion from a loosely membrane-bound to a tightly
associated enzyme, corresponds to PKC activation. [PKC translocation features also in arts. #B-3 & #C-3.- *Ed.*]

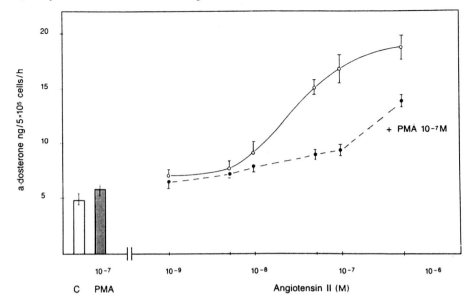

Fig. 6. As for Fig. 5, but aldosterone production by adrenal glomerulosa cells, and 100 (not 10) nM PMA. *(No ref. applicable.)*

Table 1. ANG–II-induced response in PKC-depleted cells, as % of ANG–II-induced response in untreated cells (n = 3–6).

	ANG–II	ANG–II + PMA
Prostacyclin, smooth muscle cells	27 ±4%	29 ±5%
Aldosterone, adrenal glomerulosa cells	128 ±6%	140 ±8%

Exposure of cells of either type to PMA induced a transient increase in membranous PKC activity and a persisting decrease in cytosolic activity. However, PKC activation appears to play a different role in the two cell types as judged from analysis of the effects of PMA on aldosterone or prostacyclin production.-

- In smooth muscle cells PMA induced a concentration-dependent prostacyclin production and it augmented maximal values of ANG–II-stimulated prostacyclin production by 60%. Recent work from our group showed that in these cells PMA also potentiated prostacyclin production induced by arg-vasopressin (AVP) [24]. However, in contrast with this potentiating effect of PMA, it had an inhibitory effect on phospholipase C. This inhibition by PMA was transient both for the $InsP_3$ formation and the Ca^{2+} response induced by AVP, and was inversely correlated with the membrane-associated PKC activity [24]. Various studies with different cell systems have

likewise revealed an inhibitory influence of PMA on hormone-
stimulated phospholipase C activity [25-28]. Thus, the possibility is raised that, while inhibiting ANG-II- or AVP-induced
phospholipase C activation in smooth muscle cells, PMA-induced
PKC activation increases the Ca^{2+} sensitivity of enzymes
involved in prostacyclin production [24].

- In adrenal glomerulosa cells, unlike smooth muscle cells,
PMA did not affect basal steroidogenesis. Moreover, ANG-II-
stimulated aldosterone production was greatly inhibited in
the presence of PMA. These results accord with observations
made by other groups with adrenal glomerulosa cells [17, 29].

PKC following prolonged treatment with phorbol esters

It has been shown in various systems that prolonged
exposure of cells to phorbol esters leads to a depletion
of PKC after the initial redistribution from cytosol to
membranes [30, 31]. This property of phorbol esters was
exploited in this work to evaluate the influence of PKC
on smooth muscle and adrenal glomerulosa cell responsiveness
to ANG-II.

After PMA treatment of adrenal glomerulosa cells for
24 h, membranous and cytosolic PKC activity dropped dramatically. In these PKC-depleted cells, basal and ANG-II-stimulated
aldosterone production were increased compared to untreated
cells. These results suggest that PKC activation exerts
a negative feedback modulating effect on aldosterone production
in these cells.

Compared with adrenal cells, aortic smooth muscle cells
responded very differently to prolonged treatment with PMA.
In these cells it has been shown in our laboratory that
after 24 h PMA reduces membranous and cytosolic PKC activities
by ~80%, yet membranous PKC activity can then still be
tripled (to 60% of control values) by ANG-II addition; the
24-h treatment augmented ANG-II-induced prostacyclin production
by ~60% compared with untreated cells.

We further showed that both PKC activities were suppressed
by PMA after 48 h and could no longer be activated by ANG-II.
ANG-II-stimulated prostacyclin production was decreased by
74% in these PKC-depleted smooth muscle cells compared with
untreated cells. In contrast, the ANG-II-induced rise in
cytosolic free Ca^{2+} was unaffected after pre-treatment with
PMA for 24 or 48 h.

Overall conclusions

The foregoing observations indicate that activation of
PKC, or at least of a particular isoform of PKC, plays
a positive regulatory role during stimulation of prostacyclin

production in aortic smooth muscle cells. In contrast, activation of the PKC system in adrenal glomerulosa cells appears to exert a negative feedback effect on aldosterone production. PKC isoforms might be present at different concentrations, assume several functions and exhibit different sensitivities to degradation induced by PMA [32, 33]. This might explain the discrepancy between the effects of short- and long-term treatments with PMA in aortic smooth muscle cells and in adrenal glomerulosa cells.

Acknowledgements

We thank Ms. C. Gerber-Wicht, W. Dimeck, M. Rey and M. Klein for their excellent technical assistance, and Ms. C. Gerber-Wicht for skilled secretarial help. This study was supported by Grants 3.495-0.86 and 31.227727.89 from the Swiss National Science Foundation.

References

1. Smith, J.B. (1986) Am. J. Physiol. 250, F759-F769.
2. Griendling, K.K., Rittenhouse, S.E., Brock, T.A., Ekstein, L.S., Gimbrone, M.A. Jr. & Alexander, R.W. (1986) J. Biol. Chem. 261, 5901-5906.
3. Catt, K.J., Carson, M.C., Hansdorff, W.P., Leach-Harper, C.M., Baukal, A.J., Guillemette, G., Balla, T. & Aguilera, G. (1987) J. Steroid Biochem. 27, 915-927.
4. Quinn, S.J. & Williams, G.H. (1988) Annu. Rev. Physiol. 50, 409-426.
5. Späth, A. (1988) J. Steroid Biochem. 29, 443-453.
6. Berridge, M.B. (1987) Annu. Rev. Biochem. 56, 159-193.
7. Nishizuka, Y., Takai, Y., Kishimoto, A., Kikkawa, U. & Kaibuchi, K. (1984) Rec. Progr. Hormone Res. 40, 301-345.
8. Capponi, A.M., Lew, P.D., Jornot, L. & Vallotton, M.B. (1984) J. Biol. Chem. 259, 8863-8869.
9. Rasmussen, H. (1986) New Engl. J. Med. 314, 1094-1101.
10. Capponi, A M., Lew, P.D. & Vallotton, M.B. (1985) J. Biol. Chem. 260, 7836-7842.
11. Brock, T.A., Alexander, R.W., Ekstein, L.S., Atkinson, W.J. & Gimbrone, M.A. Jr. (1985) Hypertension 7, 1105-1109.
12. Smith, J.B., Smith, L., Brown, E.R., Sabir, M.A., Davis, J.S. & Farese, R.V. (1984) Proc. Nat. Acad. Sci. 81, 7812-7816.
13. Nabika, T., Velletri, P.A., Lovenberg, W. & Beaven, M.A. (1985) J. Biol. Chem. 260, 4661 4670.
14. Lang, U. & Vallotton, M.B. (1989) Biochem. J. 259, 477-484.
15. Lang, U. & Vallotton, M.B. (1987) J. Biol. Chem. 262, 8047 8050.
16. Ishizuka, T., Miura, K., Nagao, S. & Nozawa, Y. (1988) Biochem. Biophys. Res. Comm. 155, 643-649.

17. Nakano, S., Carvallo, P., Rocco, S. & Aguilera, G. (1990) *Endocrinology 126*, 125-133.
18. Haning, R., Tait, S.A.S. & Tait, J.F. (1970) *Endocrinology 87*, 1147-1167.
19. Takai, Y., Kishimoto, A., Inoue, M. & Nishizuka, Y. (1977) *J. Biol. Chem. 252*, 7603-7609.
20. Sando, J.J. & Young, M.C. (1983) *Proc. Nat. Acad. Sci. 80*, 2642-2646.
21. Wüthrich, R.P., Loup, R., Favre, L. & Vallotton, M.B. (1986) *Am. J. Physiol. 250*, F790-F797.
22. Fredlund, P., Saltman, S. & Catt, K.J. (1975) *Endocrinology 97*, 1577-1586.
23. Hernandez-Sotomayer, S.M.T. & Garcia-Sainz, J.A. (1988) *Biochim. Biophys. Acta 968*, 138-141.
24. Chardonnens, D., Lang, U., Rossier, M., Capponi, A.M. & Vallotton, M.B. (1990) *J. Biol. Chem. 265*, 10451-10457.
25. Brock, T.A., Rittenhouse, S.E., Powers, C.W., Ekstein, L.S., Gimbrone, M.A. & Alexander, R.W. (1985) *J. Biol. Chem. 260*, 14158-14162.
26. Burch, R.M., Ma, A.L. & Axelrod, J. (1988) *J. Biol. Chem. 263*, 4764-4767.
27. Nambi, A., Nambi, P., Whitman, M., Stassen, F.L. & Crooke, S.T. (1986) *Mol. Pharmacol. 31*, 81-84.
28. Troyer, D.A., Gonzales, O.F., Douglas, J.G. & Kreiberg, J. (1988) *Biochem. J. 251*, 907-912.
29. Kojima, I., Shibata, H. & Ogata, E. (1986) *Biochem. J. 237*, 253-258.
30. Stassen, F.L., Schmidt, D.B., Papadopoulos, M. & Saran, H.M. (1989) *J. Biol. Chem. 264*, 4916-4923.
31. Stabel, S., Rodriguez-Pena, A., Young, S., Rozengurt, E. & Parker, P.J. (1987) *J. Cell Physiol. 130*, 111-117.
32. Nishizuka, Y. (1988) *Nature 334*, 661-665.
33. Ase, K., Berry, N., Kikkawa, U., Kishimoto, A. & Nishizuka, Y. (1988) *FEBS Lett. 236*, 396-400.

#B-3

PHOSPHOLIPID METABOLISM, CYTOSOLIC Ca^{2+}, AND THE REGULATION OF PROTEIN KINASE C DURING CELL STIMULATION

Joseph T. O'Flaherty

Section on Infectious Diseases, Department of Medicine, Wake Forest University Medical Center, Winston-Salem, NC 27106, U.S.A.

Cells commonly respond to stimuli by converting endogenous phospholipids into products that raise $[Ca^{2+}]_i$* and activate PKC. Elevated $[Ca^{2+}]_i$ causes PKC to translocate to plasma membrane where it binds with, and becomes activated by, phospholipid-derived DAG. However, using a DAG analogue, $[^{3}H]PDB$, to track PKC movements, we find evidence that this model does not explain all PKC translocation responses. In particular, chemotactic peptides, the leukotriene LTB_4 and PAF stimulated PKC translocation in human neutrophils that were depleted of calcium and unable to raise $[Ca^{2+}]_i$. Intracellular calcium chelators, however, were able to block these responses. Thus, some signal besides $[Ca^{2+}]_i$ per se (viz. an alternate calcium pool or an agent that requires cell calcium for its formation or action) mediates PKC translocation. This signal may regulate PKC movements in diverse cell types.

Humoral agonists stimulate cells by binding to p.m. receptors which trigger an ordered series of events. They induce G-proteins to activate PL-C, which in turn hydrolyzes p.m. GPI's to InsP's and DAG. InsP's release storage calcium and/or open Ca^{2+} channels to raise $[Ca^{2+}]_i$ while DAG stays in p.m. where it binds and activates PKC, the ubiquitous effector enzyme. Elevated $[Ca^{2+}]_i$ plays a critical role in these events: it causes cytosolic PKC to adhere with p.m. and associate with DAG. $[Ca^{2+}]_i$ and DAG thus cooperate to assemble a membranous PKC macrocomplex that actively phosphorylates response-eliciting substrate proteins [1-4]. Since direct elevators of $[Ca^{2+}]_i$ (e.g. ionophores) and PKC activators (e.g. PDB) act in synergy to stimulate cell function, the GPI/$[Ca^{2+}]_i$ axis has gained wide acceptance as an almost universally applicable explanation for PKC translocation and

---

Abbreviations (for* **aa *etc., see Fig. 2 legend).-* $[Ca^{2+}]_i$, cytosolic Ca^{2+} concentration; DAC, diacylglycerol; GPC, glycerophosphocholine(s); GPE, glycerophosphoethanolamine(s); GPI, glycerolphosphatidylinositol(s)†; InsP, inositol phosphate(s); fMLP, N-formyl-Met-Leu-Phe; PAF, platelet-activating factor, namely 1-O-alkyl-2-acetyl-GPC; PDB, phorbol dibutyrate; PKC, protein kinase C; PL(-C), phospholipase (C); p.m., plasma membrane (plasmalemma); PMN, polymorphonuclear neutrophils.

[†*but see Fig. 1 legend*

stimulus-response coupling. However, we find that this axis does not explain all instances of PKC mobilization. In particular, human PMN appear capable of bypassing $[Ca^{2+}]_i$ to effect PKC movements. This bypass likely involves some signal besides $[Ca^{2+}]_i$ that derives from, or is regulated by, phospholipid metabolites. Indeed, the pathways of stimulus-induced phospholipid metabolism produce many potential regulators of PKC.

PHOSPHOLIPID METABOLISM

In addition to cleaving GPI (Fig. 1, path c), cells form DAG from GPC's, through (i) PL-C-mediated hydrolysis of GPC (Fig. 1, path b) or (ii) PL-D-mediated hydrolysis of GPC to phosphatidic acid followed by dephosphorylation of the phosphatidate intermediate [5]. In either case, this DAG production is not coupled with InsP formation and therefore need not involve $[Ca^{2+}]_i$ rises. Furthermore, both routes of GPC hydrolysis can attack 1-O-alkyl ether GPC (Fig. 1, path a) to form 1-O-alkyl-2-acylglycerol [6, 7]. Alkyl ether glycerols have PKC-independent bioactions [8] and also inhibit DAG-induced activation of PKC [8, 9].

The G-proteins of stimulated cells activate not only PL-C but also PL-A$_2$ [10-12]. PL-A$_2$ releases **aa** from the sn-2 position of 1-O-acyl-, 1-O-alkyl- and 1-O-alkenyl-GPC and GPE's (Fig. 1, paths a-d) [13-15]. The resulting lyso phospholipids and free **aa** are bioactive. They trigger GPI turnover, raise $[Ca^{2+}]_i$, activate PKC, and elicit function in diverse cell types [16-23]. Their effects on $[Ca^{2+}]_i$ and PKC may be independent of GPI turnover. Lyso phospholipids have Ca^{2+} ionophore-like properties [17] and **aa** interacts with cell storage pools to trigger calcium extrusion [22]. Similarly, in the presence of Ca^{2+} and phospholipids, lyso GPC directly activates cell-free PKC [19]; **aa** also activates PKC but its effects are Ca^{2+}- and phospholipid-independent [20-23].

In addition to their intrinsic action lyso phospholipids and **aa** are metabolized to other bioactive products. Acetyltransferase(s) convert lyso phospholipids to the corresponding sn-2-acetyl analogues (Fig. 1, paths a-d). The best studied of these, PAF, is an exceedingly potent and broadly acting cell agonist. It operates *via* specific receptors that link to G-proteins, GPI, $[Ca^{2+}]_i$, and phospholipid deactylation [24]: PAF-challenged cells can form all of the derivatives in Fig. 1 [15]. Other acetylated phospholipids, although not yet fully examined, behave like weak PAF analogues in stimulating cells [24]. Free **aa**, on the other hand, is rapidly oxygenated to HETE's, LT's, LX's, PG's and TX's (Fig. 2).

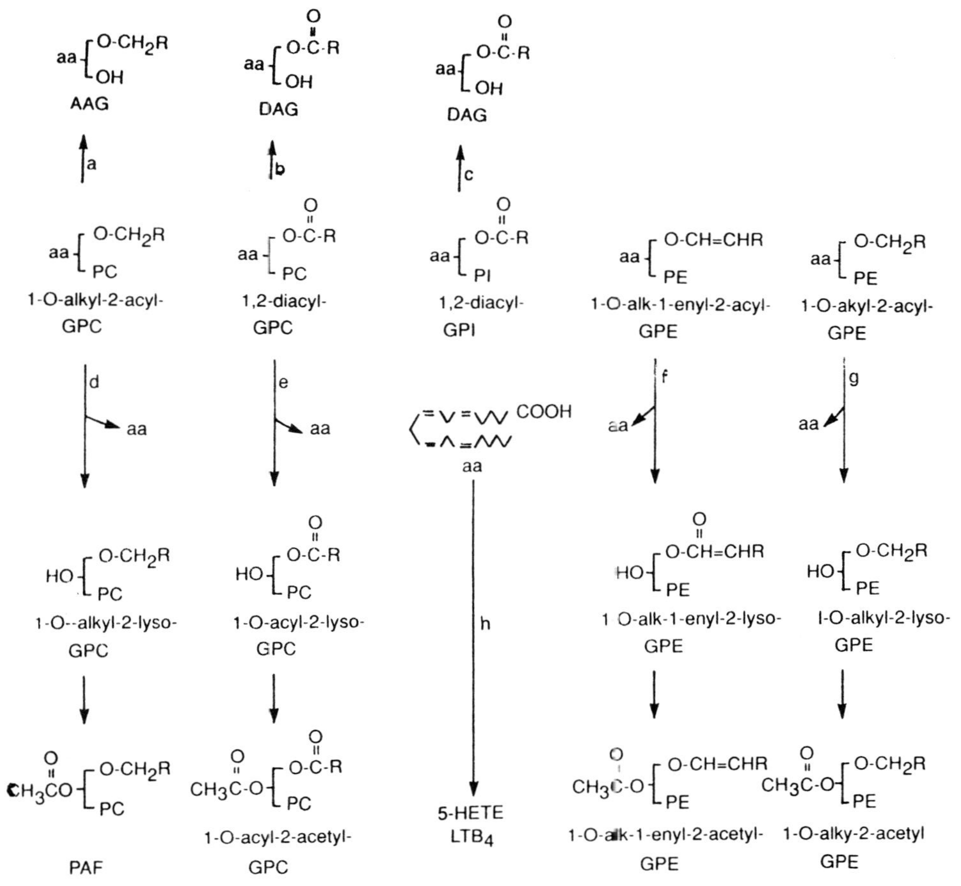

Fig. 1. Pathways of phospholipid metabolism. Stimulated cells use a PL-C to yield DAG by cleavage (path c). They may use an analogous path to form DAG from 1,2-di-acyl-GPC (b) as well as 1-O-alkyl-2-acylglycerol (AAG) from alkyl-ether GPC (a). Alternately, cells may form DAG and AAG from GPC by a 2-step pathway involving PL-D and phosphatidate phosphohydrol-ase (not shown). Finally, cells use a PL-A₂(s) to release arachi-donic ac=d (**aa**) from the *sn-2* position of various GPC, GPE and other phospholipids. Free **aa** is oxygenated to other metabolites (h; see also Fig. 2) and lyso phospho1-pids are acetylated to their corresponding 1-radyl-2-acetyl-phospholipids (paths d, e, f, and g). All of the depicted products influence cell [Ca²⁺]i, PKC and function. *Abbrevia=ions not given above are listed on the title p. or in the legend to Fig. 2. 'GPI' (as Ed. points out; see #B-6 & p. 199) is a 'treacherous' term; commonly it implies -OH at the 1 and 2 glycerol positions.*

Row 1 (DAG/AAG labels):
aa[O-CH₂R / OH] AAG — a — aa[O=C-R / OH] DAG — b — aa[O=C-R / OH] DAG — c

Row 2 (GPC/PI/GPI/PE):
aa[O-CH₂R / PC] 1-O-alkyl-2-acyl-GPC — aa[O=C-R / PC] 1,2-diacyl-GPC — aa[O=C-R / PI] 1,2-diacyl-GPI — aa[O-CH=CHR / PE] 1-O-alk-1-enyl-2-acyl-GPE — aa[O-CH₂R / PE] 1-O-akyl-2-acyl-GPE

PL-A₂ substrate: ⟨=∨=∨∨ COOH / -∧-∧∧⟩ aa

Row 3 (lyso):
HO[O-CH₂R / PC] 1-O-alkyl-2-lyso-GPC — d (aa) — HO[O=C-R / PC] 1-O-acyl-2-lyso-GPC — e (aa) — aa → h → 5-HETE LTB₄ — HO[O=C-R / O-CH=CHR / PE] 1-O-alk-1-enyl-2-lyso-GPE — f (aa) — HO[O-CH₂R / PE] 1-O-akyl-2-lyso-GPE — g (aa)

Row 4 (acetyl products):
CH₃CO[O-CH₂R / PC] 1-O-alkyl-2-acetyl-GPC PAF — CH₃C-O[O=C-R / PC] 1-O-acyl-2-acetyl-GPC — CH₃C-O[O-CH=CHR / PE] 1-O-alk-1-enyl-2-acetyl-GPE — CH₃C-O[O-CH₂R / PE] 1-O-alky-2-acetyl-GPE

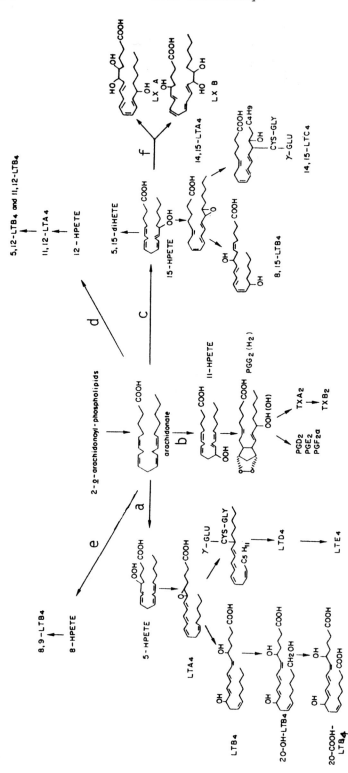

Fig. 2. Pathways of arachidonic acid (**aa**) metabolism. HETE, hydroxyeicosatetraenoates; LT, leuko-trienes; LX, lipoxenes; PG, prostaglandins; TX, thromboxanes. *See p. 367 for amplification provided by the Editor.*

Many of these eicosanoids mimic **aa** in directly activating PKC [25]. However, their most potent effects are receptor-mediated [26, 27]. Thus, LTB$_4$ stimulates PMN by binding to p.m. receptors. Like PAF, it operates through G-proteins to recruit the pathways of Fig. 1 [15, 28, 29]. PGE$_2$ and PGD$_2$, contrastingly, inhibit cell responses to diverse stimuli [27] including PAF and LTB$_4$ [30]. They use p.m. receptors [30] that couple with adenylate cyclase to elevate cAMP [26, 27], and thereby may inactivate cAMP-dependent kinase to phosphophorylate and inactivate G-proteins [31]. Finally, 5-HETE induces PMN to raise $[Ca^{2+}]_i$ by a presumed receptor mechanism. Yet it does not stimulate GPI turnover, PKC mobilization, phospholipid deacylation, or function. Rather, 5-HETE enhances the potency of PAF (but not other stimuli) up to 1000-fold in eliciting these responses [15, 32]. The product seems to act as a selective potentiator of PAF.

Evidently, then, cells use several parallel pathways to convert endogenous phospholipids into a seeming redundancy of products that have overlapping, complementary, synergistic and inhibitory actions on $[Ca^{2+}]_i$ and PKC. Moreover, $[Ca^{2+}]_i$ and PKC themselves can influence the pathways: (i) elevated $[Ca^{2+}]_i$ activates PL-C [33], PL-A$_2$ [12] and enzymes that metabolize **aa** [34, 35], and (ii) activated PKC stimulates DAG production from GPC [5], enhances the formation of PAF and of **aa** metabolites [36-38], promotes $[Ca^{2+}_i]$ extrusion [39] and depresses receptor function. Relevant to the latter point, PKC induces cells to decrease the availability of diverse receptor types including those for PAF [40], LTB$_4$ [29] and, possibly, 5-HETE [41]. It may down-regulate target receptors [42] or interfere with receptor/G-protein coupling [43, 44]. (G-proteins can enhance the ability of receptors to bind their ligands.) In any event, cells treated with PKC activators become completely unresponsive to many exogenous agonists as well as endogenously formed mediators. Thus, regulation of $[Ca^{2+}]$ and PKC by the phospholipid-metabolizing pathways may involve positive and negative feedback loops besides excitatory and inhibitory signals. These considerations are relevant to our anomalous findings on PKC translocation.

ASSAYS FOR PKC MOBILIZATION

Several assays track PKC [45]. ^{32}P-labelled cells incorporate radioactivity into PKC substrate proteins. Assays for this are lengthy and difficult to quantitate; other stimulus-activated kinases often phosphorylate PKC substrates; and phosphoproteins are rapidly dephosphorylated. Accordingly, alternate approaches have been used. Many methods measure PKC translocation to infer PKC activation. They quantitate Ca^{2+}/DAG/phospholipid-dependent histone-phosphorylating activity, [^{3}H]PDB binding, or PKC antibody binding to soluble and particulate fractions of disrupted cells. For example, Ca^{2+}

ionophores, DAG, LTB$_4$ and PAF stimulate PMN to lose soluble and gain p.m. [³H]PDB binding sites [46-49]. All of these methods, however, require lengthy cell processing, whereas PKC translocation responses may reverse within minutes. Furthermore, cell disruption exposes PKC to buffer ingredients (Ca^{2+}, Mg^{2+} and cation chelators) that can promote PKC attachment to membranes or resolubilize membrane-adherent PKC [48, 50-52]. Consequently, assays dependent on cell disruption may miss or even alter PKC movements. Dougherty & Niedel [53] devised a whole-cell [³H]PDB binding assay that avoids such problems. [³H]PDB binds with the DAG receptor on PKC. Apparently, however, the ligand does not penetrate beyond the p.m. and therefore selectively tags p.m. PKC in whole HL-60 cells [53], astrocytes [54] and PMN [55, & see below]. [³H]PDB binding briefly increases when these cells are stimulated with [Ca^{2+}]$_i$-elevating agents, but no such changes occur in cells that are blocked from raising [Ca^{2+}]$_i$. The data suggest that stimulus-induced rises in [Ca^{2+}]$_i$ cause a reversible movement of PKC from a sequestered site in cytosol to a more [³H]PDB-accessible site on surface membrane.

[³H]PDB BINDING TO PMN

Several observations in PMN validate the [³H]PDB binding assay [56, 57].
- (i) PMN incubated with 0.1-256 nM [³H]PDB rapidly incorporate radioactivity. Apparent equilibrium is attained in 1-1.5 min and persists for >60 min. The label is not metabolized during this time.
- (ii) PMN equilibrated with 125 pM [³H]PDB for 5 min release 95% of bound ³H within 2 min of exposure to 10 µM PDB or to a DAG at 100 µM. Moreover, cells pre-treated with excess PDB or a DAG take up little [³H]PDB following various stimuli.
- (iii) PMN depleted of Ca^{2+} or treated with 20 µM sphinganine (which blocks PKC/membrane adherence) show reductions of 25% and 60%, respectively, in [³H]PDB binding.
- (iv) When PMN pre-equilibrated with [³H]PDB are exposed to the Ca^{2+} ionophore ionomycin, [Ca^{2+}]$_i$ and [³H]PDB binding increase in concert (Fig. 3A). These changes do not occur in Ca^{2+}-depleted PMN (Fig. 3B).
- (v) Disrupted PMN have 10^6 accessible PDB binding sites per original cell. Whole PMN ([Ca^{2+}]$_i$ = 50 nm), however, contain only 23×10^5 accessible PDB binding sites per cell. This number falls by 30% in cells depleted of Ca^{2+} ([Ca^{2+}]$_i$ = 15 nM) and rises by 35% in PMN challenged with 10 µM ionomycin ([Ca^{2+}]$_i$ = 250 nM). Pre-treatment of the PMN with sphinganine (30 µM), PDB (10 µM) and DAG (100 µM) almost completely blocks the ionomycin effect.
Thus there is reversible and saturable binding of [³H]PDB to the DAG receptor on PKC; the tritium tags a membrane-adhered subfraction of cellular PKC, and this fraction varies with [Ca^{2+}]$_i$.

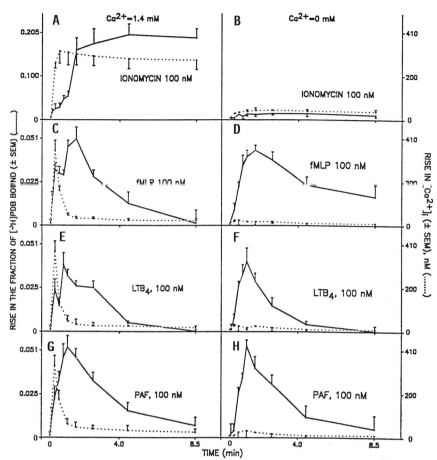

Fig. 3. Rises in [³H]PDB binding *(solid lines)* and $[Ca^{2+}]_i$ *(interrupted lines)* in control (**A, C, E & G**) and Ca^{2+}-depleted (**B, D, F & H**) PMN after challenge with the indicated stimuli. Values represent the rise in the fraction of [³H]PDB bound by 5×10^6 PMN or rise in fura 2-monitored $[Ca^{2+}]_i$. Data were corrected for the changes occurring in PMN challenged with stimulus vehicle. Each point is the mean of >7 experiments; the bar represents the S.E.M.

EFFECTS OF RECEPTOR AGONISTS

LTB_4, PAF and a chemotactic peptide fMLP stimulated PMN to increase $[Ca^{2+}]_i$ and [³H]PDB binding (C, E & G in Fig. 3). The two parameters rose in parallel and returned to control values by 5 min. To quantitate PDB receptor numbers, PMN were challenged with 100 nM of a stimulus (or stimulus vehicle) for 1.25 min, then treated with PDB-[³H]PDB (125 pM to 10 µM) for 1.25 min and centrifuged through silicone oil to isolate free and bound ligand. Scatchard analysis of binding data revealed that the stimuli (100 nM) caused 50-80% increases in binding-site numbers. With PMN stimulation for 5 min rather than 1.25 min, no such changes were seen. Furthermore, a PAF receptor antagonist blocked PAF and an fMLP antagonist blocked fMLP. Neither agonist altered responses to the opposite stimulus or to LTB$_4$. Finally, PMN pre-treated for 5 min with PAF (20 nM) had minimal increases in [^{3}H]PDB binding when challenged with 100 nM PAF, yet showed *enhanced* [^{3}H]PDB binding responses to fMLP. PAF thus desensitized PMN to itself while *priming* the cells to fMLP. The data indicate that fMLP, LTB$_4$ and PAF act through their respective receptors to stimulate PKC translocation. Their effects on PKC movements seem [Ca^{2+}]$_i$-triggered.

We tested the latter point using Ca^{2+}-depleted cells. PMN were incubated in Ca^{2+}-free buffer containing 1 µM EGTA for 30 min at 37°, twice washed in the same buffer and, if not left Ca^{2+}-depleted, rendered Ca^{2+}-repleted by treatment with 1.4 mM CaCl$_2$. Ca^{2+}-repleted PMN showed the normal increases in [Ca^{2+}]$_i$ and [^{3}H]PDB binding in response to 100 nM fMLP, LTB$_4$ or PAF. As expected, in Ca^{2+}-depleted cells these agonists failed to alter [Ca^{2+}]$_i$ appreciably; yet this challenge in these same cells caused near-normal rises in [^{3}H]PDB binding (D, F & H in Fig. 3). PDB (10 µM), a DAG (100 µM) or sphinganine (30 µM) blocked [^{3}H]PDB binding responses in Ca^{2+}-depleted as well as Ca^{2+}-repleted PMN. Moreover, Scatchard analyses of PDB binding to the Ca^{2+}-depleted cells indicated that each stimulus (100 nM) caused 35-70% increases in PDB binding site numbers at 1.25-2.5 min. Thus, receptor agonists cause near-normal PKC translocation responses in the absence of any detectable rises in [Ca^{2+}]$_i$. We note that Ca^{2+}-repleted PMN had rapid and frequently biphasic (at ~45 and ~90 sec) rises in [^{3}H]PDB binding whereas the responses in Ca^{2+}-depleted cells developed more slowly (at 90 sec) and were monophasic (Fig. 3). The receptor agonists therefore may trigger a 2-step translocation event wherein [Ca^{2+}]$_i$ causes early (at 45 sec) movements of the enzyme while some other signal acts thereafter.

EFFECT OF INTRACELLULAR Ca^{2+} CHELATION

Ca^{2+}-depleted PMN loaded with Ca^{2+} chelators (quin 2 AM, fura 2 AM, or 5',5'dimethyl BAPTA) were virtually unresponsive to fMLP, LTB and PAF (Fig. 4, interrupted lines). However, the same cells, when incubated with Ca^{2+} for 20 min, responded normally to the agonists (Fig. 4, solid lines); a weak chelator analogue, 4'-4'-difluoro BAPTA, did not inhibit [^{3}H]PDB binding responses; and chelator-loaded PMN did not leak lactate dehydrogenase or take up trypan blue.

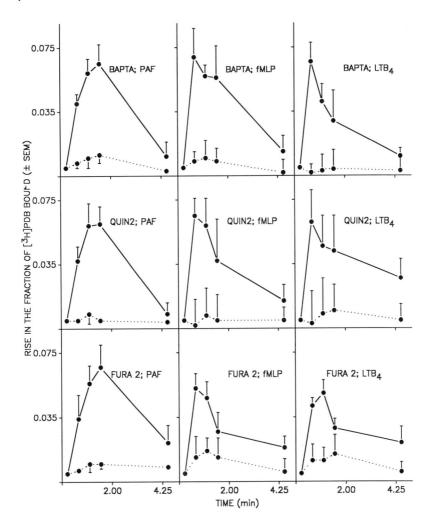

Fig. 4. Effect of intracellular Ca^{2+} chelators on $[^3H]PDB$-binding responses. PMN ($5 \times 10^7/ml$) were incubated with 10 μM 5',5'-dimethyl BAPTA, 31.6 μM quin 2, or 31.6 μM fura 2 for 30 min in the presence of 1 μM EGTA. Cells were then twice washed and suspended ($5 \times 10^6/ml$) in buffer (1 μM EGTA), incubated with (*solid lines*) or without (hence still depleted; *interrupted lines*) 1.4 mM Ca^{2+}, and challenged with agonist (100 nM). Values have a stimulus-vehicle correction as in Fig. 3; >6 experiments.
.. ..

The chelators thus appeared to inhibit PKC translocation by decreasing the bioavailability of intracellular Ca^{2+} rather than by non-specifically injuring the cells. Some intracellular pool of calcium, but not incremental $[Ca^{2+}]_i$ increases, seems required for the binding responses to the receptor agonists.

CONCLUSIONS

Our studies indicated that PMN translocate PKC using a signal that is not $[Ca^{2+}]_i$ *per se* but may be some other pool of cell calcium. Changes in p.m. or sub-p.m. Ca^{2+}, for instance, might cause the enzyme to adhere with surface membrane. Alternatively, the signal may be unrelated to Ca^{2+} but, require cell calcium to form or act. Various lipid derivatives as considered in the PHOSPHOLIPID METABOLISM section (lyso phospholipids, **aa**, **aa** metabolites), or stimulus-induced changes in membrane phospholipid composition, may arise in cells by a Ca^{2+}-enhanced mechanism and influence PKC in a manner that is relatively or completely independent of Ca^{2+}. Relevant to the latter point, it should be noted that Ca^{2+} may serve to bridge PKC and phospholipids [58] and therefore play a permissive, rather than signalling, role in $[Ca^{2+}]_i$-independent translocation responses. Finally, it seems possible that translocation responses involve stimulus-induced loss of an agent that blocks PKC mobilization (e.g. alkyl ether glycerols).

Regardless of the signal's nature or mechanism of action, however, the standard model is unable to explain: (i) PKC translocation that occurs in Ca^{2+}-depleted PMN; (ii) translocation responses that appear triggered by Ca^{2+} at submicromolar levels or that lag significantly behind changes in $[Ca^{2+}]_i$ as exemplified in Fig. 3 (PKC translocation in cell-free systems commonly requires >10 μM of the cation and occurs immediately after rises in ambient Ca^{2+}); or (iii) cellular responses to the many agents that do not raise $[Ca^{2+}]_i$. The signals detected by our studies seem better able to effect such reactions. Hence, the pathway independent of $[Ca^{2+}]_i$-transients may be involved in PKC movements under diverse conditions of cell stimulation.

Acknowledgements

Original work was supported by U.S. National Institutes of Health research grants HL-27799 and HL-26257. Jean Kimbrell is thanked for help in document preparation.

References

1. Casey, P.J. & Gilman, A.E. (1988) *J. Biol. Chem. 263*, 2575-2580.
2. Neer, E.J. & Clapham, D.E. (1988) *Nature 333*, 129-134.
3. Berridge, M.J. (1990) *J. Biol. Chem. 265*, 9583-9586.
4. Nishizuka, Y. (1988) *Nature 334*, 661-665.
5. Exton, J.H. (1990) *J. Biol. Chem. 265*, 1-4.

6. Rider, L.G., Dougherty, R.W. & Niedel, J.E. (1988) *J. Immunol. 140*, 200–207.

7. Agwu, D.E., McPhail, L.C., Chabot, M.C., Daniel, L.W., Wykle, R.L. & McCall, C.E. (1989) *J. Biol. Chem. 264*, 1405–1413.

8. Bass, D.A., McPhail, L.C., Schmitt, J.D., Morris-Natschke, S., McCall, C.E. & Wykle, R.L. (1989) *J. Biol. Chem. 264*, 19610–19617.

9. Daniel, L.W., Small, G.W. & Schmitt, J.D. (1988) *Biochem. Biophys. Res. Comm. 151*, 291–297.

10. Burch, R.M., Luini, A. & Axelrod, J. (1986) *Proc. Nat. Acad. Sci. 83*, 7201–7205.

11. Silk, S.T., Clejan, S. & Witkom, K. (1989) *J. Biol. Chem. 264*, 21466–21469.

12. Channon, J.Y. & Leslie, C.C. (1990) *J. Biol. Chem. 265*, 5409–5413.

13. Chilton, F.H. (1989) *Biochem. J. 258*, 327–333.

14. Tessner, T.G. & Wykle, R.L. (1987) *J. Biol. Chem. 262*, 12660–12664.

15. Tessner, T.G., O'Flaherty, J.T. & Wykle, R.L. (1989) *J. Biol. Chem. 264*, 4794–4799.

16. Gerrard, J.M., Kindon, S.E., Peterson, D.A., Peller, J., Krantz, K.E. & White, J.G. (1979) *Am. J. Path. 96*, 423–426.

17. Ohsako, S. & Deguchi, T. (1981) *J. Biol. Chem. 256*, 10945–10948.

18. Lapetina, E.G., Billah, M.M. & Cuatrecasas, P. (1981) *J. Biol. Chem. 256*, 11984–11987.

19. Oishi, K., Raynor, R.L., Charp, P.A. & Kuo, J.F. (1988) *J. Biol. Chem. 263*, 6865–6871.

20. Kolesnick, R.N., Musacchio, I., Thaw, C. & Gershengorn, M.C. (1984) *Am. J. Physiol. 246*, E-458–E462.

21. Kowalska, M.A., Rao, A.K. & Disa, G. (1988) *Biochem. J. 253*, 255–262.

22. Naccache, P.H., McColl, S.R., Caon, A.C. & Borgeat, P. (1989) *Br. J. Pharmacol. 97*, 461–468.

23. Chow, S.C. & Jondal, M. (1990) *J. Biol. Chem. 265*, 902–907.

24. O'Flaherty, J.T. & Wykle, R.L. (1989) in *Frontiers in Pharmacology and Therapeutics* (Barnes, P.J., Page, C.P. & Henson, P.M., eds.), Blackwell Scientific, Oxford, pp. 117–137.

25. Hansson, A., Serhan, C.N., Haeggstrom, J., Ingelman-Sundberg, M. & Samuelsson, B. (1986) *Biochem. Biophys. Res. Comm. 134*, 1215–1222.

26. Robertson, R.P. (1986) *Prostaglandins 31*, 395–411.

27. Haluska, P.V., Mais, D.E., Mayeux, P.R. & Morinelli, T.A. (1989) *Annu. Rev. Pharm. Toxicol. 10*, 213–239.

28. O'Flaherty, J.T., Redman, J.F. & Jacobson, D.P. (1990) *J. Cell. Physiol. 142*, 299–308.

29. as for 28., *J. Immunol. 144*, 1909–1913.

30. Rossi, A.G. & O'Flaherty, J.T. (1989) *Prostaglandins 37*, 641–653.

31. McAtee, P. & Dawson, G. (1990) *J. Biol. Chem. 265*, 6788-6793.

32. Rossi, A.G. & O'Flaherty, J.T. (1991) *Lipids*, in press.

33. Smith, C.D., Cox, C.C. & Synderman, R. (1986) *Science* .... *232*, 97-100.   [10988.

34. Rouzer, C.A. & Kargman, S. (1988) *J. Biol. Chem. 263*, 10980-

35. Puustinen, T., Scheffer, M.M. & Samuelsson, B. (1988) *Biochim. Biophys. Acta 960*, 261-267.

36. Parker, J., Daniel, L.W. & Waite, M. (1987) *J. Biol. Chem. 262*, 5385-5393.

37. Liles, W.C., Meier, K.E. & Henderson, W.R. (1987) *J. Immunol. 138*, 3396-3402.

38. McIntyre, T.M., Reinhold, S.L., Prescott, S.M. & .... Zimmerman, G.A. (1987) *J. Biol. Chem. 262*,15370-15376.   [1009.

39. Perlanin, A. & Synderman, R. (1989) *J. Biol. Chem. 264*,1005-

40. O'Flaherty, J.T., Jacobson, D.P. & Redman, J.F. (1989) *J. Biol. Chem. 264*, 6836-6843.

41. *as for* 40. (1988) *J. Immunol. 140*, 4323-4328.

42. Drummond, A.H. & Macintyre, D.E. (1985) *Trans. Int. Pharmacol. Soc.*, 233-234.

43. Katada, T., Gilman, A.G., Watanabe, Y., Bauer, S. & Jakobs, K.H. (1985) *FEBS Lett. 151*, 431-437.

44. Sagl-Eisenberg, R. (1989) *Trans. Int. Biochem. Soc. 14*,355-357.

45. Blackshear, P.J. (1988) *Am. J. Med. Sci. 296*, 231-240.

46. Nishihira, J. & O'Flaherty, J.T. (1985) *J. Immunol. 135*, 3439-3447.

47. Nishihara, J., McPhail, L.C. & O'Flaherty, J.T. (1986) *Biochem. Biophys. Res. Comm. 134*, 587-594.

48. O'Flaherty, J.T. & Nishihara, J. (1987) *J. Immunol. 138*, 1889-1895.

49. *as for* 48., *Biochem. Biophys. Res. Comm. 148*, 575-581.

50. Melloni, E., Pontremoli, S., Michetti, M., Sacco, O., Sparatore, B., Salamino, F. & Horecker, B.L. (1985) *Proc. Nat. Acad. Sci. 82*, 6435-6439.

51. Pontremoli, S., Melloni, E., Michetti, M., Salamino, F., Sparatore, B., Sacco, O. & Horecker, B.L. (1986) *Biochem. Biophys. Res. Comm. 136*, 228-234.

52. Phillips, W.A., Fujiki, T.,Rossi, M.W., Korchak, H.M. & Johnston, R.B. (1989) *J. Biol. Chem. 264*, 8361-8365.

53. Dougherty, R.W. & Niedel, J.E. (1986) *J. Biol. Chem. 261*, 4097-4100.

54. Trilivas, I. & Brown, J.H. (1989) *J. Biol. Chem. 264*, 3102-3107.

55. Badwey, J.A., Robinson, J.M., Horn, W., Soberman, R.J., Karnovsky, M.J. & Karnovsky, M.L. (1988) *J Biol. Chem. 263*, 2779-2786.

56. O'Flaherty, J.T., Jacobson, D.P., Redman, J.F. & Rossi, A.G. (1990) *J. Biol. Chem. 265*, 9146-9152.

57. O'Flaherty, J.T., Redman, J.F., Jacobson, D.P. & Rossi, A.G. (1990) *J. Biol. Chem. 265*, 21619-21623.

58. Ganong, B.R., Loomis, C.R., Hannun, Y.A. & Bell, R.M. (1986) *Proc. Nat. Acad. Sci. 83*, 1184-1188.

#B-4

THE ROLE OF GTP-BINDING PROTEINS IN
MEMBRANE FUSION EVENTS

J.G. Comerford and A.P. Dawson

School of Biological Sciences,
University of East Anglia, Norwich NR4 7TJ, U.K.

*With rat-liver microsomal fractions, GTP under suitable
conditions can cause fusion of small vesicles into much
larger structures. Small GTP-binding proteins as present
in our microsomes may be involved. Amongst experimental
approaches now surveyed are e.m.*, light scattering and,
especially convenient, fluorescence resonance energy transfer
between appropriate membrane probes. The membrane fusion
system has analogies to that associated with the Golgi stack.*

In earlier studies [1] on liver microsomal preparations,
which have a pronounced ATP-driven Ca^{2+} uptake system, we
found that GTP in the presence of PEG* caused slow release
of some of the accumulated Ca^{2+} and, concomitantly, $Ins(1,4,5)_3$-
induced release was enormously enhanced. Our approaches
and observations were described earlier in this book series
[2]. GTP effects have now been observed in a wide variety
of Ca^{2+}-transporting preparations of microsomes and permeabilized cells, as reviewed [3, 4]. There is strong evidence
from e.m., light scattering studies and fluorescence resonance
energy transfer data [5, 6] that the primary effect of GTP
on the microsomes is to cause fusion of small vesicles
into larger structures, offering an explanation for the effects
on Ca^{2+} transport [3]. In other systems, evidence for
permanent fusion is lacking, and Gill & co-workers [4] have
suggested that membrane junctions produced by GTP can be
transient in nature, allowing Ca^{2+} to move in a (presumably)
controlled fashion from one compartment to another. These
effects seem to require GTP hydrolysis, since the non-hydrolyzable
analogue GTP[S] is a potent inhibitor of the effects of
GTP on Ca^{2+} movements.

Recently there has been a surge of interest in membrane
fusion events inside cells. Two very clear-cut instances
where fusion is important are in secretion, where secretory
vesicles must fuse with the plasma membrane, and in packaging
of secreted material by vesicle trafficking through the Golgi

---

**Abbreviations.*- e.m., electron microscopy; e.r., endoplasmic
reticulum; GTP[S], guanosine-5-O-(3-thiotriphosphate); NEM, N-
ethylmaleimide; DTT, dithiothreitol; PEG, polyethylene glycol (M_r=
8000). For the probes R18 and F18, see text.

stack [7].   The trafficking entails both budding and fusion
events, since small vesicles which have budded off from
the e.r. can fuse with Golgi cisternae.   In secretion,
GTP seems to be strongly involved in some way, although
it is unclear whether GTP hydrolysis is required: thus some
secretory processes are stimulated by GTP[S] (non-hydrolyzable)
[8].   GTP hydrolysis seems requisite for membrane trafficking
through the Golgi stack, since non-hydrolyzable analogues
are inhibitory [7].

The latter situation being similar to that in rat liver
microsomal vesicles, possibly similar molecular mechanisms
are involved, although their nature and components are as
yet rather obscure.   A notably pertinent observation concerns
the YPT1 mutant of yeast which is deficient in membrane
trafficking [9].   Its gene product is a small GTP-binding
protein of M_r 21 kDa, one of a large family, or series
of families, of small GTP-binding proteins [10], at least
some of which are quite widely agreed to be involved in
membrane trafficking and targetting [11].   A further protein
component that has been implicated in the Golgi trafficking
system is an NEM-sensitive protein of M_r 76 kDa [12].

Microsomal membranes from rat liver contain a range
of small GTP-binding proteins, and we have some circumstantial
evidence for their involvement in membrane fusion [13].   We
now consider some of the experimental approaches we have
used to study this system, and the scope for using such
microsomes (which are easy to prepare in quantity) as a
tool to study the mechanism of GTP-induced membrane fusion.

Microsome preparation from liver.- Our Vol. 19 article
[2] gave a detailed description, with relevant literature;
hence only an outline is now given.   A Dounce homogenate
in 250 mM sucrose/1 mM dithiothreitol/1 mM EGTA buffered to
pH 7.0 is centrifuged at low **g** and then at 8000 **g** for 10 min.
The supernatant furnishes, at 36000 **g** for 20 min, a pellet
which is washed with re-centrifugation; the medium contains 10 mM
KCl (not EGTA), so removing EGTA and seemingly aiding repro-
ducible vesicle formation.   The fraction is strongly enriched
in glucose-6-phosphatase and low in cytochrome oxidase [14].
With storage overnight at 0°, or with freezing, membrane
fusion and Ca^{2+} transport activities are largely lost.

MEASUREMENT OF MEMBRANE FUSION

Methods have been developed for measuring   vesicle
transport through the Golgi stack [15].   Although quite
complex, they have the great advantage that some particular
biochemical event is being measured, e.g. attachment of *N*-acetyl-

glucosamine to vesicular stomatitis virus, which defines the particular vesicle fusion event that has occurred. The changes in Ca^{2+} transport properties accompanying vesicle fusion in liver microsomes are also well defined biochemically; but the nature of the compartments which are fusing, and the physiological function of the fusion, are not understood. Although the transport changes can be measured, it is generally more convenient (if less specific) to measure fusion by other means: e.m. and light scattering can be used, but the most straightforward, fast and generally applicable technique we have found is an adaptation of the method of Keller and co authors [16].

This method measures fluorescence resonance energy transfer between fluorescein- and rhodamine-labelled probes, viz. octadecyl rhodamine B (R18) and 5-(N-octadecanoyl)amino fluorescein (F18). Both are incorporated rapidly and essentially irreversibly into lipid bilayers, and excitation of fluorescein (at 460 nm) can result in rhodamine emission (at 595 nm) if the probes are very close together in the same bilayer. For success with the technique it is necessary to work at the correct probe/membrane ratio. Because of the potential lack of specificity in measurement of fusion in isolation, eventually a biochemical parameter must always be measured – in our case effects on Ca^{2+} transport. Under all experimental circumstances so far studied we have had complete correlation between alterations in fusion and alterations in Ca^{2+} transport.

The procedure used for probe incorporation and fluorescence measurement is essentially as published [6]. The membranes (2.5 mg protein) are suspended in 0.5 ml (total vol.) of 150 mM sucrose/50 mm KCl/10 mM HEPES–KOH pH 7.0/2 mM dithiothreitol/0.2 mM EGTA. R18 (1 µl of 3.4 mM stock solution) is added to half of the suspension, and F18 (0.6 µl of 4.1 mM stock solution) to the other half, with vigorous vortexing whilst each is added. After incubation for 5 min at 30° to allow incorporation of probes, the two halves are recombined with thorough mixing and added to the fluorimeter cell, which already contains 0.5 ml of 150 mM sucrose/50 mM KCl/10 mM HEPES-KOH pH 7.0/10% (w/v) PEG/4 mM $MgCl_2$/10 mM ATP. After mixing, vesicle fusion is initiated after 5 min by adding GTP to 40 µM. Energy transfer from F18 to R18 is measured by following the fluorescence increase at 595 nm, with excitation at 460 nm; we use a Shimadzu RF5000 fluorimeter.

Although the reaction medium contains ATP, this does not appear to be essential for the fusion reaction. Fig. 1 shows fluorescence traces in the presence and absence of ATP, where the initial rates of change of fluorescence are similar even when possible traces of endogenous ATP are

Fig. 1. Lack of an ATP requirement for membrane fusion. The fluorescence measurements were performed as in the text, except that in (**b**) Mg^{2+} was initially 0.6 mM and 0.6 mM glucose and 3.5 u./ml hexokinase were present (--- = time course if absent). Added where *arrowed*: $MgCl_2$ (2 mM)/ATP (5 mM), or GTP (0.3 mM).

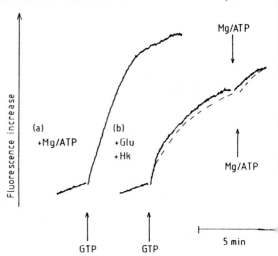

removed by glucose/hexokinase pre-treatment. However, in the absence of ATP but not in its presence there is a rapid fall-off in rate. With ATP present, the lifetime of added GTP is prolonged due both to GTP regeneration from ATP by endogenous nucleoside diphosphate kinase and to protection by ATP of GTP against hydrolysis by non-specific nucleoside phosphatases.

While we have generally included 5% PEG in our media, PEG is not a universal requirement for demonstrating effects of GTP on Ca^{2+} movements [3]. Our view has been that PEG, because of its water-structuring properties, allows membranes to come close together, thereby permitting fusion to be observed at relatively low membrane concentrations. Certainly GTP had no effect on the fluorescence trace under the conditions of Fig. 1 in the absence of PEG. In fact, this view would predict that other water-structuring agents, or very high membrane concentrations, would overcome the PEG requirement. Both these predictions hold good. Bovine serum albumin can, to a limited extent, allow GTP-dependent fusion (Fig. 2A), as can 5% polyvinylpyrrolidone (data not shown). Importantly, fusion can readily be observed in the absence of PEG if the membrane concentration is increased 10-fold (Fig. 2B).

DETECTION OF GTP-BINDING PROTEINS

It is self-evident that, since GTP hydrolysis is required for membrane fusion, a protein which can bind and hydrolyze GTP must be involved in the mechanism. However, identification of the particular species has been, and continues to be, a considerable problem.

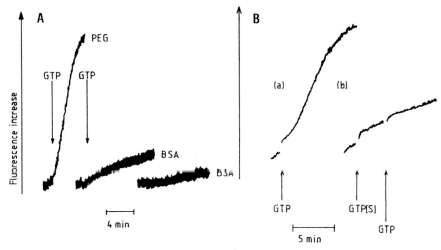

Fig. 2. Fusion in the absence of PEG. **(A)** shows a comparison between fusion in an assay medium containing 5% PEG and one where 3% BSA was substituted for PEG. Other conditions were as described in the text. Trace *on right* is a control with 3% BSA present but with no GTP addition. **(B)** shows fusion in the absence of water-structuring agents. Present in the assay medium: 18 mg/ml microsomal protein, 0.5 mM $MgCl_2$, and R18 and F18 at 10 times the standard concentrations; ATP and PEG were omitted. GTP was added *(arrows)* to 2 mM final concentration, and GTP[S] *(arrow)* to 0.2 mM final concentration.

Initially we identified two peptides, of M_r 38 and 17 kDa, which were specifically phosphorylated by the γ-phosphate of GTP under conditions of fusion [17]. However, we now think it unlikely that either of these proteins is directly involved, for two reasons. (1) As first shown by Nichitta & Williamson [18], PEG stimulates a GTPase activity in rat-liver microsomes. Stimulation of GTPase activity is to be expected if PEG allows a GTP-binding fusion protein to complete its catalytic cycle, as exemplified in Fig. 3 by data from our laboratory. However, we can find no effect of PEG on the phosphorylation level of the proteins, which would be a firm expectation if a change in steady state turnover were involved. (2) Fusion is very susceptible to proteolysis (see below), and trypsin digestion, which completely blocks fusion, does not appear to affect GTP-dependent protein phosphorylation [13].

Accordingly, we looked elsewhere. In particular, Bhullar & Haslam [19] identified a group of small (20-30 kDa) GTP-binding proteins in a variety of tissues which are capable of binding GTP after SDS-gel electrophoresis and electroblotting

Fig. 3. Stimulation of GTPase activity by PEG. Besides sucrose /KCl/buffer/DTT as in the probe studies (see text), the medium contained 2 mM adenylyl-imidodi-phosphate/1 mM ATP / 1.5 mM $MgCl_2$ /5 μM GTP/0.5 μCi [γ-^{32}P]GTP / 5 mM creatine phosphate/0.4 mg.ml^{-1} creatine kinase/1 mg.ml^{-1} microsomal protein/(when added) 5% w/v PEG. Stopping solution (0.1 ml portions of assay mixture added to 0.9 ml):- 5% charcoal/1% PEG/10 mM H_3PO_4. After vortexing and centrifuging (12000 g, 1 min), liberated ^{32}P in 0.5 ml aliquots from supernatants was scintillation-counted.

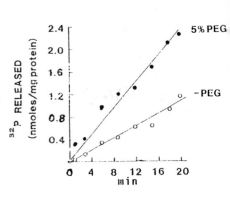

onto nitrocellulose. They defined these as 'Gn' proteins. Such proteins have now been widely described, and probably contain representatives from the *ral, rab, rap, rho* and *smg* families [10]. We have detected proteins of this sort in rat-liver microsomes [13] and, in the light of the suggestions discussed above that small GTP-binding proteins may be involved in membrane trafficking, regard them as plausible candidates for involvement in fusion. Our general methods for studying these proteins are modifications of standard procedures [19].

SDS-PAGE AND ELECTROBLOTTING OF PROTEINS

Microsomal membrane proteins are solubilized in a buffer containing 80 mM TRIS pH 6.8/100 mM DTT/2% (w/v) SDS/15% (v/v) glycerol and 0.006% (w/v) bromphenol blue. Samples (40 μg total protein) are heated at 97° for 3 min, and separated by SDS-PAGE (16% gels) as described by Drøbak and co-authors [20]. Gels obtained are soaked in transfer buffer [384 mM glycine/50 mM TRIS pH 8.5/20% (v/v) methanol] prior to electro-transfer, at room temperature, to nitrocellulose using an LKB Novablot dry-blotter apparatus (120 mA for 2 h or 100 mA for 1 h, depending on the gel surface area).

Incubation of the blots with [γ-^{32}P]GTP or [^{32}S]GTP[S], of specific activity 30 Ci/mmol (Amersham Internat[l].) and 1320 Ci/mmol (NEN-DuPont) respectively, is done essentially as we have described [21]. The blots are pre-equilibrated in the incubation buffer, 50 mM TRIS-HCl pH 7.5/0.3% (v/v) Tween 20/12 μM $MgCl_2$/1 mM DTT. After transfer to fresh incubation buffer with the addition of ATP (10 μM) and 1-2 μCi of either [γ-^{32}P]GTP or [^{35}S]GTP[S], the strips are incubated for 20 min at room temperature, and then washed 3 times with incubation buffer and air-dried. Note that DTT is required in the incubation buffer to block non-specific binding of [^{35}S]GTP[S] to the nitrocellulose, while ATP blocks non-specific binding of nucleotides. Labelled polypeptides are detected by autoradiography [21].

An important methodological point has emerged from these studies. Using [^{35}S]GTP[S] instead of [γ-^{32}P]GTP leads to dramatic improvements firstly in the resolution of bands obtainable (Fig. 4) and secondly, in cell extracts, in the number of bands that can be detected. Thus, using [^{35}S]GTP[S], *ras p21* can be identified in over-expressing cell lines while this is not the case when [γ-^{32}P]GTP is used.

INVOLVEMENT OF GTP-BINDING PROTEINS IN FUSION

Paiement and co-authors [22] found that limited proteolysis with trypsin prevented GTP-induced fusion of microsomal membranes. We have made similar observations [13], extending the study to include other proteolytic enzymes. Effects of GTP on fusion (measured by fluorescence) and Ca^{2+} transport are lost in parallel. While GTP-dependent phosphorylation of the 38 kDa polypeptide is unaffected by proteolysis, the Gn proteins show marked changes in apparent M_r under these conditions (Fig. 4). We have found a good correlation between the degree of loss of the heavier Gn proteins and loss of membrane fusion [13].

It is difficult to assess loss of the low-M_r Gn proteins, because the larger ones move into that area of the gel on trypsin treatment (Fig. 4). However, another marked effect of trypsin treatment is that the Gn proteins, which in rat-liver microsomes are totally membrane-bound, become soluble and smaller (Fig. 5). This strongly suggests that all these proteins are bound to the membrane by a relatively short (~3 kDa) stretch of polypeptide, which is joined to the cytosolic GTP-binding domain by a proteolysis-sensitive hinge region. Similar observations have been made by Lanoix & co-authors [23] and by Eide & co-authors [24], suggesting a similar susceptible hinge in G-protein subunits. The correlation between fusion and Gn proteins in susceptibility to proteolysis encourages us to the view that one or more of these proteins is directly involved in membrane fusion.

The Gn proteins apparently are integral membrane proteins and cannot be extracted by high salt concentrations or chaotropes. However, they are extractable by a variety of detergents (e.g. cholate, octylglucoside) and, as with proteolysis, fusion activity is lost in parallel with the degree of extraction of the Gn proteins as a function of detergent concentration

Fig. 4. Comparison of [^{35}S]GTP[S] and [γ-^{32}P]GTP in detection of GTP-binding proteins. **A:** duplicate autoradiograms of Western blots, to manifest GTP-binding proteins in rat-liver microsomes which, in the alternate lanes marked **+**, had been treated with trypsin (see Fig. 5 legend). [^{35}S]GTP[S] used in **B:** GTP-binding proteins shown in extracts from T15 cells (derived from NIH 3T3 cells by insertion of a plasmid containing a normal human N-*ras* gene under control of a steroid-inducible promoter [27]). Dexamethasone induction (**+**, *vs.* **−**) substantially increases GTP-binding activity in the 21 kDa region, corresponding to the appearance of an anti-*ras* antibody-reactive species (detected by immunoperoxidase reaction) in the same region (*at foot* of figure). *Adapted from [21].*

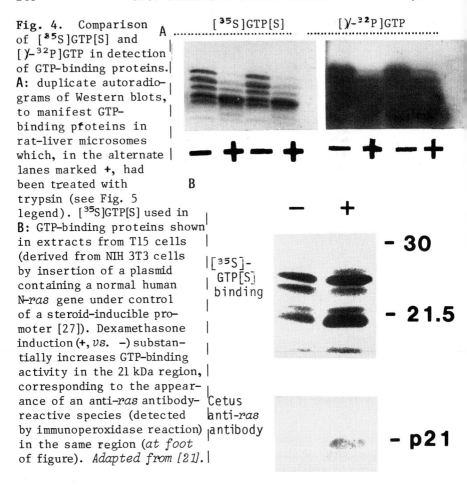

(data not shown). We have, unfortunately, not yet managed to restore fusion activity to detergent-extracted vesicles by adding back supernatant fractions.

INHIBITION BY *N*-ETHYLMALEIMIDE

NEM is known to powerfully inhibit membrane fusion in other systems. This holds also for our microsomes (Fig. 6), wherein 50% inhibition is produced by ~10 nmol NEM per mg protein. However, it is not as yet clear what is the site of NEM action. With the Golgi apparatus there is evidence that the target is a protein of M_r 76 kDa which can be extracted from the membrane fraction by ATP + Mg^{2+} [12]. With rat liver we have no evidence for the involvement of anything other than the integral membrane proteins, since extraction of the membranes with 3 M NaBr still results in vesicles which are able to fuse in the presence of ATP.

Fig. 5. Solubilization of GTP-binding proteins
by trypsin treatment. **A**: total (before trypsin);
B, pellet, and **C**, supernatant after trypsin.
The microsomes were suspended as in the probe
studies (early in the text) except that the DTT
and EGTA were 0.1 mM. Trypsin was added (10 µg/ml
final concn.) and digestion allowed to proceed
for 5 min at 30°. Trypsin inhibitor (to 40 µg/ml),
aprotinin (1 µg/ml) and pepstatin A (1 µg/ml) were
then added to stop the digestion
and prevent further non-specific
proteolysis. The sample was
centrifuged (36000 **g**, 30 min, 4°).
Pellet and supernatant samples,
and untreated microsomes, were
treated as in the text (SDS-PAGE,
electroblotting, [³⁵S]GTP[S]).

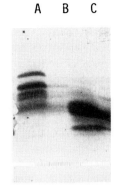

A   B   C

Fig. 6. Inhibition of fusion
by NEM. Treatment of microsomes with NEM (added to the
final assay mixture 5 min before
GTP) was as in the standard
fusion assay (see text). Note
that DTT was omitted from the
medium and preparation buffers.

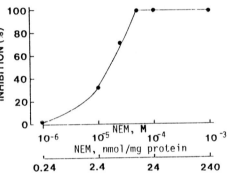

CONCLUSIONS

We consider it likely that the GTP-induced membrane
fusion in rat-liver microsomal vesicles represents a mechanism
for maintaining the spatial organization of the e.r. inside
intact cells. In particular, judging from our and other
in vitro studies [2, 3], it seems likely that it is, at
least in part, responsible for controlling the distribution
of Ca^{2+} within the e.r. This may be important in spatial
aspects of Ca^{2+} signalling [25] and in packaging of Ca^{2+}
in vesicles [26].

It also appears that fusion in rat-liver microsomes
shows sufficient similarity with other fusion systems for
it to be a reasonable model to study the hitherto totally
mysterious molecular basis of membrane fusion.

Acknowledgement
We thank the Wellcome Trust for financial support.

References
1. Dawson, A.P. (1985) *FEBS Lett. 185*, 147-150.

2. Dawson, A.P., Comerford, J.G., Cullen, P.J. & Fulton, D.V. (1989) in *Biochemical Approaches to Cellular Calcium* [Vol. 19, this series] (Reid, E., Cook, G.M.W. & Luzio, J.P., eds.), Royal Society of Chemistry, Cambridge, pp. 167-178.

3. Dawson, A.P. & Comerford, J.G. (1989) *Cell Calcium 10*, 343-350

4. Gill, D.L., Ghjosh, T.K. & Mullaney, J.M. (1989) *Cell Calcium 10*, 363-374.

5. Dawson, A.P., Hills, G. & Comerford, J.G. (1987) *Biochem. J 244*, 87-92.

6. Comerford, J.G. & Dawson, A.P. (1988) *Biochem. J. 249*, 89-93.

7. Rothman, J.E. & Orci, L. (1990) *FASEB J. 4*, 1460-1468.

8. Gomperts, B.D. (1990) *Annu. Rev. Physiol. 52*, 591-606.

9. Segev, N., Mulholland, J., & Botstein, D. (1988) *Cell 52*, 915-924.

10. Burgoyne, R.D. (1989) *Trends Biochem. Sci. 14*, 394-396.

11. Bourne, H.R. (1988) *Cell 53*, 669-671.

12. Block, M.R., Glick, B.S., Wilcox, C.A., Weiland, F.T. & Rothman, J.E. (1988) *Proc. Nat. Acad. Sci. 85*, 7852-7856.

13. Comerford, J.G. & Dawson, A.P. (1989) *Biochem. J.258*, 823-8

14. Dawson, A.P. & Irvine, R.F. (1984) *Biochem. Biophys. Res. Comm. 120*, 858-864.

15. Goda, Y. & Pfeffer, S.R. (1989) *FASEB J. 3*, 2488-2495.

16. Keller, P.M., Person, S. & Snipes, W. (1977) *J. Cell Sci. 28*, 167-177.

17. Dawson, A.P., Comerford, J.G. & Fulton, D.V. (1986) *Biochem. J. 234*, 311-315.

18. Nicchitta, C.V., Joseph, S.K. & Williamson, J.R. (1986) *FEB: Lett. 209*, 243-248.

19. Bhullar, R.P. & Haslam, R.J. (1987) *Biochem. J. 245*, 617-620.

20. Drøbak, B.K., Allan, E.F., Comerford, J.G., Roberts, K. & Dawson, A.P. (1988) *Biochem. Biophys. Res. Comm. 150*, 899-903.

21. Comerford, J.G., Gibson, J.R., Dawson, A.P., & Gibson, I. (1989) *Biochem. Biophys. Res. Comm. 159*, 1269-1274.

22. Paiement, J., Rindress, D., Smith, C.E., Poliquin, L. & Bergeron, J.J.M. (1987) *Biochim. Biophys. Acta 898*, 6-22.

23. Lanoix, J., Roy, L. & Paiement, J. (1989) *Biochem. J. 262*, 497-

24. Eide, B., Gierschik, P., Milligan, G., Mullaney, I., Unson, C Goldsmith, P. & Speigel, A. (1987) *Biochem. Biophys. Res. Comm. 148*, 1398-1405.

25. Rooney, T.A., Sass, E.J. & Thomas, A.P. (1990) *J. Biol. Chem. 265*, 10792-10796.

26. Lodish, H.F. & Kong, N. (1990) *J. Biol. Chem. 265*, 10893-1089

27. McKay, I.A., Marshall, C.J., Cales, C. & Hall, A. (1986) *EMBO J. 5*, 2617-2621.

#B-5

PROPERTIES AND REGULATION OF TWO FORMS OF THE
INOSITOL 1,4,5-TRISPHOSPHATE RECEPTOR IN RAT LIVER

F. Pietri, M. Hilly, M. Claret and J.-P. Mauger [†]

Unité de recherche de Physiologie et Pharmacologie
Cellulaire, INSERM U274, Bat. 443, UPS,
F-91405 Orsay, France

InsP$_3$[*] generated under hormonal stimulation mobilizes intracellular Ca^{2+} in a variety of cell types. Using [^{32}P]InsP$_3$, we characterized two specific binding sites in permeabilized hepatocytes or in membrane fractions prepared from liver homogenate. The low-affinity site had Kd (20-60 nM) close to the EC$_{50}$ of the InsP$_3$-induced Ca^{2+} release, which may be attributable to the low-affinity receptor since no physiological role has been ascribed to the high-affinity site (Kd = 1-2 nM). That the two sites are two forms of the same receptor was suggested by results with a membrane fraction enriched in InsP$_3$ receptors: in a low-Ca^{2+} medium InsP$_3$ bound to a single low-affinity site (Kd 30-50 nM), and Ca^{2+} addition (>1 µM) decreased the Kd of the receptor to 1-2 nM without affecting the total binding capacity.

In permeabilized hepatocytes Ca^{2+} addition also increased the proportion of the high-affinity state of the receptor and this effect was closely related to a decrease of the InsP$_3$-induced Ca^{2+} release. This suggests that the high-affinity form may be a desensitized state of the receptor.

InsP$_3$ generated under hormonal stimulation mobilizes Ca^{2+} contained in an intracellular compartment [1]. Use of [^{32}P]InsP$_3$ has allowed the identification of specific receptor(s) in membranes from brain or peripheral tissues [2, 3]. Data in the literature suggest the existence of different types of the InsP$_3$ receptor. A single low-affinity site is described in the cerebellum, with Kd 20-60 nM [3]. The high receptor density found in the cerebellum has allowed its purification and the determination of its sequence [4, 5]. In peripheral tissues some reports indicate the presence of a single high-affinity receptor with Kd 1-3 nM [2, 6, 7]; other studies suggest complex interaction between InsP$_3$ and its binding site(s) [0-10]. We studied the properties of the InsP$_3$

[†]addressee for any correspondence

[*]*Abbreviations.-* For *myo*-inositol phosphates, InsP's (*prefix* GP = glycerophospho-):- generally InsP$_3$ = Ins(1,4,5)P$_3$, and InsP$_2$ = Ins(4,5)P$_2$. CCCP, carbonyl cyanide *m*-chlorophenylhydrazone; p.m., plasma membrane.

receptor of rat liver by the use of [^{32}P]InsP$_3$. Since InsP$_3$ does not cross the p.m., we used permeabilized hepatocytes or different membrane fractions prepared from a liver homogenate.

METHODS

Hepatocytes were isolated as previously described from female Wistar rats [9] and resuspended in a cytosol-like medium containing (mM) 110 KCl, 20 NaCl, 25 Hepes-KOH (pH 7.4), 1 NaH$_2$PO$_4$, 0.2 phenylmethylsulphonyl fluoride and 1 EDTA, with 1 mg/ml serum albumin and 10 µg/ml leupeptin. Cells were permeabilized at 4° by saponin (added to 80 µg/ml). After 20 min, >98% of the cells were freely permeable to trypan blue. The fractions - crude membrane, p.m., mitochondrial and microsomal - were prepared from a rat liver homogenate [10-12]; membranes were resuspended in the cytosol-like medium (protein being the basis for wt. expressions).

[^{32}P]InsP$_3$ binding was measured by incubating permeabilized cells (1.2 × 10^6/ml) or the membrane fractions (0.1-1 mg/ml) in the cytosol-like medium during 10 min at 4° in the presence of 10,000-15,000 cpm [^{32}P]InsP$_3$ (bought from NEN/DuPont) and the indicated concentration of unlabelled InsP$_3$. Samples (0.8 ml) were then layered onto a Whatman GF/C glass fibre filter and washed with 3 ml of an ice-cold medium containing 250 mM sucrose, 10 mM NaH$_2$PO$_4$ and 4 mM EDTA (pH 7.4). The procedure was complete within 2 sec. GraphPAD software (ISI) was used to fit curves to data points by non-linear regression analysis according to models involving either a single or two InsP$_3$ binding sites. The sum-of-squares values allowed a comparison of the two fits to the same data.

InsP$_3$-induced Ca^{2+} release was measured in permeabilized hepatocytes which had been pre-incubated for 10 min in the cytosol-like medium containing no EDTA but (mM) 1 EGTA, 0.55 ^{45}Ca^{2+} (0.2 × 10^{-3} free Ca^{2+}), 1.5 Mg-ATP, 5 creatine phosphate and 1 × 10^{-2} CCCP with 5 u/ml creatine phosphokinase. InsP$_3$-induced Ca^{2+} release was then measured after inhibiting Ca^{2+} uptake by adding a mixture (10 mM glucose and 50 units/ml hexokinase) which rapidly degraded ATP. Cells were then diluted by adding 1 vol. of the ice-cold cytosol-like medium containing 2 mM EDTA and the indicated concentrations of InsP$_3$. This procedure allowed the measurement of [^{32}P]InsP$_3$ binding in the same medium.

RESULTS AND DISCUSSION

Characterization of [^{32}P]InsP$_3$ binding to permeabilized cells

We first measured binding, at 4°, to permeabilized hepatocytes, in which InsP$_3$-induced Ca^{2+} release has been well established [13]. The competitive protocol (see above;

Fig. 1. Saturation analysis of $[^{32}P]InsP_3$ binding to permeabilized hepatocytes (2.2×10^6/ml; 10 min, $4°$) in cytosol-like medium containing 17,000 cpm $[^{32}P]InsP_3$ and (concentrations indicated) unlabelled $InsP_3$; see METHODS for details. Incubations were terminated by filtration and the radioactivities retained on filters were counted in a scintillation spectrometer. The curves have been fitted to the data points according to (**A**) a single-site model or (**B**) a 2-sites model. Data are means of a triplicate determination in one typical experiment. H = high, L = low affinity.

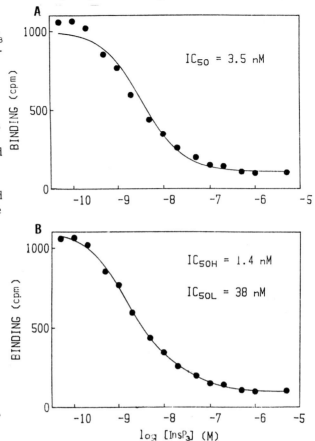

increasing concentrations of unlabelled $InsP_3$) consumed less radioligand and minimized non-specific binding as compared to a saturation binding experiment performed at constant specific radioactivity. Data shown in Fig. 1 were analyzed by non-linear regression analysis. The 'Hill' slope was <1 (0.70), suggesting a complex interaction between $InsP_3$ and its receptor(s). The curve fitted by a single-site model (Fig. 1, A) displays a systematic deviation from the data points, whereas that fitted according to a 2-sites model is in better accord with the points (Fig. 1, B). That the fit was better was confirmed statistically ($F = 25$; $P <0.001$). A high-affinity component with IC_{50} 1.4 nM and a low-affinity component with IC_{50} 38 nM were established from the data of Fig. 1.

The latter component represented 18% of the specific binding but could be under-estimated if the ligand-receptor complex had partly dissociated during the washing by filtration. Accordingly, we measured the specific binding with filtration

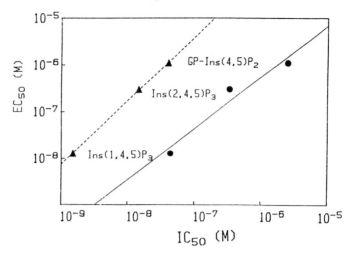

Fig. 2. Correlation between binding (IC_{50}) and Ca^{2+} release (EC_{50}), measured under identical conditions (see METHODS), as affected by $Ins(1,4,5)P_3$, $Ins(2,4,5)P_3$ and $GP\text{-}Ins(4,5)P_2$. The permeabilized hepatocytes were pre-incubated for 10 min in the presence of Mg-ATP and $0.2\ \mu M\ Ca^{2+}$ to load the intracellular Ca^{2+} compartment(s). EC_{50} is the concentration of agent that induced 50% of the maximal $^{45}Ca^{2+}$ release. IC_{50H} and IC_{50L} (▲, ●) were determined as in Fig. 1. Data are means of a triplicate determination in one typical experiment. The lines were determined from linear regression analysis.

replaced by centrifugation (5 min, 40,000 **g**). No increase in specific binding was found, which indicates that the filtration time (<2 sec) is short enough to minimize the dissociation of the ligand-receptor complex. The high- and the low-affinity sites determined from the experiment in Fig. 1 displayed respective capacities of 37 and 210 $fmol/10^6$ cells.

Specificity of the two InsP₃ binding sites

Site specificity was examined in competition experiments with different InsP's, and with heparin which inhibits $InsP_3$ binding and $InsP_3$-induced Ca^{2+} release in permeabilized hepatocytes [3, 14]. The InsP's tested inhibited the specific binding of $[^{32}P]InsP_3$ to the two sites, the potency order being $Ins(1,4,5)P_3 > Ins(2,4,5)P_3 > GP\text{-}Ins(4,5)P_2$. ['GP-Ins' here connotes deacylated phosphatidylinositol, = 'GroP-Ins'.– *Ed.*] The inhibition curves could be resolved into two components, with the different IC_{50}'s shown in Fig. 2. With heparin, the inhibition curve was adequately described by a model involving a single binding component with IC_{50} 10 nM, suggesting that heparin does not discriminate between the two $InsP_3$ binding sites.

The dose-response curves for the $^{45}Ca^{2+}$ release by the InsP's were performed under the same experimental conditions as those used for the binding experiments. The data shown in Fig. 2 indicate that the EC_{50} of $InsP_3$-induced Ca^{2+} release correlated excellently, for the high- as well as the low-affinity site, with the IC_{50} of the inhibition of $[^{32}P]InsP_3$ binding. This suggests that the two sites are closely related to Ca^{2+} release. However, the finding that the IC_{50} of the inhibition of binding to the low-affinity site was almost identical to the EC_{50} for the Ca^{2+} release suggests that this site is coupled to the opening of the Ca^{2+} channel.

Hepatic subcellular distribution of $InsP_3$ binding sites

We tried to separate the two binding sites by measuring $[^{32}P]InsP_3$ binding to different membrane fractions prepared from a liver homogenate. In each fraction tested (Fig. 3), the non-linear regression analysis of the binding data indicated that the 2-sites model better fitted the data points. A high-affinity site with Kd ranging from 1.8 to 2.7 nM and a low-affinity site with Kd 35-48 nM were found in the 4 fractions. Furthermore (Fig. 3), the proportion of the former site varied between 13% and 28% and was not significantly

Fig. 3. $InsP_3$ binding to liver subcellular membrane fractions, incubated as for Fig. 1. Non-linear regression analysis indicated that for each fraction the 2-sites model better fitted the data points. The IC_{50H} varied from 1.2 ±0.2 (S.E.M.) nM in the permeabilized hepatocytes to 2.7 ±0.3 nM in the mitochondrial fraction. The IC_{50L} varied from 35 ±15 nM in the mitochondrial fraction to 88 ±3 nM in the crude membrane fraction. B_{maxH} and B_{maxL} were determined by non-linear regression analysis. Data are means from 4 different experiments.

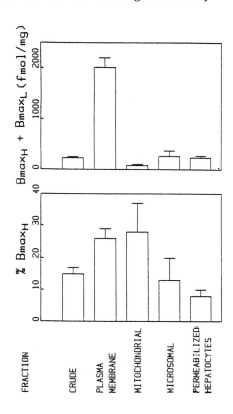

different from one fraction to another. Fig. 3 also indicates that the two InsP₃ binding sites were ~10-fold enriched in the so-called plasma membrane (p.m.) fraction which was also enriched ~10-fold in 5'-nucleotidase, an enzyme marker for p.m.   Thus, the two InsP₃ binding sites co-purified in the same membrane fractions, suggesting that they are located in the same organelle or in organelles very close in density. (Isopycnic followed differential centrifugation.)

The observation that the InsP₃ receptor was preferentially present in a p.m.-enriched fraction supports the original observation by Guillemette & co-authors [7] with a rat-liver p.m. fraction prepared by a protocol different from that now used.   Earlier [15] we had shown that the p.m. fraction partly comprised inside-out p.m. vesicles which accumulated Ca^{2+} in the presence of Mg-ATP and free Mg^{2+}.   This Ca^{2+} pool was insensitive to InsP₃.   The p.m. fraction also contains some glucose-6-phosphatase activity, indicating contamination by endoplasmic reticulum.   The latter fraction also accumulated Ca^{2+} in the presence of Mg-ATP but did not require the presence of free Mg^{2+}.   Addition of InsP₃ causes release of part of the Ca^{2+} accumulated in this compartment [10] as already described by Guillemette & co-authors [7].   These data suggest that the InsP₃ receptor probably resides on an intracellular organelle closely associated with the p.m. However, since InsP₃ released much less Ca^{2+} from the p.m. fraction than from permeabilized hepatocytes, conceivably only a small part of the InsP₃-sensitive Ca^{2+} compartment co-purified with the p.m.

Effect of Ca^{2+} on the InsP₃-binding properties

The fact that the two InsP₃-binding sites display the same specificity and the same subcellular distribution strongly suggests that the two sites are two forms of the same receptor. We had shown earlier that pre-incubating permeabilized hepatocytes at 37° before measuring InsP₃ binding at 4° led to a decrease in the number of high-affinity sites, suggesting that the two sites could be interconvertible [10, 16].   To test this hypothesis, we incubated the p.m. fraction at different temperatures before measuring the [³²P]InsP₃ binding at 4°.   Since the physiological role of InsP₃ is to increase intracellular Ca^{2+}, we have also investigated the effect of Ca^{2+} on the properties of InsP₃ binding.

The analysis of a binding-inhibition experiment performed on membranes prepared and kept at 4° throughout and incubated in a low-Ca^{2+} medium indicated the presence of the two sites, as the Scatchard plot in Fig. 4 illustrates.   Two sites with Kd 1.2 and 30 nM have respective capacities of 0.35 and 2.06 pmol/mg.   When the membranes were pre-incubated

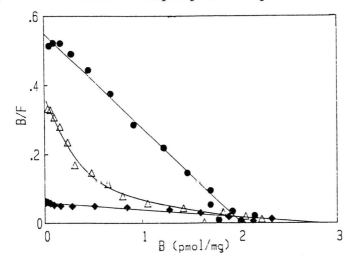

Fig. 4. Scatchard plots of $[^{32}P]InsP_3$ binding to liver p.m. fractions in the presence or absence of Ca^{2+}. The membranes were either kept at $4°$ ($\triangle$) or incubated for 15 min at $37°$ ($\blacklozenge$,$\bullet$). They were then incubated at $4°$ in the cytosol-like medium at 0.3 mg/ml as indicated in Fig. 1, in the presence of 15,000 cpm $[^{32}P]InsP_3$ and increasing concentrations of unlabelled $InsP_3$, 1 mM EDTA and either no added Ca^{2+} (free Ca^{2+} <1 nM; $\triangle$,$\blacklozenge$) or 0.97 mM Ca^{2+} (free Ca^{2+} = 1 µM; $\bullet$).

for a few minutes at $37°$ and incubated in a low-Ca^{2+} medium, a single low-affinity site having Kd 47 nM and a capacity of 2.84 pmol/mg was detected (Fig. 4). If the binding experiment was performed with ~1 µM free Ca^{2+} present, a single binding site with a high affinity (Kd = 3.6 nM) and a capacity of 2.02 pmol/mg was found (Fig. 4).

The mean values from 3 or 4 experiments indicate that the testing of three experimental conditions led to the detection of a single high-affinity site, a single low-affinity site or the two sites together in the preparation; the total binding remained constant. This strongly suggests that the two sites are different states of the same receptor. The high-affinity state is triggered by increasing Ca^{2+} to >1 µM. The small proportion of the high-affinity receptor (15-25%) found in membranes kept at $4°$ probably results from the resting cytosolic Ca^{2+} concentration found in intact cells. The reversion of the high-affinity to the low-affinity state after lowering Ca^{2+} by adding EGTA was rapid at $37°$ [16], suggesting that an equilibrium between the two states of the receptor could play some role in intact cells. Thus, during a hormonal stimulation, the increase in cytosolic Ca^{2+} should in turn increase the proportion of the high-affinity state of the receptor.

It has been demonstrated that Ca^{2+} decreases the binding of $InsP_3$ to cerebellar membranes [3]. Danoff & co-authors [17] have characterized a membrane protein, calmedin, which confers Ca^{2+}-sensitivity on $InsP_3$ receptor binding. Calmedin activity in peripheral tissues, particularly in liver, is much lower than in brain [17]. This would explain why Ca^{2+} addition did not inhibit the $InsP_3$ binding to the liver receptor but did so in the cerebellum.

Effect of Ca^{2+} on the $InsP_3$-induced Ca^{2+} release

The data shown in Fig. 2 suggest that the low-affinity state of the receptor is the better candidate for mediating $InsP_3$-induced Ca^{2+} release. Consequently, an increased proportion of the receptor's high-affinity state should be associated with a decreased $InsP_3$-induced Ca^{2+} release. To verify this prediction, we measured $InsP_3$ binding and $InsP_3$-induced Ca^{2+} release, in a low-Ca^{2+} or a high-Ca^{2+} medium. With low Ca^{2+} (<50 nM) 1 µM $InsP_3$ released ~0.7 nmol Ca^{2+}/mg, and the parallel measurement of $InsP_3$ binding revealed the presence of <5% of high-affinity site (Table 1). If Ca^{2+} were elevated above 1 µM after the loading of the intracellular Ca^{2+} pool in the presence of 0.2 µM free Ca^{2+} and before the $InsP_3$ addition (to 1 µM), then there was no longer $InsP_3$-induced Ca^{2+} release. The parallel measurement of $InsP_3$ binding indicated that the proportion of the high-affinity site increased to 40% of the total binding capacity (Table 1).

This experiment strongly suggests that the high-affinity $InsP_3$ binding site is not involved in the Ca^{2+} release from the intracellular compartment. Furthermore, the finding that the Ca^{2+}-induced transformation of the low-affinity to the high-affinity state of the receptor appears to be correlated to an inhibition of the Ca^{2+} release, suggesting that the high-affinity state could be a desensitized form of the receptor. However, it is noteworthy from the data shown in Table 1 that in contrast to what is observed in the p.m. fraction, >50% of the low-affinity receptors in permeabilized cells were resistant to Ca^{2+} addition. These receptors could represent another type of $InsP_3$ receptors, either not coupled to Ca^{2+} release or inactivated by Ca^{2+} addition without modification of their binding properties.

Other studies have also reported that Ca^{2+} inhibits $InsP_3$-induced Ca^{2+} release from microsomes or permeabilized cells [18, 19]. A feedback inhibition of the $InsP_3$ response has already been demonstrated in hepatocytes or in parotid acinar cells [20, 21].

Table 1. Effect of Ca^{2+} on $InsP_3$-induced release from permeabilized hepatocytes, pre-incubated for 10 min at 37° in a medium containing (mM) 1 EGTA, 0.55 $^{45}Ca^{2+}$, 1.5 $MgCl_2$ and 1.5 ATP. Hexokinase (50 units/ml) and glucose (10 mM) were added to stop Ca^{2+} uptake; 15 sec later, an aliquot (0.5 ml) was added to 0.5 ml of an ice-cold medium containing 2 mM EDTA with or without 2.2 mM $CaCl_2$. The final concentrations of $CaCl_2$ were 0.275 mM (low Ca^{2+}) or 1.375 mM (high Ca^{2+}). $InsP_3$ (1 µM) was added as indicated and the incubations were terminated 30 sec later. $[^{32}P]InsP_3$ binding was measured under the same conditions except that $^{45}Ca^{2+}$ was omitted.

	Low $[Ca^{2+}]$		High $[Ca^{2+}]$	
	$- InsP_3$	$+ InsP_3$	$- InsP_3$	$+ InsP_3$
Ca^{2+} content, nmol/mg	1.61	0.90	1.63	1.62
% of B_{maxH}	5	–	40	–

CONCLUSION

We have demonstrated the presence of two $InsP_3$ binding sites in rat liver, which display the specificity of the $InsP_3$ receptor. These two sites are two forms of the same receptor regulated by the Ca^{2+} concentration in the medium. In a low-Ca^{2+} medium the receptor is maintained in the low-affinity state coupled to the Ca^{2+} channel. Increasing the Ca^{2+} concentration induces transformation of the receptor to a high-affinity, unresponsive state. This mechanism could contribute to the genesis of oscillations in intracellular Ca^{2+} concentration observed in single isolated cells [22].

Acknowledgements

The work was supported by a grant from the 'Foundation pour la Recherche Médicale'.

References

1. Berridge, M.J. & Irvine, R.F. (1989) *Nature 341*, 197-205.
2. Baukal, A.J., Guillemette, G., Rubin, R., Spät, A. & Catt, K.J. (1985) *Biochem. Biophys. Res. Comm. 133*, 532-538.
3. Worley, P.F., Baraban, J.M., Supattapone, S., Wilson, V.S. & Snyder, S.H. (1987) *J. Biol. Chem. 262*, 12132-12136.
4. Snyder, S.H. & Supattapone, S. (1989) *Cell Calcium 10*, 337-342.
5. Furuichi, T., Yoshikawa, S., Miyawaki, A., Wada, K., Maeda, N. & Mikoshiba, K. (1989) *Nature 342*, 32-38.

6. Guillemette, G., Balla, T., Baukal, A.J., Spät, A.
   & Catt, K.J. (1987) *J. Biol. Chem. 262*, 1010-1015.
7. Guillemette, G., Balla, T., Baukal, A.J. & Catt, K.J.
   (1988) *J. Biol. Chem. 263*, 4541-4548.
8. Spät, A., Fabiato, A & Rubin, R.P. (1986) *Biochem. J.*
   233, 929-932.
9. Mauger, J-P., Claret, M., Pietri, F. & Hilly, M. (1989)
   J. Biol. Chem. 264, 8821-8826.
10. Pietri, F., Hilly, M., Claret, M. & Mauger, J-P. (1990)
    Cell Signalling 2, 253-263.
11. Prpic, V., Green, K.C., Blackmore, P.F. & Exton, J.H. (1984)
    J. Biol. Chem. 259, 1382-1385.
12. Dawson, A.P. & Irvine, R.F. (1984) *Biochem. Biophys. Res.*
    Comm. 120, 858-864.
13. Williamson, J.R. & Monck, J.R. (1989) *Annu. Rev. Physiol.*
    51, 107-124.
14. Cullen, P.J., Comerford, J.G. & Dawson, A.P. (1988) *FEBS*
    Lett. 228, 57-59.
15. Dargemont, C., Hilly, M., Claret, M. & Mauger, J-P. (1988)
    Biochem. J. 256, 117-124.
16. Pietri, F., Hilly, M. & Mauger, J-P. (1990) *J. Biol. Chem.*
    265, 17478-17485.
17. Danoff, S.K., Supattapone, S. & Snyder, S.H. (1988)
    Biochem. J. 254, 701-705.
18. Joseph, S.K., Rice, H.L. & Williamson, J.R. (1989)
    Biochem. J. 258, 261-265.
19. Jean, T. & Klee, C.B. (1986) *J. Biol. Chem. 261*, 16414-
    16420.
20. Ogden, D.C., Capiod, T., Walker, J.W. & Trentham, D.R. (1990)
    J. Physiol. 422, 585-602.
21. Gray, P.T.A. (1988) *J. Physiol. 406*, 35-53.
22. Berridge, M.J. & Galione, A. (1988) *FASEB J. 2*, 3074-3082.

#B-6

THE IDENTIFICATION OF INOSITOL PHOSPHATES IN [³H]-INOSITOL-LABELLED CELLS AND TISSUES

C.J. Barker[†], N.S. Wong[≠], P.J. French, R.H. Michell and C.J. Kirk

School of Biochemistry, University of Birmingham,
P.O. Box 363, Birmingham B15 2TT, U.K.

InsP's* have assumed a prominent place in our recent understanding of intracellular communication. However, only certain InsP's have defined roles in cell regulation, e.g. Ins(1,4,5)P₃ and Ins(1,3,4,5)P₄ (whose precise role in relation to Ca²⁺ is still somewhat unclear). Ins(1,3,4,5,6)P₅ and InsP₆ may have extracellular functions in the brain and immune system. Advances in HPLC and other separative techniques have revealed a large number of InsP metabolites, some of which fit into clearly defined phosphorylation/dephosphorylation pathways derived from receptor-mediated events; others such as InsP₆ may be derived directly from inositol.

Current methods for separating and identifying these compounds are discussed. These include HPLC and structural identification techniques. More rigorous structural work has led to the characterization of a number of previously unidentified InsP's. Finally some recent data relating to the interconversion of these InsP's are presented.

In the last ten years there has been an explosive growth in our understanding of inositol lipid and phosphate metabolism. This has been triggered largely by the recognition that agonist-induced breakdown of PtdIns(4,5)P₂* generates two intracellular signalling molecules: Ins(1,4,5)P₃ (*circled* in Fig. 1) [cf. Fig. 2 in art. #B-7 - *Ed.*] which mobilizes Ca²⁺ from intracellular stores [1], and DAG which activates PKC [2]. The metabolism of Ins(1,4,5)P₃ has turned out to be unexpectedly complex (Fig. 1; review: [3]).

This increasing complexity has highlighted the need to develop techniques for separating and characterizing InsP's present in the cell, some of which are probably only distantly

[†] addressee for any correspondence
[≠] NOW AT Inst. of Molecular & Cell Biology, National Univ. of Singapore, 10 Kent Ridge Crescent, Singapore 0511

Abbreviations (some changes by Ed.).- Phosphate-substituted inositols (*myoinositol* usually implied) are termed InsP's (*not* IP's; cf. 'GPI' in Fig. 3), thus InsP₃ = trisphosphate [usually Ins(1,4,5)P₃]; Ptd (prefix) = phosphatidyl ('PI' in Fig. 1 for PtdIns); PL = phospholipase (phosphoinositidase), yielding DAG = (1,2)diacylglycerol from PtdInsP₂; PKC = protein kinase C. NMR, nuclear magnetic resonance (spectroscopy); PCA/TCA, perchloric/trichloroacetic acid.

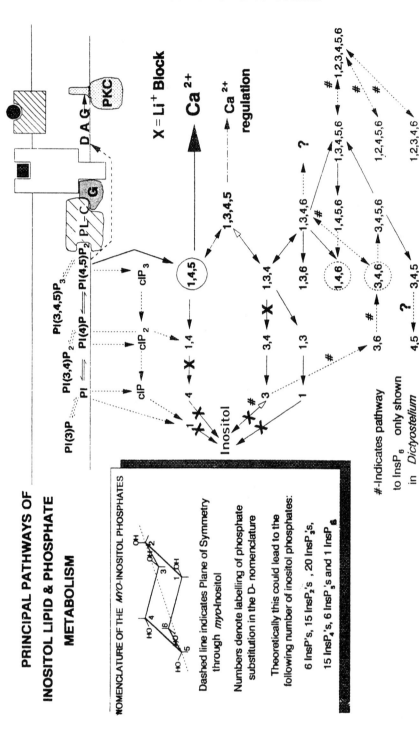

Fig. 1. Plasma-membrane (*at top*) and intracellular pathways, and (*inset*) potential numbers of isomers. G, G-protein; IP, PtdIns; cIP, Ins(1:2cyclic)P, e.g. Ins(1:2cyclic,4)P.

related to the primary second-messenger molecule. Recently, specific assays for Ins(1,4,5)P$_3$ [4, 5] and the possible co-messenger Ins(1,3,4,5)P$_4$ [6] have been developed. These permit the assay of these second messengers despite the presence (often at high concentrations) of structurally related metabolites. NMR studies have also been exploited to reveal structural information on the InsP's, but these require larger amounts of material that has to be much purer, and do not discriminate between enantiomeric forms [e.g. 7, 8]. However, most studies carried out have relied on the labelling of tissues and cells with [^{3}H]inositol, and since some of the more abundant InsP's in cells (e.g. InsP$_3$ and InsP$_6$) may have a critical role in extracellular regulation [9, 10-12], this type of protocol is likely to be important for some time to come. This article focuses on the techniques which have enabled us to identify and characterize InsP's that have been pre-labelled with [^{3}H]inositol, and attempts to assess their relative merits.

The inherent complexity of this undertaking is better understood when one considers the number of possible InsP's which can theoretically be produced when one substitutes the hydroxyl groups of the *myo*inositol ring with phosphate moieties (Fig. 1, *inset*).

The stages undertaken in the analysis of InsP's are illustrated in Scheme 1 in a mode which is the basis for the consideration now given to each stage.

LABELLING CELLS AND TISSUES WITH [^{3}H]*MYO*INOSITOL

Traditionally tissues and cells have been labelled with [2-^{3}H]*myo*inositol obtained from either Amersham International or NEN/Du Pont. Other, more expensive, variations are now available, e.g. [1,2-^{3}H]*myo*inositol marketed by NEN. [U-^{14}C]*myo*-inositol has also been used in some studies [13], but its expense precludes routine use. The position of ^{3}H on the inositol ring has been exploited in structural analysis as it is lost when certain [2-^{3}H]*myo*inositol phosphates, e.g. Ins(1,4)P$_2$, are cleaved by periodate oxidation. Radioisotopic labelling is at its most useful when isotopic equilibrium has been achieved so that radioactivity may be taken as a measure of chemical mass. With InsP's derived from rapidly turning-over pools this may occur in hours. However, many InsP's, e.g. InsP$_5$ and InsP$_6$, which do not appear to turn over rapidly cannot be labelled this quickly. Therefore many workers have, when studying InsP metabolism, focused their attentions on cells grown in culture where complete isotopic equilibrium may be achieved. In such experiments, the long-term fate of inositol in the particular cell needs to be clearly understood; but there are virtually no cases where this issue has been addressed.

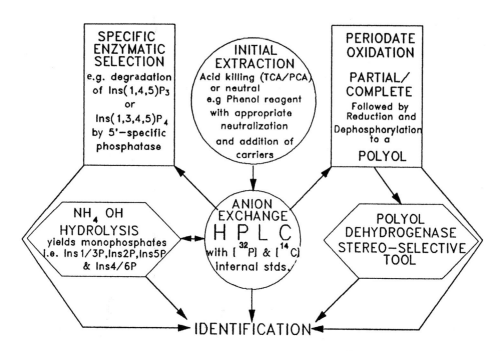

Scheme 1. Stages in the identification of InsP's.

The basic assumption made in all labelling studies is that the inositol molecule itself is immutable, i.e. it remains as *myo*-inositol and is not degraded or epimerized. There is, however, sufficient literature to suggest that such assumptions are an over-simplification. Sherman and co-workers have demonstrated the existence of *scyllo*- and *neo*-inositol in many rat tissues (as well as tissues from invertebrate sources), and epimerases exist which can inter-convert *myo*- to *scyllo*-, *neo*- and *chiro*- forms [14, 15]. When cells or tissues are labelled with [2-^{3}H]*myo*-inositol the position of the ^{3}H is such that epimerization will result in the formation of an unlabelled *scyllo*-inositol, which therefore will not complicate subsequent analyses.

The importance of taking these forms of inositol into consideration has been highlighted recently by the discovery that *chiro*-inositol forms a substantial proportion of the inositol element of the proposed insulin mediators [16]. However, Stephens & Irvine [17] have shown that none of these different inositols are substrates for the *myo*-inositol 3-kinase which forms the first step in the synthesis of InsP$_6$ in *Dictyostelium*.

EXTRACTION

The choice of extraction technique will have an important influence on subsequent analyses of the InsP's. They must ensure that the phosphates are quantitatively extracted, which may be assisted by the inclusion of 'cold' inositol phosphates (see [18]), and presented in a form suitable for subsequent analysis. In cases where addition of carrier is undesirable, glassware must be siliconized in order to quantitatively recover small amounts of InsP's, particularly those with a large number of phosphate groups, e.g. $InsP_5$ and $InsP_6$. Acidic extractants commonly used, e.g. TCA or PCA, must be neutralized to prevent either acid-catalyzed phosphate migration or removal of phosphate groups, and near-neutral extraction conditions are requisite if cyclic InsP's are to be studied (e.g. [19]). After extraction and neutralization, 0.1 M EDTA taken to pH 7.0 with NaOH is added to a concentration of 10 mM. The sample is then stored at -20°.

HPLC

HPLC itself can furnish only tentative identification of InsP's, although it is commonly used without being complemented by other techniques. To maximize information from HPLC it is important to achieve the best separation possible and to include known internal standards in the run.   As [³H] is normally used in the pre-labelling process, [³²P]- or [¹⁴C]-labelled InsP's can be used. Apart from $[5-^{32}P]Ins(1,4,5)P_3$ and $[^{14}C]Ins(3)P_1$ these standards are not readily available and therefore have to be made in the laboratory. A summary of the different methodologies can be consulted in [20] or found elsewhere [21-23].

As the HPLC columns in routine use are based on anion-exchange, even the best protocols will not separate enantiomeric forms of InsP's present in cells, e.g. $Ins(1,3,4)P_3/Ins(1,3,6)P_3$, $Ins(1,4,6)P_3/Ins(3,4,6)P_3$ and $Ins(1,4)P_2/Ins(3,6)P_2$. These limitations must be borne in mind when interpreting HPLC data.

Various anion-exchange HPLC columns are in routine use. The most commonly used ones are the 250 x 4.6 mm Whatman Partisil 10 SAX (now superseded by the 12.5 cm and 25 cm Partisil 5 SAX) and the 12.5 and 25 cm Partisphere 5 WAX columns; the Adsorbosphere 5 SAX manufactured by Altech is also used by a number of groups [24-26].   The advantage of the latter two columns is that they can produce good separation of the tetrakisphosphates. The Partisphere columns, particularly the 25 cm versions (obtainable in the U.K. from Laserchrom), are probably amongst the best in current use but a careful note must be made of the fact that the elution

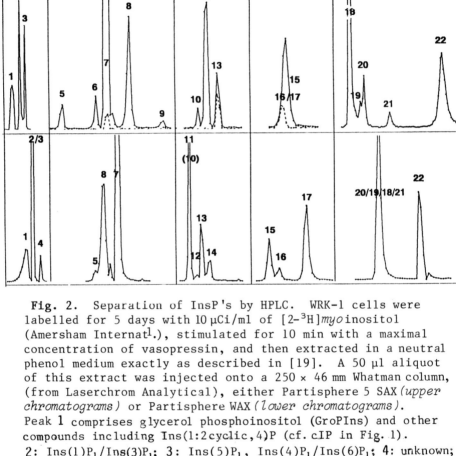

Fig. 2.   Separation of InsP's by HPLC.   WRK-1 cells were
labelled for 5 days with 10 µCi/ml of [2-³H]*myo*inositol
(Amersham Internat[1].), stimulated for 10 min with a maximal
concentration of vasopressin, and then extracted in a neutral
phenol medium exactly as described in [19].   A 50 µl aliquot
of this extract was injected onto a 250 × 46 mm Whatman column,
(from Laserchrom Analytical), either Partisphere 5 SAX *(upper
chromatograms)* or Partisphere WAX *(lower chromatograms).*
Peak 1 comprises glycerol phosphoinositol (GroPIns) and other
compounds including Ins(1:2cyclic,4)P (cf. cIP in Fig. 1).
2: Ins(1)P_1/Ins(3)P_1; 3: Ins(5)P_1, Ins(4)P_1/Ins(6)P_1; 4: unknown;
5: cInsP$_2$; 6: Ins(1,3)P_2; 7: Ins(1,4)P_2; 8: Ins(3,4)P_2; 9: Ins(4,5)P_2;
10: cInsP$_3$; 11: Ins(1,3,4)P_3; 12: Ins(1,4,6)P_3; 13: Ins(1,4,5)P_3;
14: Ins(3,4,5)P_3; 15: Ins(1,3,4,6)P_4; 16; Ins(1,3,4,5)P_4;
17: Ins(3,4,5,6)P_4; 18: Ins(1,3,4,5,6)P_5; 19: Ins(1,2,3,5,6)P_5/
Ins(1,2,3,4,5)P_5; 20: Ins(1,2,3,4,6)P_5; 21: Ins(1,2,4,5,6)P_5/
Ins(2,3,4,5,6)P_5; 22: InsP$_6$.
The dashed lines from the Partisphere SAX chromatogram
represent internal standards: [4-³²P]Ins(1,4)P_2, [4,5-³²P]-
Ins(1,4,5)P_3 and [4,5-³²P]Ins(1,3,4,5)P_4 respectively.   The
structures of the [³H]-labelled compounds were determined
by the techniques described in the text.

properties of the isomeric forms differ markedly between
the SAX and WAX columns.   This is illustrated in Fig. 2
which shows data from single runs on 25 cm Partisphere
SAX and WAX columns.

Better separations are achievable using isocratic elutions concentrating on a single class of InsP's. This is particularly true of the early part of the chromatograms illustrated. Thus, a good baseline separation of $InsP_1$'s has been obtained [23]. It is self-evident that if a reasonable separation of the majority of InsP isomers can be routinely obtained (as is the case with the newer Partisphere columns), this is preferable to many individual runs for the different classes. An example of an all-encompassing run on a 25 cm Partisphere 5 SAX column is given in ref. [27]. With increased column lifetime it may be necessary to modify gradients to maintain resolution. This is achieved by decreasing the elution gradient and/or decreasing the salt concentration in an isocratic run. Because of the very different elution properties of some InsP isomers on the SAX and WAX columns, it is possible to derive some very helpful data from the relative elution properties on the different columns.

HPLC is most useful when identifying InsP's which exist in a limited number of isomeric forms such that complete separation is more feasible, e.g. monophosphates and pentakis-phosphates although simple anion-exchange techniques will not resolve enantiomers such as $Ins(1)P_1/Ins(3)P_1$ and $Ins(1,2,3,4,5)P_5/Ins(1,2,3,5,6)P_5$.

A number of desalting methods are available for the removal of inorganic phosphate from the separated HPLC peaks. If the sample is intended for use in metabolic studies, pyrophosphate must be removed using additional purification steps [28].

PERIODATE OXIDATION

The periodate oxidation schemes originally devised by Ballou and co-workers for determining the structure of the inositol lipids form the basis of any structural identification protocol.[⊗] These methods have since been extended ([17, 21, 29-31], and N.S. Wong & C.J. Barker, unpublished work).

This technique relies on the fact that the periodate ion will cleave the inositol ring between two carbon atoms each possessing a hydroxyl group. The pattern of substitution of the phosphate groups will determine if and where the InsP molecule is cleaved. The dialdehyde produced is then reduced with $NaBH_4$ and dephosphorylated with alkaline phospha-tase. The structure of the polyol so produced will be determined by the orientation of the phosphate groups in the original molecule. This process is illustrated for Ins(1,-4,5)P_3 in Fig. 3.

[⊗] Our Vol. 19 art. [32] amplifies.

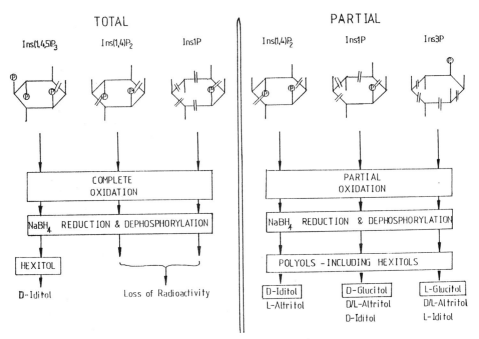

Fig. 3. Periodate oxidation strategies.

If the ring is not broken during these steps (because there are no pairs of adjacent hydroxyl groups in the InsP) the resulting polyol will be inositol. Some polyols will exist in enantiomeric form, but methods have been devised to differentiate a number of the polyols using a stereoselective enzyme polyol dehydrogenase [21, 28, 30, 31]. Thereby the stereochemistry of the original InsP can be determined. Clearly some InsP's resist attack (e.g. pentakisphosphates) and the process includes some redundancy so that identification using this technique usually requires complementary analyses to produce a definitive identification. It is most powerful when used to elucidate the structure of tris- and tetrakis-phosphates. An example of how these methods have been employed in the identification of InsP's in WRK-1 cells has been reported in a previous volume in this series [32].

A recent innovation in this technique established in our laboratory by Nai Sum Wong and independently by Len Stephens widens its scope in the identification of bis- and mono-phosphates. The difficulty with these structures is that the periodate will cleave the molecule at multiple points and, because the molecule is labelled only in the 2-position, radioactivity is often lost. Fig. 3 illustrates this in its left panel. On the right is a new variation of this technique which

uses short incubation periods and dilute periodate solutions
to generate partially degraded InsP's. The products produced
are then reduced and dephosphorylated in the normal way.
A mixture of polyols will be produced, and the hexitols
are of particular importance as they can only have been
the product of a single break in the ring. In the case
of Ins(1,4)P$_2$ the products are altritol and iditol. The
configuration of the iditol is then resolved using the polyol
dehydrogenase, and hence the location of the phosphate can
be settled unambiguously. [L-iditol is derived from Ins(3,6)P$_2$
and D-iditol from Ins(1,4)P$_2$.]

This technique can even be extended to resolve the
enantiomeric monophosphate pair Ins(1)P$_1$/Ins(3)P$_1$. Fig. 3 illustrates that a number of partially oxidized products are
produced from these compounds, this time in very much lower
yields. In this case the chirality of the sorbitol produced
is used to define which enantiomer is present. Unfortunately,
it is not possible to distinguish the Ins(4)P$_1$ and Ins(6)P$_1$
isomers in this way as the *cis* hydroxyls in the 1-, 2-
and 3-positions will always be oxidized before the *trans*
hydroxyls in the 4-, 5- and 6-positions. Methods of approaching
this problem are discussed below.

In these analyses it is assumed that the original molecule
is based on *myo*-inositol. If another inositol were present
the interpretation of the data would be erroneous. For
example, L-altritol could be derived from *myo*-Ins(1,3,4)P$_3$
and *neo*-Ins(1,3,4)P$_3$. If one finds polyol products which
do not co-elute with those which can arise from *myo*-inositol
phosphates, it is possible that the structures being analyzed
are based on another inositol.

AMMONIACAL HYDROLYSIS

Another technique which was successfully used by Ballou
and his colleagues was alkaline hydrolysis of the inositol
molecule using (NH$_4$)OH. This reaction occurs without migration
of the phosphate groups, and therefore each phosphorylated
position on an inositol phosphate will be represented by
an inositol monophosphate in the hydrolysate. The rates
of hydrolysis of the individual phosphate groups vary; hence
the relative abundances of the resulting inositol phosphates
are not proportional to the relative abundance of the phosphates
in the original molecule. Inositol phosphates with a phosphate
group in the 2-position yield more Ins(2)P$_1$ than any other
phosphate. Simple preparative gradients are then used to
resolve the Ins(2)P$_1$ from the other phosphates. The absence
of a 2-phosphate would therefore enable one to narrow the
identification of an InsP$_4$ from 15 to 5 candidate molecules,

and combined with HPLC data on a Partisphere WAX column the choice could be narrowed down to 1 of 2 candidates or a unique candidate.

A gradient has been devised recently which will separate all the *myo*-inositol monophosphates, except for the two enantiomeric pairs Ins(1)P$_1$/Ins(3)P$_1$ and Ins(4)P$_1$/Ins(6)P$_1$ [23]. If these pairs could be resolved, e.g. by chiral HPLC, then it may be possible to get an absolute configuration for an InsP without using periodate oxidation. Using such prolonged gradients to separate the InsP's has led to some anomalies: thus certain InsP's in HL-60 cells, when analyzed using these techniques, show peaks on the chromatogram which do not co-elute with internal standards. This is particularly noticeable with the monophosphates derived from the InsP$_6$:- here there is no co-elution of the [^{32}P]Ins(4)P$_1$ standard and the peak one might assume to represent this monophosphate. Whether this result is artefactual or represents a novel structure is currently under investigation.

SPECIFIC ENZYMATIC TOOLS

With the above protocols some short-cuts are feasible if a specific enzyme is available as a tool, e.g. the Ins(1,4,5)P$_3$/Ins(1,3,4,5)P$_4$ 5-phosphatase. This has been used to remove either Ins(1,4,5)P$_3$ ([33] and C.J. Barker & A. Craxton, unpublished work) or Ins(1,3,4,5)P$_4$ [29] in order to quantitate other InsP species which might co-migrate with them on HPLC.

FUTURE DIRECTIONS

The methodology used to identify InsP's is time-consuming. Efforts to reduce this might lie in several directions. The crux of the problem is the fact that many InsP's exist in enantiomeric forms; the solution would be to use chiral HPLC, but it is unlikely that a column that separated InsP$_1$'s would also resolve InsP$_5$'s due to the considerable difference in charge. A compromise solution might be to use the ammoniacal hydrolysis techniques described above to reduce any inositol polyphosphates needing study to a representative selection of monophosphates which could then be separated by chiral HPLC. Even the introduction of chiral separations at the level of the polyol products of periodate oxidation, instead of using the stereospecific polyol dehydrogenase, would reduce the complete analysis time significantly. As stated at the beginning of this review, care must be taken in the analyses to ensure that the compound is based on *myo*-inositol.

The techniques described here form the basis of our current understanding of InsP metabolism. However, structural studies such as these do not describe the interrelationships between these identified compounds. There is a link at least *in vitro* with cell homogenates between the second-messenger $Ins(1,4,5)P_3$ and $Ins(1,3,4,5,6)P_5$ [23,25,34] though this has been seriously disputed in at least one whole-cell system [28, 35]. Observations consistent with the conversion of $Ins(1,3,4,5,6)P_5$ to $InsP_6$ have recently been reported [36].

Whether the latter observations indicate a direct connection between $Ins(1,4,5)P_3$ and $InsP_6$ is debatable, as Stephens & Irvine [17] have demonstrated the stepwise synthesis of $InsP_6$ from inositol *via* a route which does not include $Ins(1,4,5)P_3$. Unpublished work by P.T. Hawkins, A. Stanley and M.R. Hanley (see [17]) has indicated a 2-OH kinase which will phosphorylate $Ins(1,3,4,5,6)P_5$ to $InsP_6$. We have made similar observations. For example, the 40% ammonium sulphate cut from rat-brain cytosol which we use in preparing labelled $Ins(1,3,4,5)P_4$ will, with longer incubation periods, apparently phosphorylate $Ins(1,3,4,5,6)P_5$ to $InsP_6$. Moreover, when $Ins(1,3,-4,5)P_4$ is used as a starting material, not only is $Ins(1,3,4,5,6)P_5$ formed but also $InsP_6$. Whether this represents a direct conversion from $Ins(1,3,4,5)P_4$ to $Ins(1,3,4)P_3$ to $Ins(1,3,4,6)P_4$ to $Ins(1,3,4,5,6)P_5$ and $InsP_6$ or occurs *via* intermediate dephosphorylation products such as inositol or $Ins(3)P_1$ is currently unclear.

Acknowledgements

Work in our laboratory was supported by the Medical Research Council and the Royal Society.

References

1. Berridge, M.J. & Irvine, R.F. (1989) *Nature 341*, 197-205.
2. Nishizuka, Y. (1989) *Annu. Rev. Biochem. 58*, 31-44.
3. Shears, S.B. (1989) *Biochem. J. 260*, 313-324.
4. Tarver, A.P., King, W.G. & Rittenhouse, S.E. (1987) *J. Biol. Chem. 262*, 17268-17271.
5. Palmer, S., Hughes, K.T., Lee, D.Y. & Wakelam, M.J.O. (1988) *Cell Signalling 1*, 147-154.
6. Donie, F. & Reiser, G.A. (1989) *FEBS Lett. 254*, 155-158.
7. Johnson, L.F. & Tate, M.E. (1969) *Can. J. Chem. 47*, 63-73.
8. Radenberg, T., Scholz, P., Bergmann, G. & Mayr, G.W. (1989) *Biochem. J. 264*, 323-333.
9. Vallejo, M., Jackson, T., Lightman, S. & Hanley, M.R. (1987) *Nature 330*, 656-658.
10. Hanley, M.R., Jackson, T.R., Cheung, W.T., Dreher, M., Gatti, A., Hawkins, P.T., Patterson, S.I., Vallejo, M., Dawson, A.P. & Thalstrup, I. (1988) *Cold Spring Harb. Symp. Quant. Biol. 103*, 435-445.

11. Nicoletti, F., Bruno, V., Fiors, L., Cavallaro, S. & Canonico, P.L. (1989) *J. Neurochem. 53*, 1026-1030.

12. Hawkins, P.T., Reynolds, D.J.M., Poyner, D.R. & Hanley, M.R. (1990) *Biochem. Biophys. Res. Comm. 167*, 819-827.

13. Maccallum, S.H., Barker, C.J., Hunt, P.H., Wong, N.S., Kirk, C.J. & Michell, R.H. (1990) *J. Endocrin. 122*, 379-389.

14. Sherman, W.R., Goodman, S.L. & Gunnell, K.D. (1971) *Biochemistry 10*, 3491-3499.

15. Sherman, W.R., Hipps, P.P., Mauck, L.A. & Rasheed, A. (1978) in *Cyclitols and Phosphoinositides* (Wells, W.W. & Eisenberg, F., J eds.), Academic Press, New York, pp. 279-295.

16. Mato, J.M., Kelly, K.L., Abler, A., Jarrett, L., Corkey, B.E., Cashel, J.A. & Zopf, D. (1987) *Biochem. Biophys. Res. Comm. 146*, 764-770.

17. Stephens, L.R. & Irvine, R.F. (1990) *Nature 386*, 580-583.

18. Wreggett, K.A., Howe, L.R., Moore, J.P. & Irvine, R.F. (1987) *Biochem. J. 245*, 933-934.

19. Wong, N.S., C.J. Barker, Shears, S.B., Kirk, C.J. & Michell, R.H. (1988) *Biochem. J. 252*, 1-5.

20. Barker, C.J. (1991) in *High Performance Liquid Chromatography in Neuroscience Research* (Homan, R.B., Cross, A.J. & Joseph, M.I eds.), Wiley, Chichester, in press.

21. Stephens, L.R., Hawkins, P.T., Morris, A.J. & Downes, C.P. (1988) *Biochem. J. 249*, 283-292.

22. Stephens, L.R., Hawkins, P.T., Carter, N., Chahwala, S.B., Morris, A.J., Whetton, A.D. & Downes, C.P. (1988) *Biochem. J. 249*, 271-282.

23. Stephens, L.R., Hawkins, P.T., Barker, C.J. & Downes, C.P. (1988) *Biochem. J. 252*, 721-733.

24. Balla, T., Guillemette, G., Baukal, A.J. & Catt, K.J. (1987) *J. Biol. Chem. 262*, 9952-9955.

25. Shears, S.B. (1989) *J. Biol. Chem. 264*, 19879-19886.

26. Hughes, P.J., Hughes, A.R., Putney, J.W. Jr. & Shears, S.B. (1989) *J. Biol. Chem. 264*, 19871-19878.

27. Staddon, J.M., Barker, C.J., Murphy, A.C., Chanter, N., Lax, A.J Michell, R.H. & Rozengurt, E.J. (1991) *J. Biol. Chem.*, in press.

28. Stephens, L.R. & Downes, C.P. (1990) *Biochem. J. 265*, 435-452

29. Barker, C.J., Morris, A.J., Kirk, C.J. Michell, R.H. (1988) *Biochem. Soc. Trans. 16*, 984-985.

30. Stephens, L.R., Hawkins, P.T. & Downes, C.P. (1989) *Biochem. J*

31. *as for* 30., *262*, 727-737.   [*259*, 267-276

32. Kirk, C.J., Michell, R.H., Morris, A.J. & Barker, C.J. (1989) in *Biochemical Approaches to Cellular Calcium* [Vol. 19, this ser (Reid, E., Cook, G.M.W. & Luzio, J.P., eds.), R. Soc. Chem., Cambridge, 191-

33. Nogimori, K., Menniti, F.S. & Putney, J.W. (1990) *Biochem.J. 2*

34. Balla, T., Hunyady, L., Baukal, A.J. & Catt, K.J.   [195-. (1989) *J. Biol. Chem. 264*, 9386-9390.   [65

35. Stephens, L.R., Berrie, C. & Irvine, R.F. (1990) *Biochem. J. 2*

36. Ji, H., Sandberg, K., Baukal, A.J. & Catt, K.J. (1989) *J. Biol. Chem. 264*, 20185-20188.

#B-7

PHOSPHOINOSITIDES AND OTHER SECOND MESSENGERS: EFFECTS OF LITHIUM

Nicholas J. Birch, Geraint M.H. Thomas[†] and Mark S. Hughes

Biomedical Research Laboratory, Wolverhampton Polytechnic, Lichfield Street, Wolverhampton WV1 1DJ, U.K.

Li^{+*} has diverse in vitro and in vivo effects, both pharmacological and biochemical (especially on phosphoinositide metabolism), of uncertain relevance to its prophylactic use in psychiatry for recurrent affective disorders. Thus, its administration to rats reduced brain inositol and increased $Ins(1)P_1$.[⊗] The Mg^{2+}-dependent enzyme InsmonoPase was inhibited in rat mammary gland by Li^+ to an extent depending on concentration. Some cell signalling process appears to be the locus of action; but Li^+ inhibition of cyclic nucleotide and phosphoinositide systems shown in vitro needs confirmation in vivo by long-term experiments on the nervous system, modelled on the therapeutic situation.

Many physiological processes are affected profoundly by small changes in concentration of simple metal ions, and these may cause subtle changes in overall biological function. The use of lithium in medicine is a very clear example of this. The diversity of Li^+ effects reflects a very fundamental biochemical level of action of the metal. Li^+ has been shown to have significant effects on InsP metabolism both *in vitro* and *in vivo* though it is at present unclear how this relates to lithium carbonate therapy for preventing the major changes in mood (affect) that characterize the affective disorders [1]. Of the two sides of the coin, mania or depression, the manic state is preferred by the patient but dreaded by their family; depression is traumatic

---

[†]Present address: Physiology Dept., Univ. Coll. London WC1E 6BT.
Editor's policy: 'lithium' in MS. replaced, where appropriate, by Li^+. *Abbreviations include:* NMR, nuclear magnetic resonance (spectroscopy); Pase, phosphatase; PKC, protein kinase C; p.m., plasma membrane.
[⊗]In the phosphoinositide area *(Editor's conventions):* phosphate-substituted inositols (*myo*inositol implied) are termed InsP's (*not* IP's); subscript indicates number of phosphate groups, e.g. $InsP_3$ = trisphosphate [usually $Ins(1,4,5)P_3$], $InsP_2$ = bisphosphate [usually $Ins(4,5)P_2$], $InsP_1$ = monophosphate [generally $Ins(1)P_1$]; *mono* or *poly* signifies an inositol phosphate family; Ptd (prefix) = phosphatidyl; PL = phospholipase, yielding DAG = (1,2)diacylglycerol from $PtdInsP_2$.

for the patient but provides some temporary relief for those close to the patient.

Lithium was introduced by Cade in 1949 [2] for treating acute mania, but nowadays is most commonly used *prophylactically* in the control of bipolar affective disorders [1]. It is used by ~40,000 patients in the U.K. alone, and the investment of ~£1.1 million in prescriptions has been reckoned to save £23 million per year in hospital in-patient costs [3]. It is always administered orally, usually as lithium carbonate at a total daily dose of 30 mmol (2 g). Blood lithium is measured regularly, 12 h after the previous dose [4]. Serum concentrations should lie in the range 0.4-0.8 mmol/L; with higher levels there may be toxic side-effects which may include tremor, dizziness, drowsiness and diarrhoea [4]. Severe intoxication leads to renal damage, escalating blood concentrations and ultimately death.

CHEMISTRY

Alkali metals (Group 1A) readily lose an electron from each atom to yield univalent cations. Most compounds of the alkali metals are almost entirely ionic, although lithium has rather more tendency to form covalent compounds. Lithium is the lightest metal of Group 1A, having the smallest ionic radius, and the largest field density at its surface. In solution, the very small diameter of Li^+ in relation to the aqueous solvent results in a large hydration sphere of uncertain size. Consequently, lithium salts do not conform to ideal solution behaviour [5]. In an aqueous environment the radius increases out of proportion to the radius of the other Group 1A elements, resulting in poor ionic mobility and low lipid solubility under physiological conditions (Fig. 1). Moreover, lithium is the least reactive of the alkali metals and its general chemical properties resemble calcium and magnesium as might be predicted from the so-called 'diagonal relationship' between lithium and magnesium in the periodic table. It has been proposed that Li^+ may interact with Mg^{2+}- and Ca^{2+}-dependent processes in physiology [6-9].

MECHANISM OF ACTION

The mode of action of Li^+ has yet to be elucidated. Many suggestions have been made although no conclusive case has been adduced for any postulated mechanism. Increases in the research activity in any particular biochemical area frequently result in the proposal of 'the definitive' mode of action of Li^+. Most of these proposals relate to a newly found analytical capability for a particular system or metabolite. That this, of all possible systems, is the

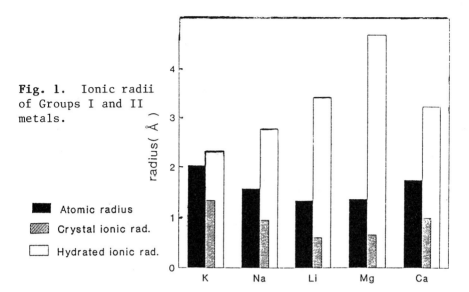

Fig. 1. Ionic radii of Groups I and II metals.

correct one is statistically improbable and intrinsically unlikely. Most recent proposals involve the cell membrane of intracellular compartments where Li^+ may interact with second messengers or other physiological control systems [10–12].

In all these proposed mechanisms it is essential to prove that the relevant conditions exist in the cell *during lithium therapy in patients*. It is important to attempt to establish the extent to which Li^+ enters the cells, or perhaps becomes closely associated with cell membranes. Recent work using NMR suggests that Li^+ uptake into cells under conditions of acute exposure is relatively low [13–15].

LITHIUM AND THE PHOSPHOINOSITIDE SIGNALLING SYSTEM

Many hormonal and neuronal signals are transduced through receptor-mediated activation of phosphoinositase C (PL-C; inositol-lipid-directed phospholipase C) which, as summarized in Fig. 2, hydrolyzes $PtdIns(4,5)P_2$ to DAG and $Ins(1,4,5)P_3$ in the p.m. Both products are second messengers: DAG stimulates PKC and $InsP_3$ releases Ca^{2+} from intracellular stores located in the endoplasmic reticulum ([16], and arts. in Vol. 19, this series). The $InsP_3$ is subsequently converted via intermediates to *myo*inositol; this is then converted to Ptd-inositol which is used to replenish $InsP_2$ stores.

The inositol lipid cycle is inhibited by Li^+ [16]. This is the basis for one unifying hypothesis proposed for Li^+ action [12]. Long-term administration of Li^+ to rats (10 mmol/kg) resulted in a reduction in brain inositol [17]

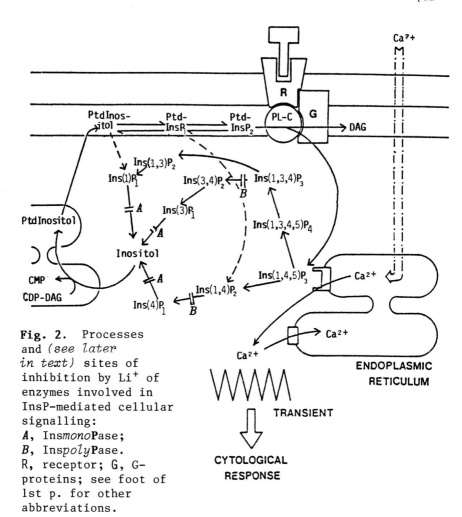

Fig. 2. Processes and *(see later in text)* sites of inhibition by Li+ of enzymes involved in InsP-mediated cellular signalling: *A*, Ins*mono*Pase; *B*, Ins*poly*Pase. R, receptor; G, G-proteins; see foot of 1st p. for other abbreviations.

and an increase in the reaction substrate $Ins(1)P_1$ [17]. The Mg^{2+}-dependent enzyme Ins*mono*Pase was totally inhibited in rat mammary gland by high concentrations of Li+ (250 mM) and partially inhibited by lower concentrations (2 mM) [18].

At clinically relevant concentrations Li+ has been shown to inhibit Ins *mono*Pase *uncompetitively* (irreversibly). This mode of inhibition is extremely unusual and can result in catastrophic metabolic consequences [19]. Models of uncompetitive inhibition of an enzyme in a metabolic pathway at steady state show that with the accumulation of reaction substrate the increase in the extent of inhibition is non-linear [19] (Fig. 3). In consequence the primary substrate at an earlier stage along the metabolic path is rapidly depleted

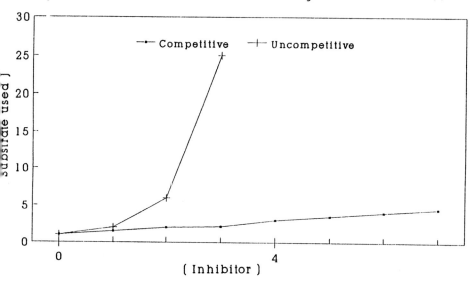

Fig. 3. Graphical simulation of substrate utilization rate
as affected   by competitive or uncompetitive inhibition.
In a concentration range near the K_i value a small change in
concentration of either substrate or uncompetitive inhibitor
has an exaggerated effect on rate of reaction and the substrate
concentration must rise to an unsustainable level to overcome
the inhibition. [Arbitrary units used.] *After Cornish-Bowden [19].*

and the   regulation of the system as a whole soon becomes
unstable and chaotic.

    Li^+ may also inhibit other enzymes involved in the
interconversion and breakdown of Ins*poly*P's, although not
always by an uncompetitive mechanism [20-22] (Fig. 2)[@]. As
a consequence either of the uncompetitive inhibition of the
*mono*Pase or of the inhibition of other enzymes, or a combination
of these, Li^+ reduces the cell concentrations of inositol, which
would otherwise be converted to Ptd-inositol. This reduction
ultimately attenuates the brain's response to external stimuli
[23].   This scheme has been suggested as the mechanism
of action of Li^+ in the affective disorders, since dietary
sources of inositol cannot cross the blood-brain barrier
and brain cells must therefore rely on endogenous supplies
which would become rate-limiting were Li^+ to act in this
manner [12].

    A second consequence of Li^+ action in receptor-activated
cells is the accumulation of DAG, which may increase or
prolong the activation of PKC [23].   It should be noted
that DAG and $Ins(1,4,5)P_3$ are separate branches of two parallel
───────────
[@]See also Fig. 1 in art. #B-6. - *Ed.*

signalling systems which initially are coherent and in phase. Metabolic interference by Li^+ to desynchronize these systems may itself be a signalling system. This is analogous to 'beat' phenomena seen in the desynchronization of biological cycles in the human by rapid movement between time zones.

A third hypothesis proposes that the effects of Li^+ may alter adrenergic-cholinergic balance [24]. Adrenergic predominance has been supposed to be associated with mania and cholinergic dominance with depression [25]. It is claimed that activation of adenylate cyclase is mainly adrenergic and that of Ptd-inositol turnover is cholinergic [26]. The key function in post-receptor information transduction in each of these systems is the G protein (GTP-binding protein), and this is postulated to be the common site for both the anti-manic and the anti-depressant effects of Li^+.

LITHIUM TRANSPORT ACROSS CELL MEMBRANES

An assumption which has been made throughout the literature is that Li^+ permeates readily through cellular membranes. We have carried out a number of studies of the mucosal mechanisms of Li^+ absorption in the GI tract [27], and it has become increasingly obvious that Li^+, and indeed many other metal ions, transfer out of the GI tract not by passage through the cell but by mechanisms of paracellular transport *via* the tight junctions and pericellular spaces. Cellular transport mechanisms and carriers identified in cells may thus exist only for the domestic requirements of the intestinal cells themselves which in turn protect their own *milieu interieur* by as far as practicable avoiding accumulation of externally derived metal ions.

The Li^+ content of intestinal cells during Li^+ absorption is low. We therefore questioned whether, in fact, Li^+ readily penetrated into other (non-intestinal) cells, bearing in mind its physicochemical properties as discussed above. Accordingly we studied a number of different cell types to investigate the uptake and transport of Li^+.

Many studies of Li^+ transport in cell membranes have used human erythrocytes because they are readily available both from normal subjects and from patients. Erythrocytes are, however, atypical cells, being enucleated and lacking many enzyme systems found in more typical cells. Cellular functions therefore might not truly reflect those of other tissues though it has been suggested that measurements of Li^+ uptake into erythrocytes may have value in the identification of those patients who are most likely to respond to treatment [28-30]. Other studies have used squid axon, hepatocytes, 3T3 fibroblast cultures and liposomes to investigate Li^+ transport across the p.m. [13, 31].

In our studies of Li^+ fluxes in cells we exploited the potential of NMR to identify atomic nuclei in different molecular environments. Thereby we could distinguish intracellular from extracellular Li^+, obtaining a measure of 'free' $[Li^+]_i$. Most previous methods of determination have relied on the washing of cell samples to remove extracellular Li^+. In brief, the cells are suspended in a medium containing a large impermeant transition metal-polyanion complex which causes a spectral shift of the metal ions with which it is in contact (i.e. extracellular Li^+). The intracellular Li^+ is unshifted and the resulting spectrum displays a doublet. We recorded 31 MHz NMR spectra using a Bruker WP80SY multinuclear spectrometer. Cells were incubated in phosphate-buffered saline containing 20% D_2O (as heteronuclear lock) and up to 5 mM dysprosium tripolyphosphate (shift reagent). Li^+ concentration was adjusted appropriately [32; 33a, b].

The results of these experiments in a variety of cells [33, 34] may be summarized thus. In erythrocytes from normal, untreated subjects, incubated for various times in a range of media containing 5-50 mM Li^+, the average Li^+ content was ~10% of external Li^+ at equilibrium [33a, b]. In other cell types $[Li^+]_i$ was somewhat less than that seen in erythrocytes. In rat hepatocytes $[Li^+]_i$ appeared to saturate at 1 h when it was <5% of external $[Li^+]$ [34]. Similarly with Swiss mouse 3T3 cells $[Li^+]_i$ at 45 min was 6% of external $[Li^+]$ [33c].

Broad confirmation of our erythrocyte data has come from other laboratories [15, 35], and we now have extensive experience of such analyses in incubated cells both from untreated normal subjects and during Li^+ treatment. In Li^+-treated subjects we have found the equilibrium concentration of Li^+ to be increased to ~30% of external, both in washed cells incubated in known Li^+ solutions and in cells separated from Li^+-containing plasma in serial blood samples obtained by phlebotomy. These data are consistent with commonly accepted values determined by atomic absorption spectrometry, and also with recent results obtained by NMR in another laboratory [35].

Patients with bipolar affective disorders may exhibit differences in their membrane transport systems, and Li^+ administration causes adaptive changes [10, 30]. $[Li^+]_i$ in erythrocytes may rise after prolonged Li^+ therapy, due to either increased influx or reduced efflux rate; efflux is eventually inhibited by ~50% in Li^+ patients [30]. An increased concentration of ankyrin, a specific erythrocyte membrane protein affecting cytoskeletal structure and function, has been found in some patients with bipolar affective disorder

[36]; this also raises the possible role of erythrocyte membrane defects in the aetiology of the disease. Slowly developing changes occurring in membrane structure, or affecting membrane-bound enzymes [37], may explain the relatively long, but variable, time period required for the beneficial effects of Li+ to become clinically apparent.

INTRACELLULAR LITHIUM AND PHOSPHOINOSITIDE SIGNALLING

These findings obviously have implications in the interpretation of Li+ effects on second-messenger systems, particularly those in which complex series of metabolic interactions occur within or close to the membrane and which may involve membrane components in their regulation. A corner-stone of the argument implicating the InsP's in the therapeutic effect of Li+ is the catastrophic consequence on InsP metabolism of the uncompetitive inhibition by Li+ of Ins(1)P$_1$-monoPase. The original calculations for this were made using experimental data obtained with 2.5 mM Li+, assumed to be consistent with the concentration prevailing in the proximity of the enzyme [38].

We have investigated, at substrate concentrations close to the K$_m$ value, the inhibition by Li+ of partially purified InsmonoPase prepared from rat liver [33c]. It is likely that InsP$_3$ is metabolized largely via Ins(4)P$_1$ in rat liver and that Li+ will therefore cause accumulation of this compound. Examples of Li+ inhibition are shown in Fig. 4, at both low [1 μM) and high (1 mM) substrate [Ins(4)P$_1$] concentrations. Fig. 4 also shows, for comparison, similar projected curves, derived from the data of Gee & co-authors [39], for Li+ inhibition of bovine brain InsmonoPase.

Calculation using both the simple equations for uncompetitive inhibition of rat liver InsmonoPase, in isolation, and for the enzyme as a member of the metabolic pathway at steady state (ss; eqn. 1) [19] leads to the conclusion that

$$[S]_{ss} = \frac{K_m}{(V_{max}/v) - 1 - \{[I]/K_i\}} \qquad \text{(equation 1)}$$

with [Li+]$_i$ <1 mM the maximum possible inhibition would be 30% depression of control activity (see Fig. 4). In practice this might be much lower because the equations assume constant metabolic flux and this might be expected to become reduced during Li+ inhibition.

In separate experiments [33c] we have demonstrated that vasopressin-stimulated PtdInsP$_1$ and PtdInsP$_2$ hydrolysis in hepatocytes was only partially inhibited by 10 mM [Li+] externally. This seems to indicate that at the intracellular

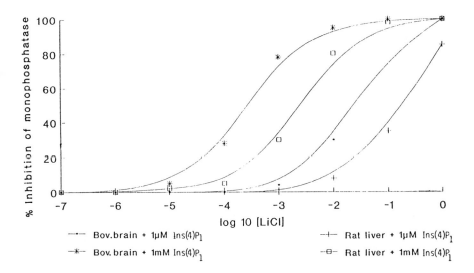

Fig. 4. A comparison of the uncompetitive inhibition of InsmonoPase from rat liver (Thomas, 1989 [33c]) and bovine brain (Gee et al., 1988 [39]) at low (1 μM) and high (1 mM) concentrations of the substrate, Ins(4)P₁.

concentrations obtained during therapy there would not be a catastrophic failure of PtdInsP metabolism, at least in peripheral tissues.

One should however be aware that $[Li^+]_i$ may be higher in excitable cells as a consequence of distribution along the potential gradient. Ehrlich & Diamond [10] have calculated from the Nernst equation that $[Li^+]$ in excitable cells should be ~10 times that of the surrounding fluid. Were this to be so, an intracellular concentration of 2 mM would be entirely consistent with normal therapeutic plasma concentrations and could, by calculation, permit an inhibition of up to 88% of control activity using the Cornish-Bowden equations [19] and assuming that the enzyme is identical to that isolated from bovine brain. This inhibition, along with the low availability of extracellular inositol in nervous tissue, might account for any neurospecific action of Li^+ on inositol-lipid-mediated signalling events. However, there is also evidence to suggest that brain cells do not specifically accumulate Li^+ to any greater extent than somatic cells [40, 41], and our studies indicate that there is significantly greater resistance to Li^+ influx through cell membranes than hitherto had been imagined.

CONCLUSION

The range of pharmacological and biochemical effects of Li^+, even at low concentrations, is enormous. There are few mammalian processes which are not affected by Li^+, and many processes in more primitive organisms are sensitive to the ion. It is likely that the locus of action resides in a cell-signalling process though the mechanism is at present unknown. We conclude that the Li^+-induced inhibition of cyclic nucleotide and InsP systems shown *in vitro* requires confirmation *in vivo* in the nervous system in long-term experiments designed to mimic the therapeutic situation.

References

1. Schou, M. (1986) *Lithium Treatment of Manic Depressive Illness: A Practical Guide*, 3rd edn., Karger, Basel, 49 pp.
2. Cade, J.F.J. (1949) *Med. J. Austral. 36*, 349.
3. McCreadie, R. (1988) in *Lithium: Pharmacology and Psychiatric Use* (Birch, N.J., ed.), IRL Press, Oxford, pp. 11-13.
4. Phillips, J.D. & Birch, N.J. (1990) in *Monovalent Cations in Biological Systems* (Pasternak, C.A., ed.), CRC Press, Boca Raton, FL, pp. 339-355.
5. Stern, K.H. & Amis, E.S. (1959) *Chem. Revs. 59*, 1-64.
6. Birch, N.J. (1976) *Br. J. Psychiat. 116*, 461.
7. Birch, N.J. (1976) *Nature 264*, 681.
8. Birch, N.J. (1987) in *Magnesium in Cellular Processes and Medicine* (Altura, B., Durlach, J. & Seelig, M., eds.), Karger, Basel, pp. 212-218.
9. Frausto da Silva, J.J.R. & Williams, R.J.P. (1976) *Nature 263*, 237-239.
10. Ehrlich, B.E. & Diamond, J.M. (1980) *J. Membr. Biol. 52*, 187-200.
11. Drummond, A.H. (1987) *Trends Pharmacol. Sci. 8*, 129-133.
12. Berridge, M.J., Downes, C.P. & Manley, M.R. (1982) *Biochem. J. 206*, 587-595.
13. Thomas, G.M.H., Hughes, M.S., Partridge, S., Olufunwa, R.I., Marr, G. & Birch, N.J. (1988) *Biochem. Soc. Trans. 16*, 208.
14. Partridge, S., Hughes, M.S., Thomas, G.M.H. & Birch, N.J. (1988) *Biochem. Soc. Trans. 16*, 205-206.
15. Espanol, M.C. & Mota de Freitas, D. (1987) *Inorg. Chem. 26*, 4356-4359.
16. Michell, R.H. (1989) *Biochem. Soc. Trans. 17*, 1-3.
17. Allison, J.H. & Stewart, M.A. (1971) *Nature New Biol. 233*, 267-268.
18. Naccarato, W.F., Ray, R.E. & Wells, W.W. (1974) *Arch. Biochem. Biophys. 164*, 194-201.
19. Cornish-Bowden, A. (1986) *FEBS Lett. 203*, 3-6.
20. Hallcher, L.M. & Sherman, W.R. (1980) *J. Biol. Chem. 255*, 10896-10901.

21. Batty, I. & Nahorski, S.R. (1985) *J. Neurochem. 45*, 1514-1521.
22. Ragan, C.I., Gee, N., Jackson, R. & Reid, G. (1988) *as for* 3., 205-207.
23. Shears, S.B. (1988) *as for* 3., 201-204.
24. Avissar, S., Schreiber, G., Danon, A. & Belmaker, R.H. (1988) *Nature 331*, 440-442.
25. Janovsky, D.S., El-Jousef, M.K. & Davis, J.M.   (1973) *Arch. Gen. Psychiatry 27*, 542-547.
26. Ebstein, R.P., Belmaker, R.H., Grunhaus, L. & Rimon, R. (1976) *Nature 259*, 411-413.
27. Davie, R.J. (1991) in *Lithium and the Cell: Inorganic Pharmacology and Biochemistry* (Birch, N.J., ed.), Academic Press, London, in press.
28. Pandey, G.N., Sarkadi, B., Hass, M., Gunn, R.B., Davis, J.M. & Tosteson, D.C. (1978) *J. Gen. Physiol. 72*, 233-247.
29. Duhm, J. (1978) *Prog. Clin. Biol. Res. 21*, 551-573.
30. Rybakowski, J., Frazer, A. & Mendels, J. (1978) *Comm. Psychopharmacol. 2*, 105-112.
31. Ehrlich, B.E. & Russell, J.M. (1984) *Brain Res. 311*, 141-143.
32. Hughes, M.S. (1991) *as for* 27., in press.
33. (a) Phillips, J.D., (b) Hughes, M.S., (c) Thomas, G.M.H. (1989) *Ph.D. Thesis*, Wolverhampton Polytechnic/CNAA.
34. Thomas, G.M.H. & Olufunwa, R.I. (1988) *as for* 3., pp. 289-291.
35. Riddell, F.G., Patel, A. & Hughes, M.S. (1991) *J. Inorg. Biochem.*, in press.
36. Zhang, Y. & Meltzer, H. (1988) *Psychiat. Res. 27*, 267-271.
37. Ehrlich, B.E., Diamond, J.M., Fry, V. & Meier, K. (1983) *J. Membr. Biol. 75*, 233-240.
38. Thomas, A.P., Alexander, J. & Williamson, J.R. (1984) *J. Biol. Chem. 259*, 5574-5594.
39. Gee, N.S., Ragan, C.I., Watling, K.J., Aspley, S., Jackson, R.G., Reid, G.G., Gani, D. & Shute, J.K. (1988) *Biochem. J. 249*, 883-889.
40. Thellier, M., Wissocq, J.C. & Heurteux, C. (1980) *Nature 283*, 299-302.
41. Renshaw, P.F., Wicklund, S. & Leigh, J.S. (1988) *as for* 3., pp. 277-278.

#ncB

NOTES and COMMENTS relating to

CYTOPLASMIC TRANSMISSION SYSTEMS, AND SOME AGONIST EFFECTS

This subsection opens with supporting ('nc') articles.

From p. 195 there is Forum discussion material.

From p. 197 there is supplementary material provided by the Editor.
- Included in this material: Nitric oxide, p. 198.

Consult the start of the main Contents list concerning the book structure and inescapable compromises in the section assignment for certain contributions whose wide-ranging subject-matter does not fall neatly into one section. In particular, some 'supplementary material' that arguably falls in topic-area #B will be found elsewhere, e.g. in subsection '#ncE'.

#ncB-1

A Note on

NEUROLEPTICS, LITHIUM, INOSITOL PHOSPHATES AND CALCIUM: A NOVEL APPROACH TO THE PHARMACOLOGY OF PSYCHOSIS

M. Adib Essali, Indrajit Das[†], Jacqueline de Belleroche and Steven R. Hirsch

Departments of Psychiatry and Biochemistry, Charing Cross and Westminster Medical School, St. Dunstan's Road, London W6 8RP, U.K.

The role of $Ins(1,4,5)P_3$* as a second messenger to mobilize Ca^{2+} from intracellular stores has been shown in various cell types including pl'ts* [1, 2]. There is abnormal production of InsP's, seemingly associated with current or past treatment with neuroleptics, in pl'ts from schizophrenic patients [3]. The effect of neuroleptics, accompanied in some experiments by Li^+, on $InsP_3$ formation and on $[Ca^{2+}]_i$ in pl'ts from healthy volunteers has now been explored *in vitro*.

Methods for InsP's.- Separated pl'ts [3] were labelled by incubation for 2 h at 37° in a pH 7.4 buffer (mM: NaCl, 134; $NaHCO_3$, 12; KCl, 2.9; NaH_2PO_4, 0.4; $MgCl_2$, 1; Hepes, 5; glucose, 5) containing 1 mM EGTA, 10 mM citrate and 40 μCi/ml $[2-^3H]myo$inositol. After washing, pl'ts were resuspended with 1 mM Ca^{2+} and 2 mg/ml albumin present and incubated at 37° for 30 min. After adjustment to 1×10^9 pl'ts/ml, activation was effected in a shaking bath at 37° after adding Tmb at different concentrations to 0.5 ml aliquots. Reactions were terminated after 1 min by adding 0.1 ml ice-cold 20% (w/v) perchloric acid. After extraction [4], $[^3H]InsP_3$ was separated by i.e.c. and counted [3]. $[^3H]$-$InsP_3$ isomers were characterized by FPLC analysis (Fig. 1).

Study of Ca^{2+} mobilization entailed measuring $[Ca^{2+}]_i$ by fura-2 ([5]; concentration equation: [6]), on pl't suspensions kept 30 min at 37° with 5 mM EGTA and 5 μM fura-2-AM present in the buffer. After washing and resuspension (3×10^8/ml), 510_{em} nm readings were taken at room temperature in a fluorimeter (MPF-3, Perkin-Elmer) with alternation between 340_{ex} and 380_{ex} nm.

[†]addressee for any correspondence (Psychiatry Dept.).
*Abbreviations (some introduced by Ed.).- $[Ca^{2+}]_i$, intra-platelet Ca^{2+} concentration; InsP, *myo*inositol phosphate, e.g. $InsP_3$ (trisphosphate, usually the 1,4,5 isomer); pl't(s), platelet(s); Tmb, thrombin. Neuroleptic drugs (prefix ± denoting optical isomer): But, butaclamol; CPZ, chlorpromazine; Flu,flupenthixol (usually c- = *cis*-); Hal, haloperidol; Rac, raclopride; Sul, sulpiride. In liquid chromatography (LC): FP = fast-protein (Pharmacia), i.e.c. = (an)ion-exchange, AmF = ammonium formate.

Fig. 1. FPLC of platelet InsP₃'s, isolated by 0.8 M AmF/0.1M formic acid elution from Dowex–formate columns: pool from 3 runs diluted and loaded onto a Mono Q HR 5/5 i.e. column (FPLC system) equilibrated with 1.7 M AmF (pH 3.7, H_3PO_4) – used as eluent 'B' ('A' = water) for the stepped gradient shown (linear increase from 0.75 to 1 M between 5 and 30 min). Flow–rate 1.2 ml/min; 0.3 ml fractions counted for ³H. Note isomer separation.

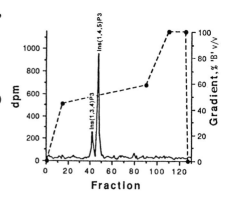

[³H]InsP₃ formation as affected by neuroleptics, alone or with Li⁺

Various neuroleptics had no effect on InsP generation in resting pl'ts, but Tmb-stimulated generation was augmented by pre-incubation for 15 min with CPZ, Hal, Flu or But, although not with Sul or Rac (Fig. 2A). Testing of individual isomers showed stereospecificity with But and Flu but not with Rac (Fig. 2B).

In resting and Tmb-stimulated (0.2 u./ml, 1 min) pl'ts, 10 mM Li⁺ in the final 30-min incubation caused a rise in total [³H]InsP's, attributable mainly to InsP₁ but also to InsP₂ and InsP₃ as obtained from Dowex-formate columns. FPLC showed the InsP₃ rise to be due mainly to the Ins(1,4,5)P₃ isomer. With Li⁺, [³H]Ins(1,4,5)P₃ was reduced. Its normalized peak area [7] without Li⁺ was ~90% of the InsP₃ fraction for resting and stimulated pl'ts. In *resting* pl'ts Li⁺ raised total peak areas by 15%, 29 ±6% (± S.E.M.) of the total being in the Ins(1,3,4)P₃ peak, and 65 ±2.4% in the Ins(1,4,5)P₃ peak which showed a 23% reduction after Li⁺.

[³H]InsP₃ accumulation in *Tmb-stimulated* pl'ts was augmented without or with Li⁺ by 100 nM CPZ (+61 ±3% and +63 ±5% respectively), and similarly by Hal or *cis*-Flu but not by *trans*-Flu (cf. Fig. 2B) or by Sul (not shown). This effect was mainly on Ins(1,4,5)P₃, which comprised 88-90% of the normalized peak-area profile with CPZ alone [see above for basal value; with CPZ Ins(1,3,4)P₃ comprised 10-12%] or, with Li⁺ besides CPZ, 65 ±8%; as mentioned above, the Li⁺ effect was also found without CPZ (71 ±4%).

Regulation of [Ca²⁺]ᵢ by thrombin (Tmb), neuroleptics and Li⁺

[Ca²⁺]ᵢ, 64.5 ±14 nM in resting pl'ts, showed a rise with CPZ, dose-dependent from 0.1 to 10 µM; at <0.1 µM CPZ was ineffective but did augment the rise caused by Tmb, while the maximum produced by 10 µM CPZ was unaffected by Tmb (Fig. 3). Hal behaved like CPZ but was less potent.

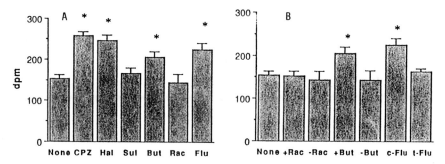

Fig. 2. [³H]InsP₃ generation(15 min incubation with 100 nM drug; see text) in Tmb-stimulated (0.05 u./ml) pl'ts. **A:** CPZ, Hal, Sul, +But, -Rac and *cis*-Flu. **B:** stereoisomer comparisons (t = *trans*). See 1st p. for drug names. Usually 3 expts. (6 for c- & t-Flu in B), each in triplicate. *P<0.05 *vs.* Tmb-alone mean.

Li⁺ added to 10 mM (37°, 1 h; no Ca²⁺) raised [Ca²⁺]ᵢ by 120% (*P* <0.00001), or 80% in pl'ts challenged with Tmb (0.03 u./ml). Basal levels rose with 100 nM Hal (30%) or CPZ (54%), and the response to Tmb alone was augmented by 14% and 23% respectively. The responses to Tmb alone or with drug fell, as % of basal [Ca²⁺]ᵢ, with Li⁺ present; but this was due to elevation of basal [Ca²⁺]ᵢ by Li⁺, which hardly affected the absolute value of the drug-induced rise.

DISCUSSION

Evidently the formation of the second messenger Ins(1,4,5)P₃ can be significantly augmented by neuroleptics of various chemical classes, with specificity insofar as Hal, CPZ and, with isomer comparisons, c-Flu and +But were effective at therapeutically relevant concentrations; only therapeutically active isomers had significant effects. Neither of the substituted benzamides tested, Sul and Rac, influenced the Tmb-induced formation of [³H]InsP₃. In contrast with the other neuroleptics tested, which may interact with dopamine D₁ and D₂ receptors

Fig. 3. [Ca²⁺]ᵢ in pl'ts challenged with CPZ (concentrations indicated) then Tmb. When the CPZ increase had peaked (fura-2 readings, giving the amount of Ca²⁺ mobilized), the further rise (not additive) after adding Tmb was monitored - note the nil rise at 10 µM CPZ. (Tmb alone, 0.03 u./ml, gave Δ = 96 ±17 nM.) Each point is mean of 2 expts. with triplication at least.

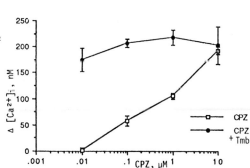

as well as with other monoamine receptors, the substituted benzamides are selective dopamine D_2 antagonists [8, 9]. Conceivably it is because pl'ts lack receptors able to bind dopamine D_2 antagonists that Sul and Rac were inactive in our system. Taken together, these findings indicated that neuroleptics may regulate InsP formation *via* some pl't-surface D_1-type (5-HT subtype [10]?) receptors linked to the phosphatidyl-InsP system. That the latter system's Ca^{2+}-mobilizing action may contribute to the therapeutic action of Li^+ ([11] & Birch, #B-6, this vol.) and neuroleptics, and to their therapeutic [e.g. 12] and toxicological [review: 13] interactions, gains support from effects of agents now studied *in vitro*, and of the neuroleptic trifluoperazine in respect of $InsP_3$ [14].

The present effects included a rise in $[Ca^{2+}]_i$ with CPZ or Hal alone or, if <100 nM, amplification of the Tmb-induced rise, not additively.- The drugs may be acting on the same Ca^{2+} pool as Tmb. The pl't Ca^{2+} pool released by Tmb-stimulated $InsP_3$ formation is probably in the dense tubular system [review: 15]. Li^+ raised basal $[Ca^{2+}]_i$. The individual or conjoint stimulating effects of CPZ, Hal and Tmb were additively augmented by Li^+, as if Li^+ and the drugs were affecting different Ca^{2+} pools. For the drug-induced rise in $InsP_3$ formation, Li^+ was not requisite but was synergistic. Such interactions at the molecular level may throw light on synergistic effects in schizophrenic or other patients.

References

1. Brass, L.F. & Joseph, S.K. (1985) *J.Biol. Chem. 260*, 15172-15179.
2. O'Rourke, F.A., Halenda, S.P., Zavoico, G.B. & Feinstein, M.B. (1985) *J. Biol. Chem. 260*, 956-962.
3. Essali, M.A., Das, I., de Belleroche, J. & Hirsch, S.R. (1990) *Biol. Psychiat. 28*, 475-487.   [238, 491-499.]
4. Palmer, S., Hawkins, P.T., Michell, R.H. & Kirk, C.J. (1986) *Biochem. J.*
5. Grynkiewics, G., Poenie, M. & Tsien, R.Y. (1985) *J. Biol. Chem. 260*, 3440-3450.
6. Daniel, J.L., Dangelmaier, C.A. & Smith, J.B. (1987) *Biochem. J. 246*, 109-114.
7. Scott, R.P.W. (1987) in *Quantitative Analysis Using Chromatographic Techniques* (Katz, E., ed.), Wiley, Chichester, pp. 63-98.
8. Hyttel, J. (1980) *Psychopharmacol. 67*, 107-109.
9. Fleminger, S., Van de Waterbeemd, H., Rupniak, N.M.J., Reavill, C., Testa, B., Jenner, P. & Marsden, C.D. (1983) *J. Pharm. Pharmacol. 35*,
10. De Keyser, J., De Waele, M., Convents, A., Ebinger, G. [363-368.] & Vauquelin, G. (1989) *Eur. J. Pharmacol. 162*, 437-445.
11. Berridge, M.J. & Irvine, R.F. (1989) *Nature 341*, 197-205.
12. Carman, J.S., Bigelow, L.B. & Wyatt, R.H. (1981) *J. Clin. Psychiat. 42*, 124-128.
13. Himmelhoch, J.M. & Neil, J.F. (1980) in *Handbook of Lithium Therapy* (Johnson, F.N., ed.), MTP Press, London, pp. 51-67.
14. Rao, G.H.R. (1987) *Biochem. Biophys. Res. Comm. 148*, 768-775.
15. Siess, W. (1989) *Physiol. Rev. 69*, 58-178.

#ncB-2

A Note on

ELUCIDATION OF THE NATURE OF THE LIVER CELL RECEPTOR-OPERATED CALCIUM CHANNEL

John N. Crofts, Bernard P. Hughes and Gregory J. Barritt[†]

Department of Medical Biochemistry, School of Medicine, Flinders University of South Australia, G.P.O. Box 2100, Adelaide, South Australia, 5001, Australia

Hormones such as α_1-adrenergic agonists and vasopressin, which use Ca^{2+} as an intracellular messenger in their actions on the liver cell, induce the release of Ca^{2+} from the endoplasmic reticulum and the inflow of Ca^{2+} across the p.m.* through a putative ROCC [1, 2]. Besides ROCC's, liver cells also possess a basal Ca^{2+} inflow system which is active in the absence of agonists.

Elucidation of the nature of the liver-cell p.m. ROCC requires a reliable assay for the measurement of Ca^{2+} inflow through the ROCC and through the basal Ca^{2+} inflow system. We felt it desirable to measure rates of Ca^{2+} inflow in intact liver cells rather than in p.m. vesicles. With the exception of the patch-clamp technique, artifacts may be introduced by the use of p.m. vesicles for measuring Ca^{2+} inflow. Here we aim to summarize and evaluate the methods currently available for measuring Ca^{2+} inflow to hepatocytes. These methods include $^{45}Ca^{2+}$ exchange [1], glycogen phosphorylase activation [2, 3], increase in quin 2 fluorescence [2, 4], Mn^{2+}-induced quenching of quin 2 [5], $^{40}Ca^{2+}$ uptake [6], patch-clamp measurement of Ca^{2+} currents, and measurement of the decrease in $[Ca^{2+}]_o$.

One of the main difficulties in measuring rates of Ca^{2+} inflow to intact hepatocytes is the presence of a variety of intracellular pathways which can remove Ca^{2+} that enters the cytoplasmic space. These pathways complicate the interpretation of data obtained using most methods employed for measuring Ca^{2+} inflow.

The use of $^{45}Ca^{2+}$ exchange involves measurement of its intitial rate under steady-state conditions with respect to Ca^{2+} [1]. The validity of this method has been verified

---

[†]addressee for any correspondence
*Abbreviations.- $[Ca^{2+}]_i$ and $[Ca^{2+}]_o$, intracellular and extracellular free Ca^{2+} concentrations, respectively; DTPA, diethylenetriaminepenta-acetic acid; p.m., plasma membrane; ROCC, receptor-operated Ca^{2+} channel.

by compartmental analysis [1].   Measurement of Ca^{2+} inflow by the activation of glycogen phosphorylase depends on the ability of Ca^{2+} to activate its kinase following Ca^{2+} addition to hepatocytes incubated in the absence of added $[Ca^{2+}]_o$ [2, 3].   The principle of measuring Ca^{2+} inflow using quin 2 is similar to that by glycogen phosphorylase except that the fluorescence of the Ca^{2+}-quin 2 complex is used to monitor the $[Ca^{2+}]_i$ increase [5].   Measurement of $^{40}Ca^{2+}$ uptake by atomic absorption spectroscopy [6] is a direct but relatively insensitive measure of Ca^{2+} inflow.

While the patch-clamp technique is potentially one of the best methods, in practice it has been extremely difficult to isolate Ca^{2+} currents in liver-cell patches. While measurement of a decrease in $[Ca^{2+}]_o$ is also direct, it requires a sensitive procedure for measuring $[Ca^{2+}]_o$ changes.

A new method for the measurement of Ca^{2+} inflow to hepatocytes is the quenching by Mn^{2+} of the fluorescence of intracellular quin 2.   Evidence has been obtained which indicates that Mn^{2+} moves through the ROCC in hepatocytes [5]. Fig. 1 shows the principle of the Mn^{2+} quench assay.

The advantages and disadvantages of the four methods which have commonly been used, or which show the greatest potential for future use, are summarized in Table 1.   Each of the tabulated assays is a somewhat indirect measure of the rate of Ca^{2+} inflow to hepatocytes.   Moreover, it is difficult to obtain an accurate estimate of Ca^{2+} inflow through the ROCC as distinct from total Ca^{2+} inflow in the presence of an agonist.   The latter parameter includes Ca^{2+} inflow through both a basal inflow system and the ROCC's.   Estimates of Ca^{2+} inflow through the ROCC's are probably best made using Mn^{2+}-induced quenching of quin 2. When a large number of routine assays are to be performed under fixed conditions the glycogen phosphorylase activation assay is probably the most useful.

Acknowledgement

The work was supported by a grant from the National Health and Medical Research Council of Australia.

References

1. Barritt, G.J., Parker, J.C. & Wadsworth, J.C. (1981)
   J. Physiol. 312, 29-55.
2. Joseph, S.K., Coll, K.E., Thomas, A.P., Rubin, R. &
   Williamson, J.R. (1985) J. Biol. Chem. 260, 12508-12515.

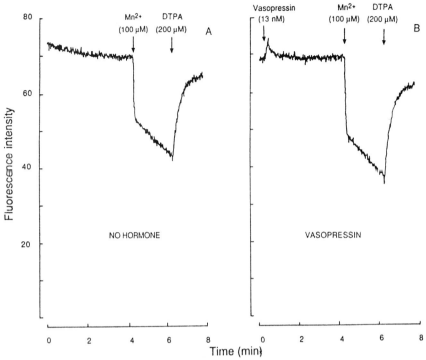

Fig. 1. The rate of quenching by Mn^{2+} of the fluorescence of intracellular quin 2. Hepatocytes loaded with quin 2 were incubated in the absence of externally added Ca^{2+}. Vasopressin (13 nM) (**B**) or vehicle (**A**) and $MnCl_2$ (100 µM) were added at the times indicated. The initial and very rapid decrease in fluorescence induced by the addition of Mn^{2+} is reversed by DTPA and is due to the quenching of extracellular quin 2. The rate of Ca^{2+} inflow is proportional to the rate of decrease in fluorescence in the second phase of quenching which follows addition of Mn^{2+}. *From [5], by permission.*

3. Binet, A., Berthon, B. & Claret, M. (1985) *Biochem. J.* *228*, 565-574.
4. Crofts, J.N. & Barritt, G.J. (1989) *Biochem. J. 264*, 61-70.
5. Crofts, J.N. & Barritt, G.J. (1990) *Biochem. J. 269*, 579-587.
6. Blackmore, P.F., Waynick, L.E., Blackman, G.E., Graham, C.W. & Sherry, R.S. (1984) *J. Biol. Chem. 259*, 12322-12325.

[Table 1 OVERLEAF

Table 1. Advantages and disadvantages of current assays for plasma membrane Ca^{2+} inflow in hepatocytes.

Assay	Advantages	Disadvantages
$^{45}Ca^{2+}$ exchange	Based on sound theory.	Reliance on compartmental analysis.
	Provides absolute values of Ca^{2+} inflow	Some technical difficulties
Glycogen phosphorylase activation	Relatively easy to perform.	Assumes that phosphorylase activity reflects $[Ca^{2+}]_i$ and $[Ca^{2+}]_i$ increase $\propto Ca^{2+}$ inflow.
	Good assay for use in multiple tests under fixed conditions	Cells are initially exposed to low $[Ca^{2+}]_o$.
		Does not readily provide an absolute value of Ca^{2+} inflow
Increase in quin 2 fluorescence	Relatively easy to perform.	Assumes that the $[Ca^{2+}]_i$ increase $\propto Ca^{2+}$ inflow.
	More direct than phosphorylase.	Cells are initially exposed to low $[Ca^{2+}]_o$.
	Provides absolute value for Ca^{2+} inflow	Leakage of quin 2 from cells
Mn^{2+}-induced quenching of quin 2	Relatively sensitive.	Assumes that Mn^{2+} acts like Ca^{2+}.
	Relatively easy to perform.	Calculation of absolute Mn^{2+} inflow rates difficult.
	Reasonably direct	Leakage of quin 2 from cells

#ncB-3

A Note on

cAMP [125I] SCINTILLATION PROXIMITY ASSAY (SPA)
- A HOMOGENEOUS RADIOIMMUNOASSAY FOR cAMP

R. Heath, B.A. Bryant, R.C. Martin and P.M. Baxendale

Research and Development - Assays,
Amersham International plc,
Forest Farm, Whitchurch, Cardiff CF4 7YT, U.K.

In an aqueous environment relatively weak β-emitters, notably ^3H and ^{125}I (Auger electrons), need to be close to scintillant molecules in order to produce light. If not, the energy is dissipated and lost in the solvent. This concept has been used to develop a range of homogeneous assays [1] by coupling specific antibodies (Ab's) onto fluomicrospheres (beads containing scintillant):-

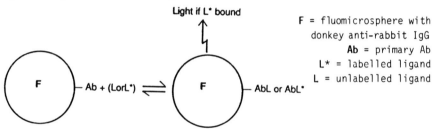

Light if L* bound

F = fluomicrosphere with donkey anti-rabbit IgG
Ab = primary Ab
L* = labelled ligand
L = unlabelled ligand

Ab + (LorL*) ⇌ AbL or AbL*

Such a system has been developed for cAMP measurement (Fig. 1).

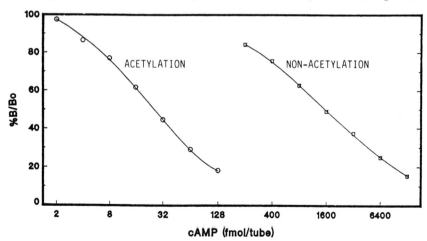

Fig. 1. Typical cAMP [^{125}I] SPA standard curves. Pre-assay acetylation of samples and standards using a freshly prepared 2:1 (by vol.) mixture of triethylamine and acetic anhydride increases sensitivity by improving the affinity of sample/ standard cAMP binding to the Ab [2].

The assay is based upon the competition between unlabelled cAMP and a limited amount of ^{125}I-labelled cAMP for binding to a limited amount of specific Ab. Ab-bound ligand is reacted with donkey anti-rabbit IgG bound to fluomicrospheres. The labelled ligand-Ab fluomicrosphere complex emits light which can be directly measured using a standard β-scintillation counter.

The assay has the advantage of being homogeneous and therefore requires no separation of bound from free radioactive ligand. All components are added at the start of the assay. This means that the assay can easily be automated and also reduces the amount of hands-on time for manual assays. Two assay ranges have been developed. Both have overnight room-temperature incubations. A high-sensitivity range including an acetylation protocol has the range of 2-128 fmol/tube (0.7-42 pg/tube) and a sensitivity of 2 fmol/tube (0.7 pg/tube). The lower sensitivity range without acetylation has a range of 0.2-12.8 pmol/tube (66-4214 pg/tube) and a sensitivity of 78 fmol/tube (26 pg/tube), as shown in Fig. 1.

Human plasma and urine have been assayed, with no pre-extraction, over dilution ranges 1:50 to 1:800 for urine (non-acetylation assay) and 1:40 to 1:320 for plasma (acetylation assay).

References

1. Bosworth, N. & Towers, P. (1989) *Nature 341*, 167-168.
2. Steiner, A.L. (1979) in *Methods of Hormone Radioimmunoassay* (Jaffe, B.M. & Behrman, H.R., eds.), Academic Press, N. York, p. 3. *Also relevant:* Product literature (Kit Code RPA 538), Amersham International plc.

Comments on #B-1: S.O. Døskeland - RESPONSE MEDIATION BY cAMP

   P-O. Berggren asked about the stability of cAMP analogues
and whether they can be electroporated into cells.   **Reply.-**
In general, electroporation is better suited for stable macro-
molecules.   Yet phosphorothioate analogues and N^6-benzoyl-cAMP
are quite stable to degradation; but possibly they may eventually be transported out of the cell.   **Answer to R. Heath:**   for
looking at effects of cAMP on cell systems, no one technique
is ideal. Use of several is preferable; different ones
may suit different cell systems. **U. Lang asked** about the time
relationships of cAMP production and bound and total cAMP
during glucagon stimulation. **Reply:** the equilibration between
bound and free cAMP is very quick, namely ~5 sec.

Comments on #B-2: U. Lang - PKC IN ANGIOTENSIN II-STIMULATED CELLS
            #B-3: J.T. O'Flaherty - CELL STIMULATION EVENTS
            #B-4: A.P. Dawson - G-PROTEINS IN MEMBRANE FUSION

   R. Bruzzone, concerning the rise in prostacyclin response
to ANG II after PMA for 24 h but not after 48 h, and the absence
of PKC-β after 24 h: what happens to PKC-α after 48 h?   **Lang:**
not yet studied.   **P-0. Berggren asked** whether, following down-
regulation of PKC, any overshoot in $[Ca^{2+}]_i$ and $InsP_3$ formation
is seen. **Reply:** not under our conditions; probably the effect
depends on the concentration of PMA used. **Another question put**
to Lang concerned the strong contrast, in the response to
PMA, between ANG II-stimulated PGI_2 production in aortic smooth
muscle cells, potentiated by PMA, and stimulated aldosterone
production in adrenal glomerulosa cells, inhibited by PMA:
taking into account that PMA blocks $PtdInsP_2$ breakdown, might
the PL-C/DAG lipase pathway be the major source of arachidonic
acid in the glomerulosa cell?   **Reply.-** Production of PGI_2 (or
PGC_2), as distinct from aldosterone, was unaffected by PMA in
the glomerulosa cells. However, it is indeed conceivable that
arachidonic acid arises from different sources in different
cell types.

   O'Flaherty, answering question by M. Maley concerning the
differences between PMA and PDB in binding and receptor interactions.- PMA is classically used, probably because it was the
first active phorbol ester to be discovered.   However, there
is at least 50% non-specific binding of PMA to the p.m., chiefly
because of the hydrophobic nature of the myristate moiety.
Such binding is much lower with PDB, which is much more potent
and, I feel, should be used in preference to PMA.   Answer to
N.T. Thompson.- The effect of 5-HETE on PAF-stimulated responses
does not imply a long-term change in cell responsiveness: the
effect is readily reversed by washing the cells. Moreover, its
effects are seen for only 0-20 min after its addition, and
neutrophils rapidly metabolize it to the inactive product 5,20-
diHETE.   The useful suggestion that an effect of 5-HETE itself
might be found by adding it after PMA has not been tried.

G. **Milligan** asked **Dawson** whether, given the inhibitory effects of GTP's, he felt that GTP hydrolysis is essential for function and that the energy of hydrolysis might be necessary for fusion. **Reply.-** Presumably energy is required for fusion and GTP hydrolysis could provide energy. O'Flaherty asked whether the fusion shows reversibility - a pre-condition for extrapolation into cellular systems? **Reply.-** Using the fluorescence assay it is not possible to measure reversibility, since once the probes have mixed they will not segregate again. Yet when measuring Ca^{2+} movements there is certainly no rapid reversal, on a time scale of minutes, of the GTP[S] blockage. On the other hand, we would certainly expect reversibility *in vivo*, in accord with reports from the groups of Gill and of Thomas. **Answer to G.J. Barritt.-** We did try *in-situ* covalent attachment of GTP to microsomal proteins; many became labelled, and we will be trying modified GTP's. **To Berggren:** The GTP effect in permeabilized cells isn't due to fusion between intracellular stores. Saponin- and electro-permeabilized cells clearly differ.

See opposite for comments on #**B-5** (J-P. Mauger)

Comments on #**B-6** : C.J. Barker - IDENTIFICATION OF InsP's
              #**B-7**: N.J. Birch - InsP's AND Ca^{2+}: LITHIUM EFFECTS
              #**ncB-2**: G.J.Barritt - LIVER-CELL PLASMA MEMBRANE ROCC
              #**ncB-3**: R. Heath - cAMP ASSAY BY SPA

Barritt asked **Barker** whether non-*myo* inositol polyphosphates had been found in cell extracts. **Reply.-** Other forms of inositol have been detected by others in very small quantities. In our studies there have been hints of chiral inositol. **Birch,** answering D.Scheller on how Li^+ crosses the blood-brain barrier: possible ways include passive diffusion and Na^+/H^+ exchange.

Dawson asked **Barritt** whether slow dissociation of quin-2-Mn explains the apparent slowness of quin-2 fluorescence recovery, compared with the initial quench rate, when DTPA is added. **Reply.-** The recovery is quite fast, but possibly the recovery may in part represent removal of intracellular Mn so that some of the initial decrease is not due to extracellular quin-2. **Bruzzone** asked why fura-2 was not used in the Ca^{2+} influx studies. **Reply.-** Indeed it would be better, but unlike other labs. we have found it difficult to load into hepatocyte suspensions. **Comment by Berggren.-** Quin-2 is actually a good general choice provided that its intracellular concentration is known - which means that every cell system has to be thoroughly characterized; too high a concentration will obscure the effects. D. **Wermelskirchen** asked whether Mn^{2+} influences intracellular Ca^{2+} binding. $[Mn^{2+}]_i$ is ~1000-fold higher than $[Ca^{2+}]_i$. Mn^{2+} supposedly competes with Ca^{2+} for the binding site on the quin-2 molecule, so should influence other Ca^{2+}-binding sites too. **Reply.-** Mn^{2+} affinity is much higher for quin-2 than for intracellular binding sites; yet the situation may be complex.

Heath, answering G. Milligan, agreed that the only limiting factor in SPA development is the availability of a primary antiserum of suitably high affinity. S.O. Døskeland enquired about non-specific counts when the beads are incubated with ^{125}I-ligand alone without Ab. **Reply:** if, say, 10^6 dpm of ^{125}I-cAMP are added, the background cpm might be 616 and 318 for non-acetylation and acetylation assays respectively, and 10726 and 6978 the maximum cpm bound.

Comments on #B-5: J-P. Mauger - HEPATIC InsP$_3$ RECEPTOR

Mauger, answering G.J. Barritt.- Our p.m. preparations show some enrichment in marker enzymes but are not pure. Tests with saponin suggest that both right-side-out and inside-out vesicles are present. Non-specific binding is assessed by computer after adding cold InsP$_3$ (5 μM). **Answer to T.R. Cheek:** it is cytosolic, not intraluminal, free Ca^{3+} that affects the affinity of the InsP$_3$ receptor; (**answering A.P. Dawson**) we don't know whether an affinity change affects specificity of binding, e.g. for Ins(2,4,5)P$_3$ or InsP$_4$: we have so far studied only Ins(1,4,5)P$_3$. **Answer to G. Milligan:** affinity changes seem not to be due to a Ca^{2+}-induced phosphorylation, as judged mainly by the lack of a temperature influence. **Replies to Døskeland.**- Preliminary results with labelled InsP$_3$ indicate that the dissociation rate constant is ~100-fold higher for the low- than for the high-affinity site. We haven't investigated whether unlabelled InsP$_3$ affects this constant. **Reply to Berggren:** our data accord with what Rubin, Putney and Spät have published.

―――――

SOME LITERATURE PERTINENT TO 'B' THEMES, noted by Senior Editor

Corbin, J.D. & Johnson, R.A., eds. (1988) *Meths. Enzymol. 159*, 792 pp.- 'Initiation and Termination of **Cyclic Nucleotide Action**'. Themes include PK activation involving cAMP (T.J. Martin), cascade systems (E. Shacter), use of analogues (S.J. Beebe, K. Purvis), and muscle studies (P. Cohen). Material in #B-1 (Døskeland) also appears.

Rozengurt, E., Murray, M., Zachary, I. & Collins, M. (1987) *Proc. Nat. Acad. Sci. 84*, 2282-2286.- PKC activation enhances **cAMP** accumulation in 3T3 cells; PT inhibits.

Farese, R.V., *et al.* (1986) *Biochem. Biophys. Res. Comm. 135*, 742-748.- **ACTH** activates InsP/Ca^{2+} *and* cyclic nucleotide systems (adrenal).

Lugnier, C. & Schini, V.B. (1990) *Biochem. Pharmacol. 39*, 75-84.- Cyclic nucleotide **PDases** from aortic endothelium (characterization).

Howlett, A.C., *et al.* (1909) *Biochem. Pharmacol. 38*, 2087-2089.- **Adenylate cyclase** regulation by cannabinoids (neuroblastoma cells).

Lai, W.S. & El-Fakahany, E.E. (1990) *Biochem. Pharmacol. 39*, 221-222.- Muscarinic receptor-mediated **cGMP formation** and binding of *N*-methylscopolamine was antagonized by the **DAG kinase inhibitor** R59 022, serving to prolong DAG elevation and hence PKC activation. Cf. R 59949: (1991) *41*, 835-838; secretion and 47 kDa protein, platelets.

NITRIC OXIDE AS A TRANSCELLULAR MESSENGER *(Outline compiled by Ed.)*

Studies on endothelium–derived relaxing factor (EDRF), which acts on vascular smooth muscle and also blocks platelet aggregation and adhesion, led to its identification as NO, biosynthesized from L–arginine [1]. A useful inhibitor in studying its biosynthesis is L–*N*–monomethyl arginine (L–NMMA) [1]. NO turns out to have various cellular sources and diverse roles, e.g. in the cytotoxicity of phagocytic cells and in neural inter–cell communication. Generally its target cells lie close to those from which it diffuses, and it acts, within seconds of its release, through stimulation of cytosolic guanylate cyclase. The activation of this cyclase involves its haem group, as amplified in a useful review [2] of signal transduction mechanisms involving NO. Differences between cell types in response to NO reflect differences in the consequences of cGMP formation. These consequences may, speculatively, comprise $[Ca^{2+}]_i$ withdrawal involving Ca^{2+}–binding proteins, or inactivation of myosin light chain (hence smooth–muscle relaxation) through phosphorylation of its kinase by a cGMP–activated PK. The likelihood that NO has a neural messenger function has been reinforced by immunohistochemical demonstration of the presence of NO synthase in particular brain neurones and in peripheral nerve cells, as well as in vascular endothelium [3]. Conceivably NO might play a role in cardiovascular disease and, not through cGMP formation, the killing of malignant or other cells by macrophages [2].

As a measure of NO formation, a bioassay based on relaxation of aortic strips has been used. Chemical methods include: measurement of absorption at 540 nm after diazotization and coupling; measurement of chemiluminescence (such as ozone may generate), the NO having been withdrawn from the test system by a continuous N_2 stream or else converted *in situ* by superoxide dismutase to NO_2^- which is finally reconverted to NO by iodide; also, with ^{15}NO, conversion to $[^{15}N]$nitrobenzene and then GS–MS with selective ion monitoring.

References

1. Moncada, S., Palmer, R.M.J. & Higgs, E.A. (1989) *Biochem. Pharmacol. 38*, 1709–1715 [& (1989) *Biochem. Soc. Trans. 17*, 642].
2. Ignarro, L.J. (1991) *Biochem. Pharmacol. 41*, 485–490.
3. Bredt, D.S., Hwang, P.M. & Snyder, S.H. (1990) *Nature 347*, 768–770.

Comments on a Forum talk (no publication text) by R.M.J. Palmer:–

Palmer, answering R. Heath.– For assaying NO, measurement of cGMP formation is widely applicable, but other assays can be useful too *(see above)*. **Replies to D. Scheller.**– (1) The half–life of NO, reckoned to be several seconds *in vivo*, depends *in vitro* on the choice of experimental conditions. (2) Vasodilation by CO_2 is mainly an autoregulatory process, not influenced by NO at the molecular level.   (3) If Arg becomes depleted, there is plenty enzyme capacity to quickly replenish the pool.   **Reply to G. Milligan.-** NO synthase has been purified, but in staining studies with antisera (cf. [3] above) some anomalies have been encountered.   **Reply to J.M. Pfeilschifter.-** The enzyme does not need Ca^{2+} for its induction, or any cofactor for its function.   For induction in macrophages the cytokines INF_8, TNF and LPS are effective.

=========

Nomenclature of glycerol phosphoinositol (cf.'GPI', in #B-6 and elsewhere): guidance kindly provided by C.J. BARKER

"This is the deacylated form of Phosphatidyl inositol. Its correct nomenclature is GroPIns; however, inositol phosphate biochemists have traditionally labelled a peak on their HPLC chromatograms GPI. Recent work (as you are no doubt aware) has seen the characterization of a series of proteins -and possible insulin mediators- anchored by a PI-glycan structure, a glycosylated phosphatidyl inositol -it too has been abreviated GPI. This leads to confusion! So when I say GPI on the HPLC Figure, I mean GroPIns.

"This is further confused by the fact that although this peak more-or-less co-elutes with a standard GroPIns we have shown it contains at least four different compounds, including cIP (correctly, Ins(1:2cyclic)P!!"

SOME FURTHER LITERATURE ON 'B' THEMES, noted by Senior Editor

Owen, C.S. (1988) *Cell Calcium 9*, 141-147.- $[Ca^{2+}]_i$ determination employing **indo 1** on lymphocyte suspensions; AM cleavage aided by Pluronic F-127, and $[Ca^{2+}]_i$ transients are buffered if $[indo 1]_i$ is high, enabling measurement of the releasable Ca pool.

Lai, F.A. & Meissner, G. (1989) *J. Bioenerg. Biomembr. 21*, 227-246.- 'The muscle ryanodine receptor and its intrinsic Ca^{2+} channel activity'; skeletal and cardiac **muscle s.r.**

Clarke, D.M., *et al.* (1989) *Nature 339*, 476-478.- In **s.r. Ca^{2+}-ATPase** the high-affinity Ca^{2+}-binding sites have been located in the transmembrane domain, using site-specific mutagenesis.

Gusovsky, F. & Daly, J.W. (1990) *Biochem. Pharmacol. 39*, 1633-1639.- Commentary on **'MTX'**: 'Maitotoxin: A unique pharmacological tool for research on calcium-dependent mechanisms' (flux effects).

Albano, E., *et al.* (1989) *Biochem. Pharmacol. 38*, 2719-2725.- CCl_4 and Ca^{2+} homeostasis (hepatocytes); toxicity not Ca^{2+}-mediated.

Barritt,G.J., *et al.* (1988) *Biochem. Pharmacol. 37*, 161-167. **Quinacrine** effects on VP-induced changes in glycogen phosphorylase activity, Ca^{2+} transport and $Ptd-InsP_2$ metabolism (hepatocytes).

Bouchelouche, P., *et al.* (1991) *Biochem. Pharmacol. 41*, 243-253.- $[Ca^{2+}]_i$ didn't influence **drug accumulation** in 'wild' or 'MDR' cells.

Michell, R.H., Drummond, A.H. & Downes, C.P., eds. (1989)
Inositol Lipids in Cell Signalling, Academic Press, N. York, 560 pp.
Palmer, S., Lee, D.Y., Hugher, K.T. & Wakelam, M.J.O. (1988)
Cell. Signalling 1, 147-156.- Radioligand **assay of Ins(1,4,5)P$_3$**;
see (1989) *Biochem. J. 260*, 593-596, for application to bovine
adrenocortical microsomes for binding-site studies.
Sasakawa, N., *et al.* (1989) *Cell. Signalling 1*, 75-84.- Cultured
chromaffin cells: Ca^{2+} uptake is needed for high-K$^+$- but not
ANG II-triggered InsP$_3$ accumulation (semi-needed for carbamylcholine).
Negishi, M. & Ito, S. (1990) *Biochem. Pharmacol. 40*, 2719-2725.-
Activation of GABA$_A$ receptor-coupled Cl$^-$ channels mediates GABA-
evoked **catecholamine release** from chromaffin cells with ouabain
present, involving Ca^{2+}-sensitive stimulation of PI metabolism. Cf.
exocytosis-model role of these cells: Schweizer *et al.* (1991) *ibid. 41*, 163-
169. Tachikawa, E., *et al.* (1990) *Biochem. Pharmacol. 40*, 1505-1513.-
PKC modulates, but is not essential for, **catecholamine release**
from chromaffin cells (digitonin-permeabilized; PKC inhibitors used).
O'Brian, C.A., *et al.* (1990) *Biochem. Pharmacol. 39*, 49-57.- A
novel **PKC inhibitor**, *N*-myristyl-Lys-Arg-Thr-Leu-Arg (affects PS too)
Froscio, M., Murray, A.W. & Hurst, N.P. (1989) *Biochem. Pharmacol.
38*, 2087-2089.- **PKC inhibition** by auranofin (interaction with -SH?)

Twomey, B.M., Clay, K. & Dale, M.M. (1991) *Biochem. Pharmacol.
41*, 1449-1454.- A putative **PKC inhibitor**, K252a, inhibits superoxide production in neutrophils activated by both Ptd-InsP$_2$-depen-
dent and -independent mechanisms for respiratory-burst activation.
(1) Giembycz, M.A. & Diamond, J., and (2) *idem* (1990) *Biochem.
Pharmacol. 39*, (1) 1297-1312, (2) 2711-2713.- The Ser-containing
peptide **kemptide** (**PKA-assay** substrate; accepts phosphate from
this cAMP-dependent PK) was evaluated (1) and used to partially
characterize PKA's in guinea-pig lung which affect contractility (2).
Khan, N.A., Quemener, V. & Moulnoux, J-Ph. (1990) *Biochem.
Pharmacol. 40*, R1-R4.- In *Xenopus laevis* oocytes, **spermidine transport** was augmented by phorbol esters without PKC activation.
Zorn, N.E. & Russell, D.H. (1990) *Biochem. Pharmacol. 40*, 2689-
2694.- **Nuclear PKC** was activated by vasoactive intestinal peptide
(**VIP**) in purified nuclei from rat splenocytes (bears on immunity).

Nahorski, S.R. & Potter, B.V.L. (1989) *Trends Pharmacol.
Sci. 10*, 139-144.- 'Molecular **recognition** of inositol polyphosphates by intracellular receptors and metabolic enzymes'.
Rhee, S.G., Suh, P-G., Ryu, S.H. & Kee S.Y. (1989) *Science 244*, 546-
550.- **Phosphoinositidases C** (Ptd-InsP$_2$-specific) & their regulation.
Biochem. Pharmacol. 41 (1991).- **Calmodulin**: (1) Elliott, M.E., *et al.*
1083-1086: involvement in ANG II stimulation of early steps of **aldosterone** synthesis; (2) Ho, A.K., *et al.*, 897-903: role in regulating
pinealocyte cAMP and **cGMP,** the noradrenaline response (accumulation
entailing a rise in [Ca^{2+}]$_i$ and activation of PKC.

*Consult other #nc subsections, especially #ncE, for other items
within the #B ambit.*

#C

HORMONE ORIGINATION AND ACTIONS, ESPECIALLY INSULIN AND GLUCAGON

#C-1

IMMUNONEUTRALIZATION: A TOOL FOR INVESTIGATING CELLULAR INTERRELATIONSHIPS, AS IN THE ISLETS OF LANGERHANS

[1]Vincent Marks[t], [1]Kim Tan, [2]Ellis Samols and [2]John Stagner

[1]Guildhay Antisera Ltd., 6 Riverside Business Park, Walnut Tree Close, Guildford GU1 4UG, U.K.

[2]Department of Medicine, University of Louisville, Kentucky 40206, U.S.A.

Transit of an agent from one cell to a nearby target cell may be via the interstitial fluid, or possibly via portal vessels where such a system exists. Such a system within the islets of Langerhans evidently links the centrally located insulin-secreting B-cells to the glucagon-secreting A-cells and the somatostatin-secreting D-cells respectively. The evidence now outlined hinges on an immunoneutralization approach using Ab's* of very high avidity, specificity and purity. There is portal transit, with regulatory implications, from B- to A- and thence to D-cells, as shown by active-hormone output measurements whilst pancreas (rat or dog) is perfused, in the forward or backward mode, with neutralizing Ig Ab's to each hormone concerned. Yet transits via the interstitial fluid are not excluded since with the present use of intact pancreas rather than slices or isolated islets, the Ab's remain within capillaries. Immunoneutralization is a powerful approach, applicable in other inter-cell contexts.

A complete understanding of the ways in which the cells of a multicellular organism influence the growth, development or behaviour of other cells, especially those adjacent to them, remains one of the major problems confronting biologists. It is of especial importance to embryologists, endocrinologists, oncologists and other investigators of cell growth and behaviour. Already partly understood ways by which cells influence one another include hormonal or endocrine mechanisms and those involving paracrine interactions between adjacent cells, i.e. those wherein a humorally active substance released into the interstitial fluid by one cell type influences another by reacting with receptors on its surface [1].

Classical endocrinology grew from knowledge acquired through the study of the anatomically discrete ductless glands, notably the thyroid, parathyroid, adrenal and pituitary glands and the gonads, all of which secrete their products into the general circulation.

[t]addressee for any correspondence (is at Univ. of Surrey, GU2 5XH)
*Abbreviations.- Ab, antibody; Ig, immunoglobulin.

A now well recognized mechanism whereby one group of cells can influence the behaviour of another more specifically than is possible by secretion of their hormones into the general circulation is through a portal system. The best known example of such a mechanism is the hypothalamico-hypophyseal portal system which was first described more than 40 years ago. Others may exist but have not yet been identified, possibly because of a lack of clear anatomical separation between the constituent parts of the total module.

We believe that there is now good anatomical and functional evidence for a portal system linking the insulin-secreting B-cells of the islets of Langerhans to the glucagon-secreting A-cells and somatostatin-secreting D-cells respectively. This article describes how we arrived at that conclusion. It also provides an opportunity for describing the general principles of immunoneutralization and its application to the investigation of hormonal and cellular interrelationships.

ANATOMY OF THE ISLETS OF LANGERHANS

Islets are scattered throughout the pancreas of all mammalian and avian species, but do not have a uniform anatomy either across or even within species [2]. They are, on the average, 250 μm in diameter and contain ~3000 cells. Altogether the islets account for ~1.0-1.5% of the volume of the pancreas but 10% of the blood flow.

Microanatomy

Mammalian islets contain at least 4 distinct types of cell which can be distinguished from one another on the basis of their histological, morphological, staining, immuno-cytochemical and functional properties. They are generally referred to as A, B, D and PP cells (PP = pancreatic polypeptide), though various other names have been used to describe them in the past. The islets were long considered to be no more than simple collections of cells, albeit important ones, and to consist mainly of A- and B-cells that secreted glucagon and insulin respectively and functioned independently of one another.

The mammalian islet of Langerhans is in fact a remarkably sophisticated micro-organ, each islet having its own individual morphology and vasculature. The islets of most mammalian species have certain features in common insofar as they consist of a central core of more-or-less pure B-cells surrounded by a mantle of D-cells and either A- or PP-cells, depending upon whether they are of dorsal or ventral pancreatic lobe origin.

In the human and rat islet the mantle of A- and D-cells is perforated by one or more arterioles carrying blood from the pancreatic artery into the islets. This splits up into capillaries which traverse the islet tissue and eventually emerge as a venule which drains into the pancreaic vein.

It was originally believed that blood in the intra-islet blood vessels flowed in the direction mantle-to-core, but this view was challenged by Bonner-Weir & Orci [3].   They suggested that, in the rat at least, the blood flow was core-to-mantle; in other words, that the core of B-cells was perfused before the mantle of A- and D-cells whose vascular relationship to one another, if indeed there was one, could not be determined by the purely morphological techniques available to them.

THE PARACRINE HYPOTHESIS

Our discovery that exogenous glucagon stimulated insulin secretion by direct action on the B-cells and, conversely, that insulin inhibited glucagon secretion by the A-cells, led us to suggest that, far from being random collections of cells, the islets were complex organelles consisting of interactive cells whose main function was to secrete exactly the right amount of insulin in response to ingestion of a meal [4-6].

We originally proposed that A- and B-cells might communicate with one another by diffusion of their respective hormonal secretions in the interstitial fluid in which both types of cell are bathed [7].   This hypothesis, which postulated a paracrine interrelationship between A- and B-cells, was slow to gain acceptance [5], and indeed for many years was frankly disbelieved.   Once accepted it had to be extended to include the D-cells [8, 9] after these were shown to be the exclusive source of pancreatic somatostatin, a potent inhibitor of both insulin and glucagon.   The revised paracrine hypothesis proposed that A-, B- and D-cells regulate one another's secretion through the intra-islet interstitial fluid they all share.

Vascular *vs.* paracrine communication

An extensive literature has accumulated to support the hypothesis of a complex intra-islet relationship between the different islet cell types. Recent observations on the structure and function of islet vasculature have, however, made it less likely than formerly that the paracrine hypothesis is capable of providing a complete and totally satisfactory explanation of all the major intra-islet cellular relationships [9-12].

Taking into account that, compared with the rest of the pancreas and with other organs, the islets have strikingly high vascularization [13], and that the islet capillaries are impervious to Ig's [14], one of us (E.S.) questioned the total correctness of the 'paracrine' theory of intra-islet interaction which had been strongly supported by numerous experiments using pancreas slices, isolated islets and even individual cells *in vitro* with and without Ab's in the incubation medium.

On the basis of the new information on the microvasculature of the islets of Langerhans, we proposed the existence of an intra-islet portal system wherein blood perfusing the central core of B-cells picks up insulin which is then delivered downstream to the A- and D-cells, constituting the mantle, where it serves to regulate the secretion of glucagon and somatostatin respectively. This hypothesis has been subjected to rigorous investigation, the results of which have consistently served to substantiate it. The proposition that the vasculature of individual islets does indeed constitute a tiny intra-islet portal system and that blood flows progressively past B-, A- and finally D-cells is consistent with all observations made to date [15-17].

The investigative system developed by Samols & co-workers to test which of the two views of intra-islet interrelationships, i.e. paracrine *vs.* endocrine, represents the true situation utilizes the perfused isolated pancreas preparation with or without concomitant immunoneutralization. A novel twist was the decision to perfuse the preparation sequentially, with randomized order, in two modes - *anterograde* (through the pancreatic artery) and *retrograde* (through the pancreatic vein) [18]. No changes in the hormonal secretion pattern would be expected in relation to the direction of flow if intra-islet cellular interactions occurred solely, or mainly, as a result of changes in interstitial fluid composition, whereas they might be expected to occur if intra-islet cellular interactions were mediated *via* an intra-islet 'endocrine' mechanism.

Although the effects, both stimulatory and inhibitory, of insulin, glucagon and somatostatin upon the secretory function of the various pancreatic cell types can be deduced from observations made *in vivo*, and *in vitro*, it is often difficult - and in some cases impossible - to distinguish direct from indirect effects mediated by nearby or contiguous cells. Mechanically separated islet cells, and those grown in monoculture, have been used to try and overcome some of the problems, but techniques employing these preparations are not entirely free from difficulties since cells do not necessarily behave normally under such unphysiological circumstances.

IMMUNONEUTRALIZATION

Immunoneutralization techniques have also been used to overcome some of the difficulties of studying intercellular relationships, but in so doing they have created others [19-22]. Details of the methods employed in the immunoperfusion experiments employed by us are described elsewhere and will not be repeated. Suffice it to say that only high-avidity, very specific and purified Ig's with neutralizing Ab activity are employed [15-18, 21, 22] since they alone have the ability to bind with hormones sufficiently rapidly and completely to render them biologically inert. Typically the antisera were from sheep or rabbits, and $(NH_4)_2SO_4$ was used as a precipitant.

The importance, in immunoneutralization experiments – especially those involving immunoperfusion – of using only purified, high-affinity and specific Ab's cannot be over-emphasized. These requirements have not always been appreciated by earlier authors whose results must therefore be considered with some degree of circumspection.

PERFUSION AND IMMUNONEUTRALIZATION

Islet capillaries are impermeable during perfusion experiments to introduced Ig's and are excluded from the interstitial fluid [14]. Consequently Ab's to insulin, glucagon and somatostatin added to the perfusion fluid remain exclusively within the intra-islet vasculature, and any effect they exert must be through their ability to react with and neutralize, after entering the blood vessels, the hormones against which they have been raised.

Ig's therefore provide an excellent and indeed unique means of distinguishing effects that are attributable to events happening within the vasculature rather than as a consequence of inter-cell reactions that are mediated through the interstitial fluid or by mere cellular juxtaposition. Thus whilst insulin in the circulation is completely neutralized during a perfusion experiment in which insulin Ab's are present in the perfusion fluid, insulin that is already in, or is subsequently released into, the interstitium remains unneutralized and hence still available for interaction with any A- or D-cells with which it comes into contact.

Under 'ordinary' *in vitro* conditions, as when isolated pancreatic islets or slices are used, exclusion of Ig's from the interstitium does not occur; indeed if Ig penetration of the interstitium did not occur the experiments would be worthless. Preparations of this type, therefore, provide pharmacological rather than physiological information, and results thereby obtained should be interpreted accordingly.

From the known pharmacological effects of the various islet endocrine cells it can be surmised that:
- (i) if intra-islet interactions are mediated largely or entirely through paracrine mechanisms, there would be no difference in the amount of hormones secreted in response to perfusion regardless of whether the perfusion was made in the anterograde or the retrograde direction;
- (ii) profound differences, depending on whether the perfusion was in one direction or the other, would be observed if interactions between islet cells were mediated by intra-islet capillaries carrying hormones from one islet cell type to another through a micro-portal system.

RESULTS

Results of the perfusion experiments performed by Samols and his co-workers using Ab's to neutralize the various endogenous pancreatic hormones, and radioimmunoassays to estimate hormone output, can be summarized as follows.

(i) Anterograde perfusion of the isolated dog or rat pancreas with anti-insulin Ig's produced a prompt rise in glucagon secretion. No such rise occurred when the perfusion was made in the reverse (retrograde) direction through the pancreatic vein [16]. These observations are what would be expected if insulin, secreted by core B-cells, exerted a tonic inhibition on glucagon secretion from which it was released as soon as the intravascular insulin was neutralized.

(ii) Retrograde infusion of anti-insulin Ig's had no effect upon somatostatin secretion whereas anterograde perfusion produced an increase [11]. A small increase might have been anticipated as a result of release of the D-cells from tonic inhibition by insulin secreted by the core B-cells; the very large increase that was seen, however, was unexpected and raised the possibility that blood circulation through the mantle is not entirely random, but first goes past the A-cells - where it picks up glucagon - and then past the D-cells which respond by increasing their own secretion of somatostatin.

(iii) When perfused in the retrograde direction, anti-glucagon Ig's did not affect somatostatin output by the isolated rat pancreas but somewhat reduced its insulin output. In the anterograde direction, on the other hand, they caused a moderate but significant fall in somatostatin secretion without affecting insulin output [17].

(iv) Anti-somatostatin Ig's infused in the anterograde direction had no effect upon insulin or glucagon output. Retrograde infusion, on the other hand, produced a significant rise in glucagon and a very much larger rise in insulin output.

This would be expected if glucagon secretion by pancreatic A-cells had been released from tonic inhibition by somatostatin secreted by the D-cells and had then become available to stimulate the core B-cells to secrete insulin [16].

From these experiments it is possible to conclude that within the islets of Langerhans, particularly in the rat but probably in man also, there is an orderly sequence of perfusion from B- to A- and then to D-cells. The value and importance of immunoneutralization in arriving at these conclusions cannot be overemphasized. They could not, we believe, have been reached with the same degree of confidence in any other way known to us at the present time.

DISCUSSION

Immunoneutralization is clearly an extremely powerful tool in the elucidation of many physiological questions, but must be combined with other physiological techniques in order to realize its full potential. Although the data obtained using pancreatic perfusion, disclosing vascular pathways, by their nature do not reinforce our earlier hypothesis of important paracrine inter-islet relationships, the data do not disprove the large body of evidence that already exists in favour of the hypothesis.

We have chosen to discuss the endocrine pancreas as a model of the usefulness of immunoneutralization as a research tool. Work recently carried out in our laboratories, employing Ab's raised against thyroid hormones and used *in vivo* to study thyroid-pituitary relationships, throw considerable doubt upon the correctness of some seemingly well established concepts in exactly the same way as our work on the endocrine pancreas.

This work has been made possible only because of our increasing ability to make and purify large amounts of very specific, highly avid Ab's against small haptenic as well as the larger proteinaceous hormones and growth factors, at a cost that makes their use in immunoneutralization experiments both realistic and practicable.

References

1. Franchimont, P., ed. (1986) *Paracrine Control* issue of *Clinics in Endocrinology & Metabolism*, 15, 1-207.
2. Hellman, B. & Lernmark, A. (1970) in *The Structure and Metabolism of the Pancreatic Islets: a Centennial of Paul Langerhans' Discovery* (Falkmer, S., Mellan, B. & Taljedal, I.B., eds.), Pergamon, Oxford, pp. 453-462.

3. Bonner-Weir, S. & Orci, L. (1982) *Diabetes 31*, 833-839.
4. Marks, V. & Samols, E. (1968) in *Recent Advances in Endocrinology*, 8th edn. (James, V.H.T., ed.), Churchill, pp. 111-138.
5. Samols, E., Tyler, J. & Marks, V. (1972) in *Glucagon. Molecular Physiology. Clinical and Therapeutic Implications* (Lefebvre, P.J., ed.), Springer, Berlin, pp. 151-173.
6. Morgan, L.M., Flatt, P.R. & Marks, V. (1988) *Nutr. Res. Rev. 1*, 79-97.
7. Samols, E., Marri, G. & Marks, V. (1966) *Diabetes 15*, 855-866.
8. Reichlin, S. (1983) *New Engl. J. Med. 309*, 1495-1501 & 1556-1563.
9. Samols, E., Weir, G.C. & Bonner-Weir, S. (1983) in *Handbook of Experimental Pharmacology*, Vol. 66/II (Lefebvre, P.J., ed.), Springer, Berlin, pp. 13-173.
10. Samols, E. (1983) *as for* 9. but Vol. 66/I, pp. 485-518.
11. Samols, E. & Stagner, J.I. (1988) *Am. J. Med. 85*, 31-35.
12. Marks, V., Tan, K.S., Stagner, J.I. & Samols, E. (1990) *Biochem. Soc. Trans. 18*, 103-104.
13. Lifson, N., Kraminger, K.G., Mayrand, R.R. & Lander, E.J. (1980) *Gastroenterology 79*, 466-473.
14. Kvietys, P.R., Perry, M.A. & Granger, D.N. (1983) *Am. J. Physiol. (Gastrointest. Liver Physiol. 8)*, G519-G524.
15. Stagner, J.I., Samols, E. & Bonner-Weir, S. (1988) *Diabetes 37*, 1715-1721.
16. Samols, E., Stagner, J.I., Ewart, R.B.L. & Marks, V. (1988) *J. Clin. Invest. 82*, 350-354.
17. Stagner, J.I., Samols, E. & Marks, V. (1989) *Diabetologia 32*, 203-206.
18. Stagner, J.I. & Samols, E. (1986) *J. Clin. Invest. 77*, 1035-1037
19. Schatz, H. & Kullek, U. (1980) *FEBS Lett. 122*, 207-210.
20. Fujimoto, W.Y., Kawazu, S., Ikeuchi, M. & Kanazawa, Y. (1983) *Life Sci. 32*, 1873-1876.
21. Tan, K.S., Tsiolakis, D. & Marks, V. (1985) *Diabetologia 28*, 435-440.
22. Tan, K.S., Atabani, G. & Marks, V. (1985) *Diabetologia 28*, 441-444.

#C-2

REGULATION OF SECRETORY-GRANULE PROTEIN BIOSYNTHESIS IN THE PANCREATIC B-CELL, STUDIED BY TWO-DIMENSIONAL GEL ELECTROPHORESIS

P.C. Guest and J.C. Hutton[†]

Department of Clinical Biochemistry,
University of Cambridge, Addenbrooke's Hospital,
Hills Road, Cambridge CB2 2QR, U.K.

The effect of glucose on the biosynthesis of pancreatic islet proteins was investigated by 2-D gel electrophoresis, combined with fluorographic analysis. Subcellular fractionation of [35S]methionine-labelled rat islets showed that the biosynthesis of the majority of i.s.g.* proteins is increased 15- to 30-fold by glucose stimulation. By contrast, only a minority of cytosolic proteins and no p.m. constituents showed a response of similar magnitude. The findings indicate the possibility of multiple mechanisms for stimulation of biosynthesis in the islets of Langerhans.

Insulin biosynthesis is regulated by circulating nutrients, neurotransmitters and hormones, principally through control of the translation of pre-formed mRNA. Since subsequent storage and secretion of the hormone depends upon its packaging into secretory granules, the question arises as to how the synthesis of other granule proteins is coordinated in response to stimulation of the cell. This point is underscored by the presence of >100 different polypeptides in the granule besides insulin [1]. These include: the proteolytic enzymes involved in conversion of proinsulin to insulin [2, 3], minor co-secreted peptides [4], membrane proteins involved in secretion, and ion pumps involved in regulation of the intragranular environment [5].

We have analyzed subcellular fractions prepared from pulse-chase-labelled rat islets by using 2-D gel electrophoresis and fluorography, to examine the effects of glucose on the synthesis of i.s.g.* proteins. This approach posed a number of technical difficulties including the need for rapid fractionation of radiolabelled tissue homogenates, identification of proteins subjected to post-translational processing and assignment of specific proteins to subcellular compartments. Ways to overcome these problems are now discussed.

[†]addressee for any correspondence

*Abbreviations.- BSA, bovine serum albumin; e.r., endoplasmic reticulum; i.s.g., insulin secretory granule; p.m., plasma membrane; MES, potassium 2-(N-morpholino)ethanesulphonic acid.

EFFECT OF GLUCOSE ON THE BIOSYNTHESIS OF TOTAL ISLET PROTEINS

Islets of Langerhans were isolated from the pancreatic tissue of New England Deaconess Hospital (NEDH) rats by a collagenase digestion technique [6], and maintained before use in an incubation medium consisting of modified Krebs bicarbonate buffer (120 mM NaCl/5 mM KCl/1 mM $MgSO_4$/2.5 mM $CaCl_2$/24 mM $NaHCO_3$), containing 20 mM Hepes (pH 7.4), 0.1% BSA and 2.8 mM glucose.

Preliminary studies [6, 7] showed that islets incubated in the presence of high glucose (16.7 mM) for 20 min incorporated 2.8 ±0.3 (S.E.M.; n = 8) times as much [^{35}S]methionine into total protein compared to islets exposed to low glucose (2.8 mM). Two-dimensional (2-D) electrophoretic analysis of ^{35}S-labelled islets [7] showed that among the 260 spots amenable to analysis the majority exhibited a 2- to 4-fold biosynthetic induction in response to high glucose, and ~10 spots showed an increase of 10- to 20-fold (Fig. 1).

Although this method proved adequate for detection of the biosynthetic responses of major islet proteins, it may not account for that of minor constituents such as secretory granules and p.m. proteins. In order to increase the sensitivity and resolution of detection and, at the same time, determine the localization of glucose-stimulated proteins ^{35}S-labelled islets were subjected to subcellular fractionation prior to electrophoretic analysis.

SUBCELLULAR LOCALIZATION OF GLUCOSE-STIMULATED PROTEINS

Previous investigations have shown that newly synthesized i.s.g. proteins are delivered to nascent granules after a delay of 30-90 min, following their intracellular passage through the e.r. and Golgi complex [8, 9]. Accordingly, to allow newly synthesized secretory proteins sufficient time to reach their cellular destination and ensure completion of post-translational modifications prior to fractionation, islets were incubated with [^{35}S]methionine for 1 h followed by a further 3 h incubation in non-radioactive media [7].

The subcellular fractionation procedure [7] is outlined in Fig. 2. In order to facilitate efficient homogenization and recovery of fractions, it was necessary to combine radiolabelled islets with unlabelled rat insulinoma as carrier tissue, prior to fractionation. Thus, immediately before completion of the islet incubation, a post-nuclear supernatant fraction was prepared from freshly excised rat insulinoma tissue (propagated in NEDH rats by subcutaneous injection of an insulinoma cell suspension) according to our described method [1].

The homogenization medium consisted of 10 mM MES (pH 6.5) containing 0.3 M sucrose, 1 mM $MgSO_4$ and 1 mM EGTA.   The combined tissues, after 10 strokes in a Potter homogenizer, were centrifuged to remove unbroken cells and nuclei.   The supernatants were subjected to discontinuous Nycodenz density gradient centrifugation.   The resulting interface material (fractions A, B & C in order of increasing density) and pellets were washed twice in homogenization medium and stored at -70°.   The material not entering the gradient (soluble fraction) was reduced in volume using a Centricon 10 concentrator (Amicon, Danmere, MD, U.S.A.) and stored at -70°.

Marker protein analyses [7] (Fig. 3) showed that the soluble fraction contained cytosolic constituents, fraction A was enriched in p.m., fraction B was a mixture of p.m. and secretory granules, fraction C was enriched in secretory granules, and the pellet was a mixture of lysosomes, mitochondria and e.r.

Analysis by 2-D electrophoresis of ^{35}S-labelled proteins in i.s.g.-enriched fraction C [7] showed that high glucose stimulated the incorporation of radioactivity into >60 out of 240 polypeptides by 15- to 30-fold, compared to islets incubated in low glucose (Fig. 4).   Most of the remaining proteins were either stimulated 2- to 4-fold by high glucose or showed no response.   Fig. 5 shows diagrammatically the predominant radioactive spots in Fraction C [7].   Of 80 designated major spots, 32 were localized predominantly in fraction C and co-migrated with proteins in highly purified insulinoma granules (Table 1 & Fig. 6).   Of this number, 25 proteins (78%) were stimulated 15- to 30-fold by high glucose, while the remaining 7 proteins showed moderate or low responses.

None of ~40 proteins detected in p.m.-enriched fraction A [7] were stimulated by high glucose [Fig. 7].   In contrast, incorporation into 3 proteins was decreased markedly.   Most of the remaining proteins showed no response.   Comparison of the arrays of spots in each fraction showed that none of the p.m. components were contaminants from other fractions (data not shown).

Out of 160 proteins in the cytosolic protein-enriched soluble fraction [7], ~15 were stimulated by high glucose (Fig. 8). However, 7 of these were found in significant amounts in the pellet, indicating that the soluble fraction may contain significant quantities of lumenal e.r. proteins released during homogenization. Therefore, at most 8 of the glucose-stimulated constituents are likely to be of cytosolic origin.   Also present in this fraction was a small cluster of proteins which were decreased 10- to 20-fold by high glucose, while

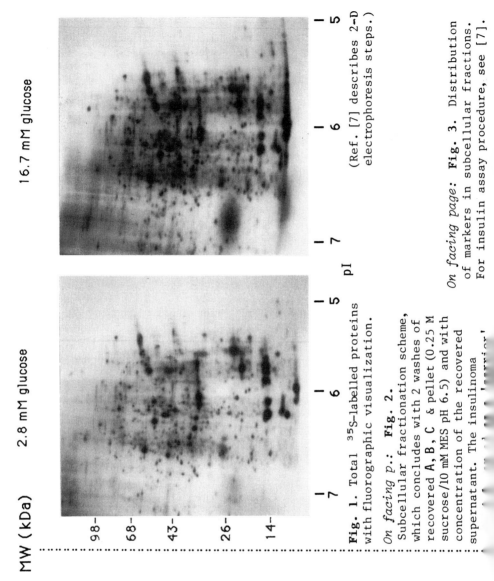

16.7 mM glucose

2.8 mM glucose

MW (kDa)

pI

Fig. 1. Total ^{35}S-labelled proteins with fluorographic visualization. (Ref. [7] describes 2-D electrophoresis steps.)

On facing p.: **Fig. 2.** Subcellular fractionation scheme, which concludes with 2 washes of recovered **A, B, C** & pellet (0.25 M sucrose/10 mM MES pH 6.5) and with concentration of the recovered supernatant. The insulinoma

On facing page: **Fig. 3.** Distribution of markers in subcellular fractions. For insulin assay procedure, see [7].

the remaining constituents were stimulated 2- to 4-fold or showed no response.

The mixed nature of fraction B and the pellet material [7] did not permit the unequivocal assignment of spots to these fractions or the identification of glucose-responsive proteins. For these reasons, the data for these fractions are not presented.

[Text (DISCUSSION) continues on p.219.

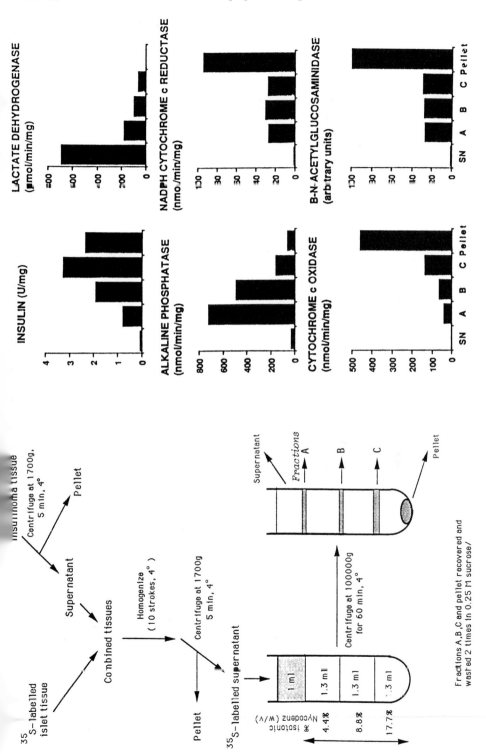

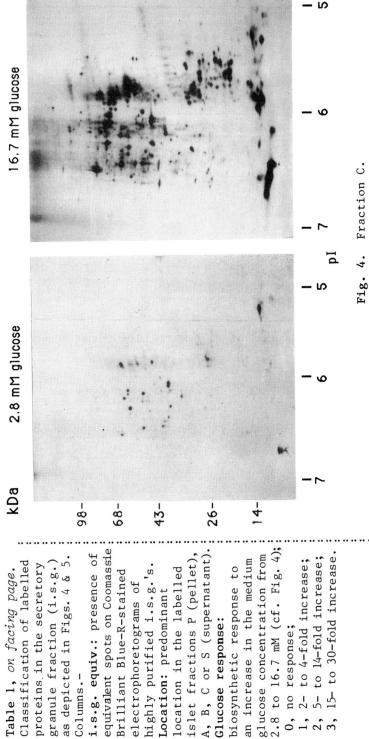

Fig. 4.   Fraction C.

Table 1, *on facing page.*
Classification of labelled
proteins in the secretory
granule fraction (i.s.g.)
as depicted in Figs. 4 & 5.
Columns.-

i.s.g. equiv.: presence of
equivalent spots on Coomassie
Brilliant Blue-R-stained
electrophoretograms of
highly purified i.s.g.'s.
Location: predominant
location in the labelled
islet fractions P (pellet),
A, B, C or S (supernatant).
Glucose response:
biosynthetic response to
an increase in the medium
glucose concentration from
2.8 to 16.7 mM (cf. Fig. 4);
    0, no response;
    1, 2- to 4-fold increase;
    2, 5- to 14-fold increase;
    3, 15- to 30-fold increase.

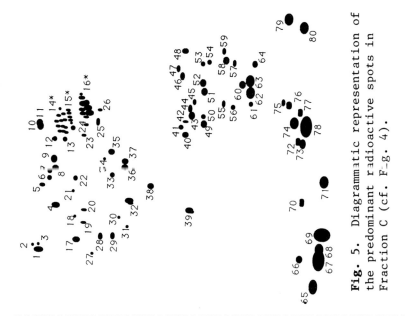

Fig. 5. Diagrammatic representation of the predominant radioactive spots in Fraction C (cf. F-g. 4).

Spot no.	i.s.g. equiv.	Location	Glucose response
1	–	P	2
2	–	C	3
3	–	P	3
4	+	S,P	1
5	+	P	1
6	+	C	2
7	+	S,P	2
8	+	C	3
9	+	P	3
10	+	P	2
11	+	P	2
12	–	C	3
13	–	C	3
14	+	C	2
15	+	E	3
16	+	C	3
17	+	C	2
18	+	E	3
19	+	C	3
20	+	C	2
21	+	C	1
22	+	E	2
23	–	C	2
24	–	E	1
25	+	C	3
26	–	C	3

Spot no.	i.s.g. equiv.	Location	Glucose response
27	–	C	2
28	–	C	2
29	–	C	1
30	+	C	3
31	+	P	3
32	+	C	2
33	+	C	1
34	–	P	2
35	+	B	1
36	+	B	1
37	+	C	2
38	+	C	3
39	–	C	2
40	–	C	2
41	+	C	3
42	+	C	3
43	+	C	3
44	–	P	2
45	–	C	2
46	+	C	3
47	–	C	3
48	+	C	3
49	–	C	3
50	+	C	3
51	–	C	3
52	–	C	3
53	+	C	3

Spot no.	i.s.g. equiv.	Location	Glucose response
54	+	C	3
55	–	C	3
56	–	C	3
57	–	C	3
58	–	C	3
59	–	C	2
60	+	C	3
61	+	C	3
62	+	C	3
63	+	P	3
64	+	C	3
65	+	C	3
66	?	C	3
67	?	C	3
68	+	C	3
69	+	C	3
70	?	C	3
71	–	C	3
72	+	C	3
73	+	C	3
74	–	C	3
75	+	C	3
76	+	C	3
77	–	C	3
78	+	C	3
79	–	P	0
80	–	C	3

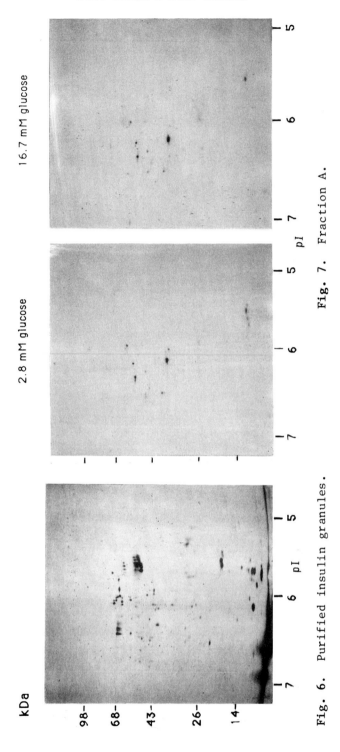

Fig. 7. Fraction A.

Fig. 6. Purified insulin granules.

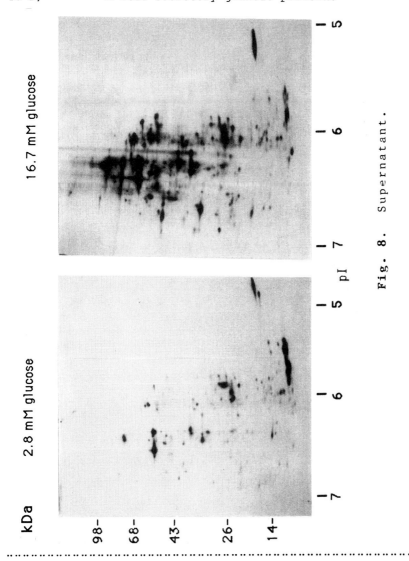

Fig. 8. Supernatant.

DISCUSSION

Essential to this study of i.s.g. biogenesis was a technique which allowed the rapid separation and quantitative recovery of islet subcellular fractions. For these reasons, the Nycodenz density gradient centrifugation technique was adopted in preference to the Percoll method described previously [1]*. One drawback to the Nycodenz method was that the i.s.g. purity was less than half that of Percoll-purified granules, as determined by insulin specific activity measurements [7].

*See also art. by Hutton *et al.* (on Ca^{2+} and prohormone proteolysis) in Vol. 19 (1989; same eds. & publisher).- *Ed.*

Also relevant is source heterogeneity: besides B-cells (insulin-secreting; 64–74% of the rat islet cell number), islets also contain A-cells (glucagon; 2–28%), D-cells (somatostatin; 4%) and PP-cells (pancreatic polypeptides; 2–20%) [10]*. However, comparison of the 2-D gel profiles of i.s.g.'s with highly purified Percoll i.s.g.'s made it relatively easy to distinguish genuine i.s.g. components from contaminants. Also, the co-migration of the labelled proteins under study with the unlabelled insulinoma proteins indicated that they had undergone complete post-translational modification.

Since the biosynthetic responses of the glucose-stimulated granule proteins were comparable to the insulin responses in size and in their short time scale [6, 7], translational control may be operating; perhaps the mRNA's encoding these proteins contain common structural elements selectively recognizable by the cell's translational machinery, but such elements are not obvious for the DNA sequence of insulin *vs.* that of chromogranin A [11], likewise a granule constituent that responds similarly [6]. Possibly, too, as proposed for insulin [12], signal-recognition-particle-mediated translocation of nascent peptides into the e.r. lumen may be accelerated; whilst this mechanism may operate, it is insufficient to account for all of the observed biosynthetic responses – since not all p.m. and granule components show equivalent responses to glucose stimulation – nor for the small number of glucose-stimulated cytoplasmic proteins since these are synthesized presumably on free ribosomes.

References (BJ = Biochem. J.)

1. Hutton, J.C., Penn, E.J. & Peshavaria, M. (1982) *Diabetologia 23*, 365–373.
2. Davidson, H.W., Rhodes, C.J. & Hutton, J.C. (1988) *Nature 333*, 93–96.
3. Davidson, H.W. & Hutton, J.C. (1987) *BJ 245*, 575–582.
4. Sopwith, A.M., Hales, C.N. & Hutton, J.C. (1984) *Biochim. Biophys. Acta 803*, 342–345.
5. Hutton, J.C. & Peshavaria, M. (1982) *BJ 204*, 161–170.
6. Guest, P.C., Rhodes, C.J. & Hutton, J.C. (1989) *BJ 257*, 431–437.
7. Guest, P.C., Bailyes, E.M., Rutherford, N.G. & Hutton, J.C. (1991) *BJ 274*, 73–78.
8. Sorenson, R.L., Steffes, M.W. & Lindall, A.W. (1970) *Endocrinology 86*, 88–96.
9. Orci, L., Ravazzola, M., Amherdt, M., Madsen, O., Vassali, J-D. & Perrelet, A. (1985) *Cell 42*, 671–681.
10. Baetens, D., Malaisse-Lagau, F., Perrelet, A. & Orci, L. (1979) *Science 206*, 1323–1325.
11. Hutton, J.C., Bailyes, E.M., Rhodes, C.J., Rutherford, N.G., Arden, S.A. & Guest, P.C. (1990) *Biochem. Soc. Trans. 18*, 122–124.
12. Welsh, M., Scherberg, N., Gilmore, R. & Steiner, D.F. (1986) *BJ 235*, 459–467.

*Note by Ed.– The contribution by V. Marks (#C-1) is pertinent.

#C-3

STIMULUS-RESPONSE COUPLING IN PANCREATIC B-CELLS: THE ROLE OF PROTEIN KINASE C

S.L. Howell, S.J. Persaud and P.M. Jones

Biomedical Sciences Division,
King's College London, Campden Hill Road,
Kensington, London W8 7AH, U.K.

PKC*, a family of Ca^{2+}/phospholipid-dependent isoenzymes, has been identified in islets of Langerhans and insulin-secreting tumour cells. Its precise role in islet function has not yet been defined although there is accumulating evidence that it is involved in the regulation of insulin secretion. We have used several experimental approaches to investigate the involvement of PKC in the rat B-cell secretory response to nutrient secretagogues (e.g. glucose) and receptor-mediated secretagogues (e.g. CCh). These include the measurement of insulin release from normal and PKC-depleted islets, correlation of secretion with the activation and redistribution of islet PKC, and assessment of PKC-mediated phosphorylation events in electrically permeabilized islets. Thereby we have obtained evidence supporting a role for PKC in sustaining cholinergic potentiation of glucose-stimulated insulin secretion. However, our data from these investigations suggest that PKC is not essential for nutrient stimulation of insulin secretion.

The hydrophobic product of the phosphodiesterase-mediated cleavage of inositol phospholipids, DAG*, is thought to have a second-messenger function through the activation of PKC [1]. PKC is now known to be a family of Ca^{2+}/phospholipid-dependent isoenzymes which have been identified in many tissues [2]. To date, at least 7 isoforms of PKC (designated α, βI, βII, γ, δ, ε, ζ) have been identified by their chromatographic separation or by nucleic acid sequencing [2-4]. It is not yet certain that all these isoforms are normally expressed in mammalian tissues, but there is convincing evidence that different tissues express different PKC isoenzymes [2]. At physiological Ca^{2+} concentrations, PKC is activated by DAG or by tumour-promoting phorbol esters such as PMA, which can substitute for DAG. PKC (in particular the γ type) may also be activated by arachidonic acid, an unsaturated fatty acid produced by phospholipase A_2-mediated cleavage of membrane phospholipids [2].

**Abbreviations.*- CCh, carbachol; DAG, diacylglycerol; Pi, inorganic phosphate; PKC, protein kinase C; PMA, 4β-phorbol myristate acetate.

PKC is present in islets and insulin-secreting tumour cells, and its sensitivity to Ca^{2+}, DAG and phosphatidylserine appears to be the same as those of PKC in other tissues [5-7]. The expression of PKC isoenzymes in B-cells has not yet been extensively studied, although two recent reports suggest that pancreatic B-cells contain both α [8] and βII [9] isoforms.

There are several lines of evidence that PKC may be involved in the control of insulin secretion from pancreatic B-cells. Insulin release can be stimulated by activation of PKC with DAG analogues [10] or phorbol esters [11, 12] and this activation is associated with phosphorylation of several endogenous substrates [13-16]. Insulin secretion can be inhibited by a number of PKC inhibitors [17-19]; but since these may show poor specificity for PKC and exert effects on other kinases, conclusions from this experimental approach are questionable [20, 21]. We have therefore used alternative approaches to investigate the functional role of PKC in the regulation of insulin secretion from rat isolated islets of Langerhans.

ACTIVATION OF PROTEIN KINASE C (PKC)

One fundamental approach towards identifying an involvement of PKC in insulin secretion is to measure the activation of the enzyme by insulin secretagogues. Glucose, which is the major physiological secretagogue, is thought to stimulate insulin secretion as a consequence of its metabolism within the B-cell [22], while the cholinergic agonist CCh enhances insulin secretion through interaction with B-cell muscarinic receptors [23]. Early changes in DAG formation in response to both glucose and CCh have been determined in rat islets [24, 25]. The increase in the DAG amount in response to both secretagogues was maximal by 60 sec and maintained for at least 5 min [24]. However, CCh-generated DAG was predominantly of a stearoyl-arachidonyl configuration, consistent with its being a product of inositol phospholipid hydrolysis, while the DAG within glucose-stimulated islets contained mainly palmitic acid, indicating that it has been produced *de novo*. Thus, the different fatty acid species of the DAG's produced by muscarinic agonists and glucose may provide a means of differential control of PKC activity by these stimuli. Activation of PKC is thought to be accompanied by its translocation, involving a redistribution of enzyme activity from a predominantly cytosolic form to a membrane-associated form [26]. The translocation of PKC may, therefore, offer an indication of the activation of PKC in response to specific stimuli. (#B-2 and several other arts. deal with this translocation; consult 'PKC' in the Index.- *Ed*.)

Table 1.[⊗] PMA-induced translocation of PKC activity. Ion-exchange purified (DE-52) PKC activity was determined in memb-rane and cytosolic fractions after 5 min exposure to 500 nM PMA or the inactive phorbol ester 4αPMA (controls), by measuring incorporation of ^{32}P from $[\gamma^{32}P]$ATP into histone type III-S [7].

Test material; activity units (& no. of obs.)		Cytosol	Membrane
Rat islets;	control	256 ±11	5 ±2
fmol/islet per min (4)	+ PMA	201 ±13*	29 ±1[†]
Rat anterior pituitary;	control	10.7 ±0.6	1.2 ±0.6
pmol/μg protein per min (4)	+ PMA	7.0 ±0.4[†]	6.0 ±0.6[†]
Human luteal cells[φ];	control	296 ±6	7 ±3
pmol/10^6 cells per min (3)	+ PMA	254 ±7[†]	38 ±3[†]
Human granulosa cells[φ];	control	265 ±5	47 ±1
pmol/10^6 cells per min (5)	+ PMA	142 ±16[†]	85 ±3[†]

[⊗]*see foot of p.* [φ] studied in collaboration with Dr. D.R.E. Abayasekara
*$P < 0.02$; [†]$P < 0.01$ (*vs.* appropriate controls; unpaired t-tests)

We have developed methods for measuring PKC activity in islet membrane and cytosolic fractions after extraction and partial purification of the enzyme by ion-exchange chromato-graphy [7]. In the unstimulated state (2 mM glucose), islet PKC activity was mainly associated with the cytosolic fraction with little detectable activity in the membrane fraction (Table 1). PMA caused a decrease in cytosolic activity which was accompanied by increased activity in the membrane fraction (Table 1). PMA stimulated translocation of PKC in several other secretory tissues, although these differed in extent of translocation (Table 1). [Arts. #B-2 & #B-3 are relevant.- *Ed.*]

CCh stimulated translocation of PKC activity from cytosolic to membrane fractions at both 2 mM and 20 mM glucose (Fig. 1a[⊗]), consistent with its reported effects on inositol phospholipid hydrolysis [24, 25, 27]. Despite the ability of CCh to stimulate PKC translocation at 2 mM glucose, it has no effect on insulin release in the absence of a stimulatory concentration of glucose [27], suggesting that activation and translocation of PKC are not sufficient for the initiation of insulin secretion. In parallel experiments under the same conditions we were consistently unable to demonstrate an effect of glucose on PKC translocation (Fig. 1b). Further investigation of the effect of glucose on distribution of PKC activity revealed that over a time-course of 1-10 min 20 mM glucose was unable to stimulate translocation of this enzyme from the cytosol to membranes (Fig. 2).

[⊗]Values in Table 1 and Figs. are means and (± or, in Figs., error bars) S.E.M.; in Fig. legends, n = no. of observations.

Fig. 1. Effects of CCh and
glucose on PKC translocation
in rat islets. (a) Incubating
islets with 500 µM CCh for
5 min in the presence of
2 mM glucose resulted in
a translocation of PKC
activity from cytosol
(open bars) to membrane
(hatched bars). Re-drawn
from data in [7].
(b) CCh stimulated an
increase in membrane-
associated PKC activity
at both 2 and 20 mM
glucose, but 20 mM
glucose alone had no
effect.
*P <0.05,** P <0.01
(n = 4-6) vs. appropri-
ate PKC activity of
unstimulated islets
(2 mM glucose).

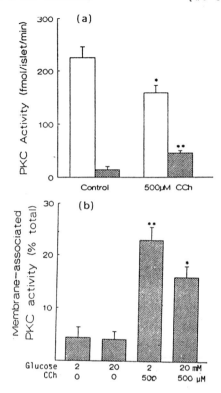

        Similar results demonstrating stimulatory effects of PMA,
DAG analogues or CCh on PKC redistribution have been reported
in islets [28] and insulin-secreting tumour cells [29-31].
The lack of effect of glucose on islet PKC translocation
in our studies [7] has been confirmed recently [28], although
conflicting results have been obtained using insulin-secreting
cell lines [30, 31].   The tumour cell line RINm5F is not
sensitive to glucose, but responds to the glycolytic intermediate glyceraldehyde with an increased generation of DAG
and enhanced insulin secretion [32].   It has been reported
that, contrary to the effects of glucose in normal islets,
glyceraldehyde produces a rapid redistribution of PKC activity
[30].   In contrast, there has been a preliminary report
[31] that glucose does not cause PKC translocation in a
glucose-sensitive tumour cell line (HIT).   It is clear from
these observations that there are differences between the
responses of normal islets and insulin-secreting cell lines.

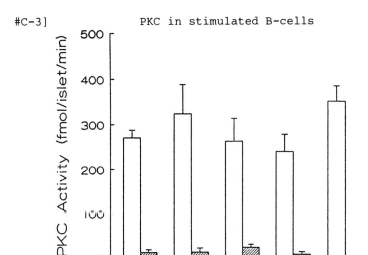

Fig. 2. Time-course of glucose effects on islet PKC distribution. Incubation of islets with 20 mM glucose for 1-10 min
caused no significant translocation of PKC activity (*open bars*,
cytosolic; *hatched*, membrane-associated) *vs.* distribution at 2 mM
glucose (t = 0). Bars represent 2-6 expts. (error bar = range, not
S.E.M., where appropriate) each covering 3-6 time-points.

This may be a consequence of the expression of different
PKC subtypes in islets and the cell lines [8, 9], and perhaps
questions the validity of using cell lines for studies of
the role of PKC in normal B-cell function.

DOWN-REGULATION OF PROTEIN KINASE C IN ISOLATED ISLETS

Measurements of PKC translocation provide some evidence
that this enzyme may be involved in cholinergic signalling
mechanisms in islets. Furthermore, the lack of effect of
glucose might suggest that PKC is not involved in glucose-stimulated secretion. However, it is possible that glucose activates
a minor isoform of PKC which does not translocate to membranes
upon activation, or whose translocation is not detectable
in present assays. Another experimental strategy is therefore
required to resolve the involvement of PKC in B-cell signalling.

In other cell types prolonged exposure to phorbol esters
results in depletion of cellular PKC activity [33], reflecting
an increased rate of degradation of the enzyme without any
change in its rate of synthesis [34]. In accord with this,
20-24 h treatment of islets with 200 nM PMA, with either
2 or 11 mM glucose present, resulted in a dramatic reduction
in islet PKC activity (Fig. 3), consistent with PMA-induced
translocation of PKC to membranes followed by progressive

Fig. 3. PMA-induced down-
regulation of islet PKC
activity. Treatment of
islets with 200 nM PMA for
>20 h *(hatched bars)* in the
presence of either 2 mM (**a**)
or 11 mM (**b**) glucose resulted
in a dramatic loss of islet
PKC activity (n = 3 or 4).
Data in **a** *obtained in colla-*
boration with D.J. Slee.
Data in [27] redrawn for **b**.

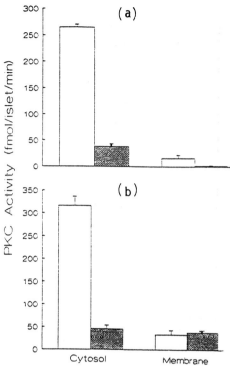

cleavage of the enzyme. In these experiments islet insulin
content was not affected by the prolonged exposure to PMA
(controls, 35.4 ±7.1 ng/islet; PMA-treated, 29.2 ±4.9; n = 3
experiments, each 7-9 observations; *P* >0.2).

To complement our studies of PKC activation by insulin
secretagogues, we have measured insulin secretion and protein
phosphorylation in normal and PKC-depleted islets.

i) Insulin secretion

As might be expected, islets in which PKC had been
depleted by prolonged treatment with PMA did not respond
to a subsequent exposure to this phorbol ester. Conversely,
islets which had been treated overnight with the inactive
phorbol ester, 4αPMA, responded to PMA with a significant
potentiation of 20 mM glucose-induced insulin secretion (Fig. 4).
While islets in which PKC had been down-regulated were unresponsive to PMA, 20 mM glucose-stimulated insulin release, measured
over 1 h from batch-incubated islets, was not significantly
affected (Fig. 4). These results, together with our previous
observations that glucose did not translocate PKC (Fig. 2),
strongly suggest that PKC is not essential for glucose-induced
insulin secretion from normal rat islets – as borne out by
later investigations [35, 36] likewise with PKC-depleted islets.

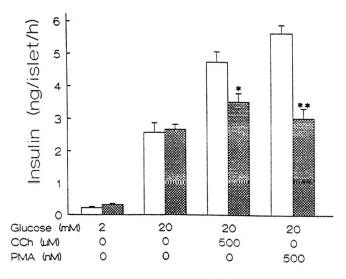

Fig. 4. Insulin secretion from PKC-depleted islets. Down-regulation of PKC activity *(hatched bars)* had no effect on glucose-induced (20 mM) insulin secretion, but totally abolished the secretory response to PMA and significantly, but not completely, inhibited the response to CCh (500 µM) at 20 mM glucose. Control islets (4αPMA-treated; *open bars*) responded to both PMA and CCh with a significant potentiation of 20 mM glucose-induced insulin release.   *$P<0.05$, **$P<0.01$ (n = 8), *vs.* control islets. *Some of the data redrawn from [37].*

However, glucose-induced insulin secretion is a biphasic process, and it has recently been reported that the second phase of secretion is inhibited in perifused PKC-depleted mouse islets [38]. We have also performed experiments using a perifusion system to investigate the dynamics of insulin release from rat isolated islets: these demonstrated that neither phase of glucose-stimulated insulin secretion is inhibited by PKC down-regulation (Fig. 5).   Indeed, we found that the first phase of the secretory response to glucose showed a slight, but significant, enhancement in PKC-depleted islets (Fig. 5). The reasons for these discrepancies are unclear, but it is interesting to note that another group has shown no inhibition of glucose-stimulated insulin release from PMA-pretreated peri-fused mouse islets [36].

PKC depletion by prolonged PMA treatment has also been applied to insulin-secreting tumour cells.   In contrast to glucose-induced insulin secretion from normal islets, glyceral-dehyde-stimulated secretion from RIN cells was almost totally abolished following PKC down-regulation [30].   These results support the attribution of an important role to PKC in nutrient

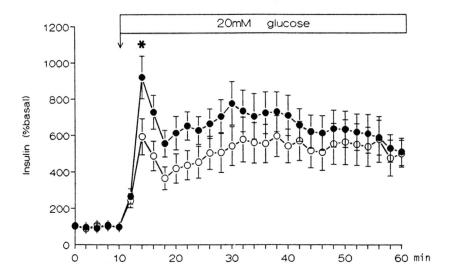

Fig. 5. Glucose-induced insulin secretion from PKC-depleted islets. Groups of islets were perifused (1 ml/min) with buffers supplemented with glucose, 2 mM and then *(arrow)* 20 mM. Insulin secretion is expressed as % increase over basal: o denotes secretion from control (4αPMA-treated islets) and ● that from PKC-depleted islets (n = 6; 6 separate expts., each 2-4 obs.).

stimulation of insulin secretion from this cell type [32], and are consistent with translocation of PKC by glyceraldehyde in RIN cells [30].

Although our studies using PKC-depleted islets do not support a role for PKC in nutrient-induced insulin secretion from normal islets, this experimental approach has provided evidence that CCh potentiates insulin secretion in part through PKC activation. Enhancement of insulin secretion by CCh was significantly reduced, but not abolished, after PMA-induced down-regulation of PKC activity (Fig. 4). This reduced responsiveness was not due to general inhibition of secretion since secretory responses to glucose were undiminished. Furthermore, the inability of CCh to fully potentiate secretion could not be accounted for by impaired coupling of muscarinic receptors to phospholipase C, nor by direct inhibition of this activity, since CCh-stimulated [^{3}H]inositol phosphate production was similar in control and PKC-depleted islets [27].

These results, together with our observations that CCh stimulates PKC translocation, suggest that PKC is one important mechanism through which CCh potentiates glucose-induced insulin secretion from normal B-cells. Again, conflicting results

have been reported in studies using insulin-secreting tumour
cells. In a preliminary study using RINm5F cells, PKC-depletion
enhanced cholinergic stimulation of insulin release [39].
However, the recent report of an inhibition of acetylcholine-
stimulated insulin secretion from PKC-depleted HIT-T15 cells
[40] is more consistent with the observations in rat islets.

ii) Protein phosphorylation

The mechanisms involved in the stages of the secretory
response after PKC activation are not well understood. Activators of PKC have been shown to increase ^{32}P incorporation
into specific proteins in islets or insulin-secreting cells
in studies using either intact cells or homogenates [e.g.
13, 41, 42]. Both of these experimental approaches have
certain technical disadvantages. With intact cells, intracellular ATP is radiolabelled by prolonged incubation of the cells
with [^{32}P]Pi, but the results of such experiments are difficult
to interpret because many insulin secretagogues stimulate
the turnover of ATP [43] and fluxes of Pi [44]. With homogenates
or subcellular fractions, kinases may phosphorylate many proteins
which are not their substrates *in situ*.

We have attempted to overcome some of these problems
by measuring PKC-dependent protein phosphorylation in rat
islets in which the plasma membranes of the islet cells
have been permeabilized by exposure to a high-intensity electric
field [16]. This procedure permits the introduction of small
molecules into the intracellular compartment, while maintaining
the cellular structure and preventing leakage of molecules
>5 kDa. The ^{32}P required to monitor protein phosphorylation
can be supplied as [$\gamma^{32}P$]ATP, which rapidly equilibrates across
the permeabilized plasma membranes to form a constant specific-
radioactivity pool of radiolabelled ATP within the cells,
thus avoiding the problems associated with radiolabelling
intact cells with [^{32}P]Pi. Furthermore, since the secretory
apparatus remains intact and functional, insulin secretion
from electrically permeabilized islets is stimulated by activators of PKC [12, 45]. Effects of experimental manipulations
on protein phosphorylation can therefore be correlated with
effects on insulin secretion.

In this experimental model, PKC activation by PMA stimulated incorporation of ^{32}P into a number of islet proteins,
notably those of apparent mol. wts. 32, 40, 45 and 50 kDa
(Fig. 6). It seems reasonable to conclude that some, or
all, of these modifications in protein phosphorylation are
involved in the secretory response of B-cells to PMA (Fig. 4).
This conclusion is supported by our measurements of protein

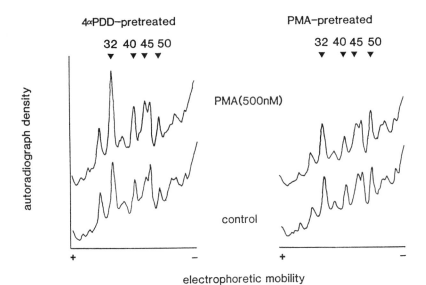

Fig. 6. PKC-dependent protein phosphorylation in electrically
permeabilized islets. Control (4αPMA-treated) and PKC-depleted
(PMA-treated) rat islets were electrically permeabilized by 5
exposures to a 3.4 kV/cm field, then incubated for 15 min in
the presence of $[\gamma^{32}P]ATP$ (0.8 Ci/mmol) and in the absence or
presence of PMA (500 nM). Proteins were separated by SDS-PAGE
on 7-17% gradient gels. ^{32}P incorporation was assessed by autoradiography and quantitated by scanning densitometry of the
autoradiographs. In control islets, PMA stimulated ^{32}P incorporation into several proteins *(arrows)*; in PKC-depleted islets
PMA had no detectable effect. M_r's (kDa) were estimated from
the migration position of protein standards of known M_r.

phosphorylation in PKC-depleted islets, in which PMA-dependent
^{32}P incorporation into these proteins was abolished (Fig. 6),
concomitant with abolition of PMA-induced insulin secretion
(Fig. 4).

These studies in electrically permeabilized islets suggest
that PKC-mediated protein phosphorylation is involved in the
control of insulin secretion. However, since PKC-dependent
phosphorylation events do not occur in PKC-depleted islets
which show an appropriate secretory response to glucose
(Fig. 5), we must conclude that the PKC-dependent protein
phosphorylation is not essential for glucose-induced secretion.

SUMMARY AND CONCLUSIONS

There is little doubt that PKC is involved in the regulation
of insulin secretion from pancreatic B-cells. PKC is present
in B-cells, and PKC activators are generated in response to

insulin secretagogues. Current evidence suggests an important role for PKC in the insulin secretory response to cholinergic neurotransmitters, entailing PKC activation by DAG generated from the hydrolysis of membrane inositol phospholipids. Whether PKC is involved in B-cell secretory responses to nutrient secretagogues is less certain. The available evidence suggests that PKC activation is not obligatory for glucose-induced insulin secretion.

Much work remains to be done. Immunological and biochemical characterization of the isoforms of PKC which are expressed in normal B cells is essential. The activators of these isoforms within the cell must be identified, and likewise the mechanisms controlling the generation of the activators. Finally, the identity of the protein substrates for PKC in B-cells, and their roles in the secretory process, remain to be determined.

Acknowledgements

Financial support from the British Diabetic Association and the Medical Research Council is gratefully acknowledged. S.J.P. is a Wolfson Foundation Research Fellow. P.M.J. is an MRC Senior Research Fellow (Non-Clinical).

References

1. Nishizuka, Y. (1984) *Nature 308*, 693–698.
2. Nishizuka, Y. (1988) *Nature 334*, 661–665.
3. Parker, P.J., Kour, G., Marais, R.M., Mitchell, F., Pears, C., Schaap, D., Stabel, S. & Webster, C. (1989) *Mol. Cell Endocr. 65*, 1–11.
4. Schaap, D., Parker, P.J., Bristol, A., Kriz, R. & Knopf, J. (1989) *FEBS Lett. 243*, 351–357.
5. Tanigawa, K.,Kuzuya, H., Imura, H., Taniguchi, H., Baba, S., Takai, Y. & Nishizuka, Y. (1982) *FEBS Lett. 138*, 183–186.
6. Lord, J.M. & Ashcroft, S.J.H. (1984) *Biochem. J. 219*, 547–551.
7. Persaud, S.J., Jones, P.M., Sugden, D. & Howell, S.L. (1989) *FEBS Lett. 245*, 80–84.
8. Onoda, K., Hagiwara, M., Hachiya, T., Usuda, N., Nagata, T. & Hidaka, H. (1990) *Endocrinology 126*, 1235–1240.
9. Ito, A., Saito, N., Taniguchi, H., Chiba, T., Kikkawa, U., Saitoh, Y. & Tanaka, C. (1989) *Diabetes 38*, 1005–1011.
10. Malaisse, W.J., Dunlop, M.E., Mathias, P.C.F., Malaisse-Lagae, F. & Sener, A. (1985) *Eur. J. Biochem. 149*, 23–27.
11. Hubinont, C.J., Best, L. & Malaisse, W.J. (1984) *FEBS Lett. 170*, 247–253.
12. Jones, P.M., Stutchfield, J. & Howell, S.L. (1985) *FEBS Lett. 191*, 102–106.
13. Harrison, D.E., Ashcroft, S.J.H., Christie, M.R. & Lord, J.M. (1984) *Experientia 40*, 1075–1084.
14. Thams, P., Capito, K. & Hedeskov, C.J. (1984) *Biochem. J. 221*, 247–253.

15. Hughes, S.J. & Ashcroft, S.J.H. (1988) *Biochem. J. 249*, 825–830.
16. Jones, P.M., Salmon, D.M.W. & Howell, S.L. (1988) *Biochem. J. 254*, 397–403.
17. Malaisse, W.J. & Sener, A. (1985) *IRCS Med. Sci. 13*, 1239–1240.
18. Thams, P., Capito, K. & Hedeskov, C.J. (1986) *Biochem. J. 237*, 131–138,
19. Stutchfield, J., Jones, P.M. & Howell, S.L. (1986) *Biochem. Biophys. Res. Comm. 136*, 1001–1006.
20. Garland, L.G., Bonser, R.W. & Thompson, N.T. (1987) *Trends Pharmacol. Sci. 8*, 334. *[Cf. arts.#A-6 & A-7, this book.-Ed.]*
21. Rüegg, U.T. & Burgess, G.M. (1989) *Trends Pharmacol. Sci. 10*, 218–220.
22. Hedeskov, C.J. (1980) *Physiol. Rev. 60*, 442–509.
23. Henquin, J.C. & Nenquin, M. (1988) *FEBS Lett. 236*, 89–92.
24. Peter-Riesch, B., Fathi, M., Schlegel, W. & Wollheim, C.B. (1988) *J. Clin. Invest. 81*, 1154–1161.
25. Wolf, B.A., Easom, R.A., Hughes, J.H., McDaniel, M.L. & Turk, J. (1989) *Biochemistry 28*, 4291–4301.
26. Kraft, A.S. & Anderson, W.B. (1983) *Nature 301*, 621–623.
27. Persaud, S.J., Jones, P.M., Sugden, D. & Howell, S.L. (1989) *Biochem. J. 264*, 753–758.
28. Easom, R.A., Hughes, J.H., Landt, M., Wolf, B.A.,Turk, J. & McDaniel, M.L. (1989) *Biochem. J. 264*, 27–33.
29. Yamatani, T., Chiba, T., Kadowaki,S., Hishikawa, R., Yamaguchi, A., Inui, T., Fujita, T. & Kawazu, S. (1988) *Endocrinology 122*, 2826–2832.
30. Thomas, T.P., Ellis, T.R. & Pek, S.B. (1989) *Diabetes 38*, 1371–1376.
31. Regazzi, R. & Wollheim, C.B. (1988) *Diabetologia 31*, 534A.
32. Wollheim, C.B., Dunne, M.J., Peter-Riesch, B., Bruzzone, R., Pozzan, T. & Petersen, O.H. (1988) *EMBO J. 7*, 2443–2449.
33. Murray, A.W., Fournier, A. & Hardy, S.J. (1987) *Trends Biochem. Sci. 12*, 53–54.
34. Young, S., Parker, P.J., Ullrich, A. & Stabel, S. (1987) *Biochem. J. 244*, 775–779.
35. Metz, S.A. (1988) *Diabetes 37*, 3–7.
36. Arkhammar, P., Nilsson, T., Welsh, M., Welsh, N. & Berggren, P-O. (1989) *Biochem. J. 264*, 207–215. *[Cf. #C-4. - Ed.]*
37. Jones, P.M. & Howell, S.L. (1988) *Biochem. Soc. Trans. 17*, 62–63.
38. Thams, P., Capito, K., Hedeskov, C.J. & Kofod, H. (1990) *Biochem. J. 265*, 777–787.
39. Li, G-D., Pralong, W., Regazzi, R., Ullrich, S. & Wollheim, C.B. (1989) *Diabetologia 32*, 510A.   [227-232.
40. Hughes, S.J., Chalk, J.G. & Ashcroft, S.J.H. (1990)*Biochem. J. 267*,
41. Dunlop, M.E. & Larkins, R.G. (1986) *Arch. Biochem. Biophys. 248*, 562–56
42. Brocklehurst, K.W. & Hutton, J.C. (1984) *Biochem. J. 220*, 283–290.
43. Christie, M.R. & Ashcroft, S.J. H. (1984)*Biochem. J. 218*, 87–99.
44. Freinkel, N., El Younsi, C., Bonnar, J. & Dawson, R.M.C. (1974) *J. Clin. Invest. 55*, 1179–1189.
45. Jones, P.M., Persaud, S.J. & Howell, S.L. (1989) *Biochem. Biophys. Res. Comm. 162*, 998–1003.

#C-4

REGULATION OF INSULIN RELEASE: STUDIES EMPLOYING THE PATCH-CLAMP TECHNIQUE AND FLUORESCENT PROBES

[1]Per-Olof Berggren[†] and [2]Patrik Rorsman

[1]Department of Endocrinology Karolinska Institute, Box 60500, Karolinska Hospital, S-104 01 Stockholm, Sweden

[2]Department of Medical Physics, Gothenburg University, Box 33031, S-40033 Gothenburg, Sweden

This article focuses on techniques for studying the mechanisms involved in the regulation of insulin release (measured with a column perifusion system), using normal pancreatic cells from obese hyperglycaemic mice. For electrical activity and membrane potential changes, variants of the patch-clamp technique were used, and bisoxonol as a fluorescent indicator; similarly fura-2, indo-1, quin-2 and quene-1 were used to estimate $[Ca^{2+}]_i$* and pH_i.

In resting β-cells a K^+ channel regulated by intracellular ATP predominated. Modulation of its activity was parallelled by changes in membrane potential, $[Ca^{2+}]_i$ and insulin release, implying that closure of these channels is the link between glucose stimulation and insulin release. Glucose-induced changes in $[Ca^{2+}]_i$ were parallelled by changes in insulin release. Similar effects on $[Ca^{2+}]_i$ were seen in quin-2 and fura-2 loaded cell suspensions and in individual fura-2 loaded cell aggregates. Glucose increased fura-2 and indo-1 efflux, an effect unrelated to insulin release. Probenecid (1 mM) reduced both basal and glucose-induced efflux of the indicators whilst not affecting glucose-induced increases in $[Ca^{2+}]_i$ and insulin release. Glucose slightly increased β-cell pH_i, whereas it was decreased by glibenclamide, K^+ stimulation and notably by glyceraldehyde. After glucose stimulation a rapid pH_i increase promoted re-polarization and a decrease in $[Ca^{2+}]_i$. Tolbutamide prevented these effects, suggesting involvement of ATP-regulated K^+ channels.

Insulin release from the pancreatic β-cell represents a complex process involving a variety of mechanisms acting in concert. Such mechanisms regulate membrane potential and thereby the voltage-activated Ca^{2+} channels, $[Ca^{2+}]_i$, pH_i

---

[†]addressee for any correspondence

[*]*Abbreviations.*- $[Ca^{2+}]_i$, free cytoplasmic Ca^{2+} concentration; pH_i, intracellular pH; DAG, diacylglycerol; DMO, dimethyloxazolidine; PKC, protein kinase C; p.m., plasma membrane; TEA, tetraethylammonium. AM, acetoxymethyl ester (of 'probes', e.g. fura-2). For fluorescence, ex = excitation, em = emission (nm, as set).

and the exocytotic machinery directly responsible for the transport of insulin secretory granules and their fusion with the p.m.

Glucose-stimulation of the β-cell leads, by closure of specific ATP-regulated K^+-channels [1-4], to depolarization of the p.m. and thereby opening of voltage-activated Ca^{2+} channels. Ca^{2+} flows into the β-cell along its electrochemical gradient, and the subsequent increase in $[Ca^{2+}]_i$ is believed to be the ultimate trigger for insulin release [5]. In addition, changes in pH_i [6], due to metabolism of the sugar, may be involved in regulation of the secretory event.

Although glucose is the most important insulin secretagogue under physiological conditions, the secretory response can be modified by many additional factors including various other nutrients, hormones and neurotransmitters. These modulators exert their actions by being metabolized or by activating receptors in the p.m.

With the development of new techniques for measuring $[Ca^{2+}]_i$, pH_i, membrane potential and single-channel currents, we have been provided with powerful tools for further elucidation of not only β-cell physiology but also pathophysiology. In this article some aspects of the regulation of insulin release are discussed in the light of the application of these techniques.

METHODOLOGICAL CONSIDERATIONS

Animals and preparation of islet cell aggregates

Adult obese hyperglycaemic mice (*ob/ob*) of both sexes were taken from a local non-inbred colony and starved overnight [7]. The mice were decapitated, and islets, containing >90% β-cells, were isolated by a collagenase technique. To prepare cell aggregates, isolated islets were shaken vigorously in a Ca^{2+}- and Mg^{2+}-deficient medium [8]. The disrupted islets were suspended in RPMI 1640 tissue culture medium supplemented with 10% (v/v) NU-Serum™, 100 µg/ml streptomycin, 100 I.U./ml penicillin and 60 µg/ml gentamycin, and cultured overnight. Throughout culture and incubations, cell attachment to the culture flasks was prevented by shaking the flasks gently.

The cell suspensions comprised some single cells, but mostly small cell aggregates (5-10 cells). In some experiments, freshly prepared islet cell aggregates were seeded onto plastic coverslips, allowed to attach and cultured for 1-3 days prior to the measurements. For the patch-clamp experiments, cells were plated in petri dishes and kept in culture until the measurements were performed.

Media

The medium for islet isolation as well as all experimental procedures was, unless otherwise stated, a pH 7.4 Hepes buffer containing (mM) 125 NaCl, 5.9 KCl, 1.28 $CaCl_2$, 1.2 $MgCl_2$ and 25 Hepes-NaOH [9]. It was supplemented with 1 mg/ml bovine serum albumin (but *see below*, Membrane potential measurements; albumin omitted to obviate interference with the potential-specific dye). In the patch-clamp recordings the cells were initially immersed in 'solution A', containing (mM) 140 NaCl, 5.6 KCl, 1.2 $MgCl_2$, 2.6 $CaCl_2$ and 5-10 Hepes-NaOH, pH 7.4. In the cell attached patch experiments the pipette solution, adjacent to the extracellular side of the p.m., was either as described above or (solution 'B') had KCl increased to 146 mM. In the normal membrane potential range, A will give outwardly directed K^+ currents, whereas with B they are inwardly directed. Identical media were used for outside-out patch recordings, and B was used to fill the pipette also in the inside-out recordings. In the latter configuration the intracellular side of the membrane is exposed to the bath solution, which accordingly should mimic the composition of the cytoplasm. This solution contained (mM) 125 KCl, 30 KOH, 1 $MgCl_2$, 6 NaCl, 2 $CaCl_2$, 10 EGTA and 5 Hepes (pH 7.15, $[Ca^{2+}]$ 0.06 μM). Supplemented with 0.3 mM Mg-ATP, this solution was used to fill the pipette in the outside-out recordings. Published binding constants [10] were used to calculate $[Ca^{2+}]_i$.

Patch–clamp experiments[*]

The cell-attached, inside-out and outside-out configurations of the patch-clamp technique were employed [11]. Only with this version of the technique do the intracellular milieu and thus cell metabolism remain intact. Pipettes were pulled from borosilicate or aluminosilicate glass, heat-polished at the tip and coated with silicone rubber; their resistances were ~5 mega-ohms. The pipette was connected to a patch-clamp amplifier, and data were stored on video tape or in a computer pending further analysis. At the time of analysis, the current signal was filtered at 0.5-4 kHz (-3 dB-point) with a 4-pole Bessel low-pass filter, and single-channel currents were measured directly from a digital oscilloscope. All patch-clamp experiments were performed at room temperature.

Measurements using fluorescent probes

Fluorescence measurements (normally at 37°) using cell suspensions or cells attached to cover slips were made in an Aminco-Bowman or a Perkin-Elmer LS5 spectrofluorimeter.

---

[*] This approach and those that follow appear in Fig. 1 at end of article.

Each spectrofluorimeter was equipped with thermostatically controlled cuvette holders and had been modified to allow constant stirring of the suspensions or medium. Generally the buffer contained $1-3 \times 10^6$ cells/ml or, in the measurements of membrane potential, $\sim 0.3-0.5 \times 10^6$/ml. Where cells were attached to coverslips, they were either placed at a 30° angle to the excitation beam, using a specially made holder placed in the ordinary cuvette, or fixed in a small chamber apt for microscopic work. In the microfluorimetric measurement, a Nikon TMD microscope, equipped with an epifluorescence device and a temperature control box, was used.

Measurements of $[Ca^{2+}]_i$

Cell suspensions were incubated with 2.5-12.5 μM quin-2/AM, 1-2 μM fura-2/AM or 1 μM indo-1/AM for 30-45 min. After two washes in experimental buffer and resuspension, fluorescence was recorded with ex/em settings respectively 340/490, 340,380/510 and 335/400 [12-14]. For calibration in quin-2 experiments cells were lysed with Triton X-100, giving F_{max}, followed by Mn^{2+} addition, which quenches quin-2 fluorescence and gives F_{Mn}. Approximate levels of $[Ca^{2+}]_i$ were calculated using the formula: $[Ca^{2+}]_i = Kd \, (F - F_{min})/(F_{max} - F)$ where a Kd of 115 nM [12] was used and $F_{min} = F_{Mn} + 0.16(F_{max} - F_{Mn})$ [15]. The values were corrected for fading and leakage of indicator, assuming near-linear dynamics for both parameters.

In the fura-2 measurements, $[Ca^{2+}]_i$ was calculated by the formula: $[Ca^{2+}]_i = Kd \, (R - R_{min})/[R_{max} - R(S_{f2}/S_{b2})]$ where Kd was assumed to be 224 nM, and fluorescence ratio values (R) were obtained by dividing values of the emitted fluorescence at 340 nm excitation with that obtained at 380 nm excitation [13]. The factor S_{f2}/S_{b2} is the ratio of the fluorescence of fura-2 in the absence of Ca^{2+} (f) to that in the presence of saturating Ca^{2+} (b) at 380 nm excitation. In the indo-1 measurements, calibration of the experiments in terms of absolute values of $[Ca^{2+}]_i$ were unnecessary; hence all recordings were presented as values of relative fluorescence only.

To assess leakage of indicator, samples of the cell suspension were taken at the start and end of each experiment. After pelleting the cells, the fluorescence in the supernatant was measured and the pellet was freeze-dried. In both the fura-2 and the indo-1 experiments the leakage of indicator became non-linear when glucose was added, which evidently hampered continuous estimations of $[Ca^{2+}]_i$. As discussed above, this was not a big problem in the indo-1 experiments. However, in the fura-2 measurements it became a problem and therefore only basal values of $[Ca^{2+}]_i$ were calculated, after obtaining a reliable measure of extracellular fura-2 at that time point. The freeze-dried pellets were weighed to determine the number of cells in each experiment, assuming that 1 mg dry wt. corresponds to 3.6×10^6 cells [8].

To determine the indicator content of the cells the the F_{max} values, corrected for background fluorescence, were compared with known standards of the fluorescent indicators in experimental buffer. Loading values were subsequently calculated by dividing the amount of indicator by the number of cells, and were typically 1-4 nmol quin-2/10^6 cells and 0.5-1 nmol fura-2/10^6 cells. Given that the volume of 10^6 cells is ~1 µl [8], the loading values correspond to indicator concentrations in the mM range for quin-2 and 10-fold lower for fura-2. Although these estimations were not performed with indo-1, it should be kept in mind that this indicator too is a representative of the second generation of fluorescent Ca^{2+} indicators and is closely related to fura-2 [13].

Measurements of membrane potential and cell permeability

The fluorescent dye bisoxonol, bis(1,3-diethylthiobarbiturate)-trimethineoxonol [16], at 150 nM final concentration was used to monitor qualitative changes in membrane potential (ex, 540; em, 580). Bisoxonol is a lipophilic anion which increases its fluorescence upon association with lipids and proteins [16]. The presence of bovine serum albumin in the experimental buffer will therefore increase the non-cell-associated fluorescence; hence the measurements were performed with albumin omitted. When dye-diffusion equilibrium is attained, depolarization of the p.m. will increase the fraction of the dye in the cell and hence total fluorescence. It should be noted that the use of this dye in cell suspensions does not permit resolution of the temporal changes in the β-cell electrical activity, but instead gives a qualitative measure of the average changes in membrane potential in a cell population.

Cell permeability was assessed with 25µM ethidium bromide (ex, 365; em, 590) [17], which fluoresces when in contact with nucleic acids.

Measurements of pH_i

Cell suspensions were incubated with 25 µM quene-1/AM [18] for 40 min (ex, 390; em, 530) and then washed twice in experimental buffer before resuspension in the cuvette. For calibration in each experiment the cells were lysed with Triton X-100 and the medium titrated (pH electrode) with NaOH and HCl additions; thereby a graph of fluorescence vs. pH was obtained. Since mM levels of Ca^{2+} and Mg^{2+} are known to quench quene-1 fluorescence, EGTA and EDTA

were added prior to cell lysis [18]. The small increase
in fluorescence obtained after these additions was assumed
to represent the contribution of extracellular quene-1 and
thus corrected for in the estimations of pH_i.

Measurements of insulin release and indicator efflux

The dynamics of insulin release and efflux of fura-2
and indo-1 were studied by perifusing islet-cell aggregates
in an 0.5 ml column at 37° [19]. The flow rate was 0.1-0.3 ml/min,
and 0.5-2 min fractions were collected using crystalline rat
insulin for reference. When indicator efflux was investigated
the fluorescence in the fractions was measured with appropriate
ex and em, and corrected for background fluorescence. In
accordance with the gel's exclusion limit (4000 Da), fura-2
was delayed ~1 min by the column compared to insulin.

RESULTS AND DISCUSSION

ATP-regulated K⁺ channels

Glucose promotes closure of specific ATP-regulated K^+
channels in the pancreatic β-cell, suggesting that an increase
in intracellular ATP is the factor coupling sugar metabolism
to activation of the insulin secretory process [20]. As
further support for the notion that these channels serve
as regulators of the β-cell stimulus-secretion coupling, they
are closed by sulphonylurea compounds, which stimulate insulin
release and are used in treating non-insulin dependent diabetes
mellitus (NIDDM) [5, 21, 22]. In contrast, the sulphonamide
diazoxide, which is an inhibitor of secretion, increases
their activity [5, 24].

The ATP-regulated K^+-channels were first characterized
in rat and mouse β-cells [1-3], but later demonstrated also
in human insulinoma and normal β-cells [23, 24]. For these
channels in β-cells from *ob/ob* mice we have found characteristics similar to those reported for rat and NMRI-mouse β-cells
[2, 3, 5]. In the absence of glucose the channels displayed
spontaneous activity, with openings grouped in bursts; channel
activity was markedly reduced with the sugar at 5 mM, a
concentration near the threshold for insulin release [25].
In addition, other investigations have demonstrated that
channel activity in a cell-attached patch was reduced by
87% in response to a similar increase in glucose concentration
[26]. The largest elevation in intracellular ATP is observed
between 0 and 5 mM glucose [27], and there is a marked
reduction in activity of the ATP-regulated K^+-channels in
response to such an increase in sugar level. Nevertheless,
the size of the current flowing through the channels remaining
active was unaffected, implying that the cell was still
maintained in a polarized state. Hence only marginal increases

in membrane potential and $[Ca^{2+}]_i$ were induced by 5 mM glucose [5]. Further increases in sugar concentration have been reported to slightly elevate the cellular ATP level [27, 28], suggesting that such elevations cause depolarization by closing the remaining active channels.

As shown with the patch-clamp technique in the inside-out configuration, ~15 µM ATP applied to the cytosolic side of the membrane is enough to close 50% of the ATP-regulated channels [1]. Since cytoplasmic ATP is in the mM range [28], the fact that the channel is active under physiological conditions is difficult to reconcile with a direct regulatory role for the nucleotide [29]. However, ATP is not the sole factor whereby glucose regulates channel activity. Thus, it has been demonstrated that the presence of ADP reduces the sensitivity of the channel to ATP, suggesting that the ATP/ADP ratio rather than the absolute concentration of ATP regulates channel activity [30, 31]. Cook and co-authors [32] showed that ~99% inhibition of channel activity could be expected at physiological ATP levels. If the channels operate near this fully closed state, even small increases in glucose concentration and thereby ATP levels will suffice to close the remaining channels and cause depolarization. This implies that the system will operate at very low net current levels, thus minimizing ion fluxes and energy expended to maintain ion gradients.

By raising glucose to 20 mM or by combining 5 mM glucose with 100 µM tolbutamide, it is possible to substantially depolarize the β-cell [5]. Under these conditions addition of diazoxide restored channel activity and repolarized the cells. The effects on single-channel activity were paralled by changes in membrane potential, $[Ca^{2+}]_i$ and insulin release. Measurements at room temperature revealed similar effects on membrane potential and $[Ca^{2+}]_i$ [5], justifying the comparison between these parameters and channel activity. At room temperature a transient increase in the number of active channels, resulting in a small hyperpolarization, could be observed after raising the glucose concentration to 20 mM [5].

Similar effects have been found in other studies [22, 33], and stimulation with glucose has been found to transiently increase β-cell input conductance, indicating an opening of K^+-channels [34]. This might reflect increased consumption of ATP in the initial phosphorylation of glucose, leading to a transient drop in ATP level [22, 35]. Since changes in β-cell membrane potential and $[Ca^{2+}]_i$ reflect changes in the activity of the ATP-regulated K^+-channels, the closure of these channels is the link between glucose stimulation and insulin secretion. Furthermore, the ATP-regulated K^+-channels can be regarded as sensors of the cytoplasmic ATP concentrations in β-cells.

Cytoplasmic free Ca²⁺ concentration

With quin-2, the first generation fluorescent Ca^{2+} indicator, in β-cells, glucose stimulation of insulin release was shown to be accompanied by an increase in $[Ca^{2+}]i$ [5, 36, 37]. Although monitoring changes in $[Ca^{2+}]_i$ with quin-2 has large advantages compared to earlier methods, this indicator suffers from experimental drawbacks related mainly to its weak fluorescence. Hence there is a demand for high intracellular concentrations of indicator in the measurements, resulting in a substantial buffering of $[Ca^{2+}]_i$ [38]. This is less of a problem with the second generation of indicators, such as fura-2 and indo-1, which display a much stronger fluorescence [13]. Fura-2 has been successfully employed for measuring $[Ca^{2+}]_i$ in cell suspensions as well as in individual cells, and has also enabled Ca^{2+} gradients to be revealed within single cells by applying digital image analysis [39, 40]. With two different ex wavelengths, taking the ratio as the measure of $[Ca^{2+}]_i$, error sources such as photobleaching and indicator leakage are neutralized [40]. However, it has been demonstrated that fura-2 can be unevenly distributed within cells [40-42] and sometimes the ester form is incompletely hydrolyzed [43, 44]. Moreover, leakage of fura-2 when measuring $[Ca^{2+}]_i$ in cell suspensions is commonly encountered. [*Note by Ed.*- The foregoing points are amplified and exemplified in various arts. in this book and in Vol. 19 on cell calcium (1989; same Eds.).]

Suspended fura-2-loaded β-cells displayed glucose-induced changes in $[Ca^{2+}]_i$ that resembled those previously demonstrated with quin-2 [5, 36] and were paralleled by changes in insulin release. Thus an increase in glucose concentration from 0 to 20 mM induced an initial decrease in both $[Ca^{2+}]i$ and insulin release that was followed by an increase in both parameters. At higher concentrations of indicator the glucose-induced rise in $[Ca^{2+}]_i$ was less pronounced and there was no initial decrease. Similar effects on insulin release were obtained in cells loaded with low concentrations of quin-2 or fura-2, demonstrating that concentrations of indicator needed to perform reliable measurements of $[Ca^{2+}]_i$ did not impair β-cell function. Whereas the $[Ca^{2+}]_i$ increase following glucose stimulation serves to trigger insulin release the initial reduction in $[Ca^{2+}]_i$ most likely reflects build-up of intracellular ATP, which in turn stimulates Ca^{2+} removal from the cytoplasm [29]. This process is progressively marked when the basal sugar concentration is raised, suggesting that the β-cell does not normally experience a lowering in $[Ca^{2+}]_i$ in response to a rise in glucose concentration within the physiological range [45].

$[Ca^{2+}]_i$ measurements in individual islet-cell aggregates revealed glucose effects similar to those in cell suspensions although varying somewhat between different aggregates [46]. Such variability, as yet unexplained, has been encountered by other investigators too [47, 48]. The reason for the accord in sugar-induced effects on $[Ca^{2+}]_i$ among the different cuvette experiments is that responses from numerous aggregates yielded an average.

Permeabilization of fura-2-loaded β-cells demonstrated a substantial amount of indicator bound intracellularly, but there was no evidence for uneven distribution, as judged from digital image analysis, or for incomplete hydrolysis of the ester [46, 49]. In measuring $[Ca^{2+}]_i$ with fura-2 in cell suspensions, glucose addition seemed to increase the rate of indicator leakage [46], as confirmed by directly monitoring the efflux of indicator from perifused fura-2-loaded β-cells [46, 50]; the basic efflux rate showed a 2-fold increase in response to the sugar, unrelated to insulin release. Whereas an increase in indicator efflux was observed also when indo-1-loaded cells were challenged with glucose, no efflux of quin-2 was detectable under similar conditions; yet it might have escaped detection in the efflux experiments due to its weaker fluorescence, and the possibility of glucose-enhanced leakage is not excluded. Accumulation of extracellular indicator might seriously interfere with estimations of $[Ca^{2+}]_i$ made on cell suspensions in finite volumes. If leakage is linear with time it can be corrected for. However, when it is altered in response to additions of various agonists, continuous estimations of extracellular fluorescence must be made so as to establish $[Ca^{2+}]_i$ changes.

The employment of perifusion is one way to abolish the contribution from extracellular indicator. Lowering the measurement temperature can help reduce the leakage rate [51], but is not a useful approach since it inhibits insulin release [52]. We found that probenicid (at 1 mM), an organic-anion transport blocker, both reduced basal efflux of fura-2 and indo-1 and prevented the increase promoted by glucose ([50] - studies in which P. Arkhammar and T. Nilsson participated). At this concentration there was no interference of the drug with the glucose-induced rise in either $[Ca^{2+}]_i$ or insulin release [50]. Consequently, probenecid can be used to improve the quality of the $[Ca^{2+}]_i$ measurements made with fura-2 or indo-1.

Cytoplasmic pH

As evident from studies employing [^{14}C]DMO and fluorescein for measuring intracellular pH, glucose stimulation is accompanied by intracellular alkalinization [53, 54] - which has been found to evoke insulin secretion, suggesting that an increase in pH might be part of the mechanisms regulating glucose-stimulated insulin release [55]. The fact that [^{14}C]DMO and fluorescein accumulate also in cellular compartments other than the cytosol implies that results obtained using these compounds might be of limited value for investigations of the role of pH$_i$ in the regulation of the insulin secretory process. The fluorescence H$^+$ indicators quene-1 and BCECF [bis(carboxyethylcarboxyfluorescein)] primarily signal changes in pH$_i$, since they are trapped in the cytoplasm by the same mechanism as quin-2, i.e. the cell-permeant ester form is hydrolyzed to the free acid by cytoplasmic esterases [18, 56].

By using quene-1 we have found that 20 mM glucose increased pH$_i$. The alkalinization has been suggested to reflect Na$^+$/H$^+$ and HCO$_3^-$/Cl$^-$ exchange processes [54]. In the present investigation, as previously [53], an increase in pH$_i$ was observed with HCO$_3^-$ either absent or present, suggesting that the underlying mechanisms reflect activation of Na$^+$/H$^+$ — rather than HCO$_3^-$/Cl$^-$ — exchange. Since glucose stimulation leads to the production of DAG [57], conceivably the resulting activation of PKC may be responsible for an accelerated Na$^+$/H$^+$ exchange [58]. Moreover, in normal β-cells it has proved possible to obtain a similar increase in pH$_i$ by stimulating PKC with the phorbol ester TPA (L. Juntti-Berggren, P. Arkhammar, T. Nilsson, P-O. Berggren & P. Rorsman, unpublished).

In contrast to glucose, glyceraldehyde decreased pH$_i$, an effect persisting even in the presence of D-600 and therefore not the result of Ca^{2+} influx through voltage-activated Ca^{2+}-channels. Glibenclamide and, at high levels, K$^+$ likewise decreased pH$_i$; D-600 caused complete reversal, suggesting that changes in pH$_i$ at least under some circumstances may reflect changes in [Ca^{2+}]$_i$. Our results do not favour the idea that changes in pH$_i$ are directly involved in the regulation of insulin secretion. This is supported by observations in the clonal insulin-producing RINm5F cells, where both glyceraldehyde and lactate induced cytoplasmic acidification, but only the former compound stimulated insulin release [59].

Since manipulations of pH$_i$ affect membrane potential as well as ion fluxes [60, 61], conceivably pH$_i$ may modulate ion-channel activity and thus affect insulin release indirectly. Due to increased sensitivity to the prevailing [Ca^{2+}]$_i$, more Ca^{2+}-activated K$^+$-channels are indeed opened subsequent to alkalinization of the cytoplasm [62]. In addition, a rapid increase in pH$_i$, in glucose-stimulated β-cells, induced repolarization, maybe reflecting opening of Ca^{2+}-activated K$^+$-channels [63]. Likewise the activity of the ATP-regulated K$^+$-channels was recently shown to be modulated by pH$_i$ if ATP were present on the cytoplasmic side of the channel [64]. In glucose-

stimulated β-cells, addition of NH_4Cl to 10 mM rapidly increased
pH_i, whereafter it gradually decreased towards the resting
level.   The alkalinization was accompanied by a decrease
not only in membrane potential and $[Ca^{2+}]_i$ but also in insulin
release (L. Juntti-Berggren, P. Arkhammar, T. Nilsson, P. Rorsman
& P-O. Berggren, unpublished).

Possible involvement of Ca^{2+}-activated or ATP-regulated
K^+-channels in the observed response was investigated with
blockers of the respective channels [22, 65, 66]; 10 mM TEA had only
a small effect, but tolbutamide prevented most of the decrease
in $[Ca^{2+}]_i$, suggesting that an opening of the ATP-regulated
K^+-channels was responsible for the repolarization and thereby
lowering of $[Ca^{2+}]_i$ and of insulin release.   Hence a modulatory
effect of protons on the activity of these channels could
bear on the regulation of β-cell electrical activity and
thereby the insulin secretory process.   Since glyceraldehyde
lowers pH_i, such a proton-induced modulation of channel activity
may account for part of the depolarization evoked by this
secretagogue in normal β-cells [6].

CONCLUDING REMARKS

This article has focused on aspects of stimulus-secretion
coupling in the pancreatic β-cell and on types of biochemical
and biophysical techniques that are applicable. Fig. 1 outlines
the different mechanisms that have been discussed as well
as the techniques employed.   The ready uptake and metabolism
of glucose by the β-cell results in generation of ATP,
which closes the ATP-regulated channels. The cell depolarizes,
which leads to opening of voltage-activated Ca^{2+}-channels, an
increase in $[Ca^{2+}]_i$ and thereby insulin release.   In addition,
glucose stimulation promotes alkalinization of the β-cell.

Acknowledgements

The foregoing results were obtained through financial
support from the Swedish Medical Research Council (19X-00034,
19P-9434 and 12X-08647) and the Bank of Sweden Tercentenary
Foundaton. Katarina Breitholtz, Christina Bremer-Jonsson and
Susanne Rydstedt gave valued secretarial help.

References (BBRC = Biochem. Biophys. Res. Comm.; JBC = J. Biol. Chem.)

1. Cook, D.L. & Hales, N. (1984) Nature 311, 271-273.   [446-448.
2. Ashcroft, F. M., Harrison, D.E. & Ashcroft, S.J.H. (1984) Nature 312,
3. Rorsman, P. & Trube, G. (1985) Pflügers Arch. 405, 305-309.
4. Findlay, I., Dunne, M.J. & Petersen, O.H. (1985) J. Membr.
   Biol. 88, 165-172.
5. Arkhammar, P., Nilsson, T., Rorsman, P. & Berggren, P-O. (1987)
   JBC 262, 5448-5454.
6. Arkhammar, P., Berggren, P-O. & Rorsman, P. (1986) Biosci. Rep. 6,
7. Hellman, B. (1965) Ann. N.Y. Acad. Sci. 131, 541-558.   [355-361.
8. Lernmark, A. (1974) Diabetologia 10, 431-438.

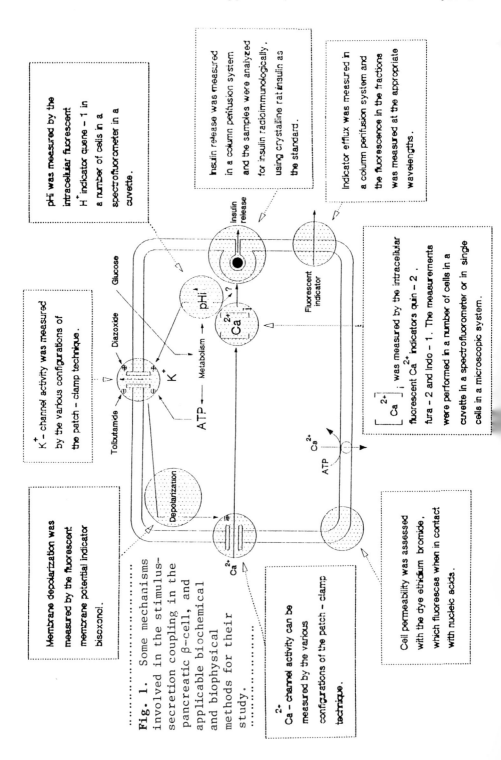

Fig. 1. Some mechanisms involved in the stimulus-secretion coupling in the pancreatic β-cell, and applicable biochemical and biophysical methods for their study.

pHi was measured by the intracellular fluorescent H⁺ indicator quene – 1 in a number of cells in a spectrofluorometer in a cuvette.

Insulin release was measured in a column perifusion system and the samples were analyzed for insulin radioimmunologically, using crystalline rat insulin as the standard.

Indicator efflux was measured in a column perifusion system and the fluorescence in the fractions was measured at the appropriate wavelengths.

K⁺ – channel activity was measured by the various configurations of the patch – clamp technique.

$[Ca^{2+}]_i$ was measured by the intracellular fluorescent Ca^{2+} indicators quin – 2, fura – 2 and Indo – 1. The measurements were performed in a number of cells in a cuvette in a spectrofluorometer or in single cells in a microscopic system.

Membrane depolarization was measured by the fluorescent membrane potential indicator bisoxonol.

Ca^{2+} – channel activity can be measured by the various configurations of the patch – clamp technique.

Cell permeability was assessed with the dye ethidium bromide, which fluoresces when in contact with nucleic acids.

9. Hellman, B. (1975) *Endocrinology* 97, 392-398.
10. Martell, A.E. & Smith, R.M., eds. (1974) *Critical Stability Constants*, Vols. 1 & 2, Plenum, New York.
11. Hamill, O.P., Marty, A., Neher, E., Sakmann, B. & Sigworth, F.J. (1981) *Pflügers Arch.* 391, 85-100.
12. Tsien, R.Y., Pozzan, T. & Rink, T.J. (1982) *J. Cell Biol.* 94, 325-334.
13. Grynkiewicz, G., Poenie, M. & Tsien, R.Y. (1985) *JBC* 260, 3440-
14. Lückhoff, A. (1986) *Cell Calcium* 7, 233-248.   [3450.
15. Hesketh, T.R., Smith, G.A., Moore, J.P., Taylor, M.V. & Metcalfe, J.C. (1983) *JBC* 258, 4876-4882.
16. Rink, T.J., Montecucco, C., Hesketh, T.R. & Tsien, R.Y. (1980) *Biochim. Biophys. Acta* 595, 15-30.
17. Gomperts, B.D. (1983) *Nature* 306, 64-66.
18. Rogers, J., Hesketh, T.R., Smith, G.A. & Metcalfe, J.C. (1983) *JBC* 258, 5994-5997.
19. Kanatsuna, T., Lernmark, Å., Rubenstein, A.H. & Steiner, D.F. (1981) *Diabetes* 30, 231-234.
20. Rorsman, P. & Trube, G. (1990) in *Potassium Channels: Structure, Function and Therapeutic Potential* (Cook, N.S., ed.), Horwood, Chichester, pp. 96-117.
21. Sturgess, N.C., Ashford, M.L.J., Cook, D.L. & Hales, C.N. (1985) *Lancet II*, 474-475.
22. Trube, G., Rorsman, P. & Ohno-Shosaku, T. (1986) *Pflügers Arch.* 407, 493-499.
23. Sturgess, N.C., Carrington, C.A., Hales, C.N. & Ashford, M.L.J. (1987) *Pflügers Arch.* 410, 169-172.
24. Ashcroft, F.M., Kakei, M., Kelly, R.P. & Sutton, R. (1987) *FEBS Lett.* 215, 9-12.
25. Hedeskov, C.J. (1980) *Physiol. Rev.* 60, 442-509.
26. Ashcroft, F.M., Ashcroft, S.J.H. & Harrison, D.E. (1988) *J. Physiol.* 400, 501-527.
27. Ashcroft, S.J.H., Weerasinghe, L.C.C. & Randle, P.J. (1973) *Biochem. J.* 132, 223-231.
28. Malaisse, W.J. & Sener, A. (1987) *Biochim. Biophys. Acta* 927, 190-195.
29. Wollheim, C.B. & Biden, T.J. (1986) *Ann. N.Y. Acad. Sci.* 488, 317-333.
30. Kakei, M., Kelly, R.P., Ashcroft, S.J.H. & Ashcroft, F.M. (1986) *FEBS Lett.* 208, 63-66.
31. Bokvist, K., Berggren, P-O. & Rorsman, P. (1988) *Diabetologia* 31, 473A.
32. Cook, D.L., Satin, L.S., Ashford, M.L.J. & Hales, C.N. (1988) *Diabetes* 37, 495-498.
33. Ribalet, B., Eddlestone, G.T., & Ciani, S. (1988) *J. Gen. Physiol.* 92, 219-237.
34. Atwater, I., Ribalet, B. & Rojas, E. (1978) *J. Physiol.* 278, 117-139.
35. Malaisse, W.J., Hutton, J.C., Kawazu, S., Herchuelz, A., Valverde, I. & Sener, A. (1979) *Diabetologia* 16, 331-341.
36. Rorsman, P., Abrahamsson, H., Gylfe, E. & Hellman, B. (1984) *FEBS Lett.* 170, 196-200.
37. Prentki, M. & Wollheim, C.B. (1984) *Experientia* 40, 1052-1060.
38. Rink, T.J. & Pozzan, T. (1985) *Cell Calcium* 6, 133-144.
39. Poenie, M., Alderton, J., Steinhardt, R. & Tsien, R.Y. (1986) *Science* 233, 886-889.

40. Tsien, R.Y. & Poeni, M. (1986) *Trends Biochem. Sci. 11*, 450-455.
41. Almers, W. & Neher, E. (1985) *FEBS Lett. 192*, 13-18.
42. Ratan, R.R., Shelanski, M.L. & Maxfield, F.R. (1986) *Proc.*
    Nat. Acad. Sci. 83, 5136-5140.   [1162.
43. Highsmith, S., Bloebaum, P. & Snowdowne, K.W. (1986)*BBRC 138*, 1153–
44. Scanlon, M., Williams, D.A. & Fay, F.S. (1987) *JBC 262*, 6308-6312.
45. Nilsson, T., Arkhammar, P. & Berggren, P-O. (1988) *BBRC 153*,
    984-991.   [17-27.
46. Arkhammar, P., Nilsson, T. & Berggren, P-O. (1989)*Cell Calcium 10*,
47. Meissner, H.P. & Atwater, I.J. (1976)*Horm. Metab.Res. 8*, 11-16.
48. Grapengiesser, E., Gylfe, E. & Hellman, B. (1988) *BBRC 150*,
    419-425.
49. Arkhammar, P., Hallman, H., Nilsson, T. & Berggren, P-O. (1988)
    Diabetologia 31, 465A.
50. Arkhammar, P., Nilsson, T. & Berggren, P-O.* (1989)*BBRC 159*, 223–228.
51. Prentki, M., Glennon, M.C., Geschwind, J-F., Matschinsky, F.M.
    & Corkey, B.E. (1987) *FEBS Lett. 220*, 103-107.
52. Wollheim, C.B. & Sharp, G.W.G. (1981) *Physiol. Rev. 61*, 914-973.
53. Lindström, P. & Sehlin, J. (1984) *Biochem. J. 218*, 887-892.
54. Deleers, M., Lebrun, P. & Malaisse, W.J. (1985) *Horm.*
    Metab. Res. 17, 391-395.
55. Lindström, P. & Sehlin, J. (1986) *Biochem. J. 239*, 199-204.
56. Rink, T.J.,Tsien, R.Y. & Pozzan, T. (1982) *J. Cell Biol. 95*, 189-196.
57. Peter-Riesch, B., Fathi, M., Schleger, W. & Wollheim, C.B.
    (1988) *J. Clin. Invest. 81*, 1154-1161.
58. Madshus, I.H. (1988) *Biochem. J. 250*, 1-8.
59. Wollheim, C.B., Ullrich, S.J., Salomon, D.C., Halban, P.A.,
    Tsien, R.Y. & Pozzan, T. (1984) *Diabetologia 27*, 347A.
60. Henquin, J-C. (1981) *Mol. Cell Endocrin. 21*, 119-128.
61. Pace, C.S. Tarvin, J.T. & Smith,J.S. (1983) *Am. J. Physiol. 244*, E3-E18.
62. Cook, D.L., Masathoshi, I. & Fujimoto, W.Y. (1984) *Nature 311*,
    269-271.
63. Rosario, L.M. & Rojas, E. (1986) *FEBS Lett. 200*, 203-209.
64. Misler, S., Gillis, K. & Tabcharani, J. (1989) *J. Membr. Biol.*
    109, 135-143.
65. Petersen, O.H. & Findlay, I. (1987) *Physiol. Rev. 67*, 1054-1116.
66. Bokvist, K., Rorsman, P. & Smith, P.A. (1990) *J. Physiol. 423*, 327-
    342.   [*also (1990) *FEBS Lett. 273*, 182-184.

Note by Ed.:

#For *'BBRC'* and *'JBC'* see start of Refs. list.
#Vol. 19, *Biochemical Approaches to Cellular Calcium* (eds. and
publisher as for present book; 1989), and this Vol. (21),
contain articles by many of the authors cited, including:-
Vol. 19: J.C. Hutton, F.R. Maxfield, T. Pozzan, T.J. Rink (whose
art., and that of A.K. Campbell, survey approaches and probes
as well as concepts), and A. Lückhoff;
this vol.: J.C. Hutton and I.H. Madshus.

#C-5

INVESTIGATION OF THE EFFECTS OF INSULIN ON SUBCELLULAR DISTRIBUTION AND PHOSPHORYLATION OF INSULIN-REGULATED PROTEINS

[1]H.G. Joost, [1]A. Schürmann, [1]C. Schmitz-Salue and [2]T.M. Weber

[1]Institut für Pharmakologie und Toxikologie,
Universität Göttingen, Robert-Koch-Strasse 40,
D-3400 Göttingen, Germany

[2]National Institutes of Health, LBM/NIDDK,
Bldg. 10, Rm. 9B06, Bethesda, MD 20892, U.S.A.

Insulin binds to a membrane receptor that has TK* activity at its intracellular domain. There ensues, as a cascade, phosphorylation and dephosphorylation of target enzymes. Then, with an unknown link to these reactions, glucose transport is activated by translocation of transporters from an intracellular compartment into the p.m.[*] Such subcellular redistribution of regulated proteins in response to insulin can be studied with isolated adipocytes from which p.m. and low-density microsomal fractions can be prepared. Adipocytes express two species of glucose transporter, glcT₁ and glcT₄, detectable with specific antisera, by photolabelling and by binding of specific ligands, e.g. cytochalasin B and forskolin. Both species are translocated to the p.m. in response to insulin. G-proteins, which may participate in glucose transport regulation, are detectable in the intracellular membrane fraction by immunoblotting and GTP binding.

Target protein phosphorylation in response to insulin has been studied in adipocytes labelled with [³²P]phosphate. Phosphoproteins can be isolated in the presence of phosphatase inhibitors from membrane fractions or lysates of whole cells. In SDS-gels, membranes show insulin-dependent phosphorylation of, e.g., ribosomal protein S6 and ATP-citrate lyase. Less abundant phosphoproteins, e.g. the insulin receptor or glucose transporters, must be isolated by immunoprecipitation with specific antisera. Insulin receptors are autophosphorylated, internalized into low-density microsomes and dephosphorylated; this cycle might be the signal for cellular translocation of the receptor. The phosphorylation state of glcT₁ and glcT₄ being unaltered by insulin, their translocation may be signalled not by their own phosphorylation state but by that of other proteins localized in intracellular vesicles.

Abbreviations.- Ab, antibody; glcT₁ & glcT₄, glucose transporters; IAPS-, 3-iodo-4-azidophenethylamido-7-*O*-succinyldeacetyl-; p.m., plasma membrane; Tyr, tyrosine; TK, Tyr kinase.
[*]Art. #F-1 (C.A. Pasternak) is relevant.- *Ed.*

Insulin produces a rapid stimulation of glucose transport activity in adipose and muscle tissue through redistribution of transporters from an intracellular pool to the p.m. [1-3]. Since other proteins are similarly translocated in response to insulin, e.g. the insulin-like growth factor II/mannose-6-phosphate receptor [4] and the transferrin receptor [5], this translocation mechanism appears to represent a general mechanism of insulin action. While the phenomenon as such has been studied thoroughly [6, 7], the mechanisms transmitting the signal to the intracellular proteins and the forces which drive the translocation of these proteins are obscure.

The finding that insulin produces characteristic changes in the phosphorylation state of several cellular proteins has led to the hypothesis that a phosphorylation cascade mediates and amplifies the intracellular actions of insulin [8, 9]. Subsequently this hypothesis gained considerable support when Kasuga and co-workers [10] discovered the protein kinase activity of the insulin receptor which stimulates phosphorylation of its own β-subunit on Tyr residues. However, the signalling chain which starts with autophosphorylation of the receptor and produces rapid changes at intracellular target proteins, e.g. the glucose transporter, is still poorly understood. Hence current research efforts are aimed at identifying a link between insulin-controlled phosphorylation reactions and the translocated proteins. This article describes experimental strategies and methods employed in order to approach this goal, and summarizes some recent results.

PREPARATION AND CHARACTERIZATION OF MEMBRANE FRACTIONS FROM BASAL AND INSULIN-STIMULATED ADIPOCYTES

The stimulatory effects of insulin on glucose transport in the adipocyte persists after homogenization of cells and can be demonstrated in an isolated p.m. preparation [11, 12]. This so-called fossil effect has facilitated the study of insulin action and led to the finding that adipocytes contain a large intracellular pool of glucose transporters from which transporters are translocated to the p.m. when cells are exposed to insulin [1, 2, 12]. Adipocytes represent an ideal model for the study of glucose transport regulation because of the large effect of insulin, the homogeneity of the cell preparation and the relative ease of the procedures for obtaining well-defined membrane fractions which retain marked responsiveness to insulin [13-15]. A murine fibroblast cell line, the 3T3-L1 pre-adipocytes [16], can be used if long-term culture of insulin-sensitive cells is desired.

The methods currently used for isolating adipocytes and preparing membranes have been published in detail [15]. The most important factors affecting the hormonal response appear

Table 1. Characteristics of p.m. and low-density microsomes (1-d.) from basal (B) and insulin-stimulated (I) adipocytes, fractionated [15] after incubation with or without insulin. Values are means ±S.E.M. (n >3) and, for rows 1-3, represent pmol, nmol and fmol respectively per mg protein, per sec for rows 1 & 2; % values in rows 5-7 are % of p.m. activity: the units for 5 & 6 are nmol/mg protein, per h (row 5) or per 10 min (6).

Parameter[@]	B, p.m.	I, p.m.	B, 1-d.	I, 1-d.
glcT act., vesicles	5.6 ±0.9	76.2 ±15	–	–
ditto, reconstituted	0.25 ±0.1	0.76 ±0.2	2.64 ±0.5	1.1 ±0.1
[^{125}I]IAPS into glcT's	10.4 ±1.9	34.2 ±5.8	43.2 ±7.3	23.3 ±2.84
Protein yield, µg/rat	443 ±66	509 ±121	162 ±13	162 ±21
5'-Nucleotidase act.	263 ±42 (100%)	254 ±43 (100%)	49 ±11 (19%)	34 ±13 (13%)
Adenylate cyclase act.	64 ±12 (100%)	61 ±14 (100%)	11 ±3 (17%)	7 ±1 (12%)
K-ras immunoreactiv.	100%	100%	4.0 ±0 6%	3.7 ±1.2%

[@] glcT, D-glucose transport(er); act., activity; IAPS is linked to forskolin for incorporation measurement. Rat body wt. ~170 g.

to be the materials used for isolating cells (vials, albumin, collagenase) and the conditions of homogenization (grinder clearance, temperature, cell concentration). Glucose transport activity can be assayed in the intact p.m. vesicles as an indicator of the quality of the membrane preparation [12]; we routinely observe a 10- to 20-fold effect of insulin (Table 1). Other parameters for quantitating the effect of insulin, e.g. the number of glucose-inhibitable cytochalasin B binding sites [12], the amount of the photolabel IAPS-forskolin [17] incorporated into the transporter (Fig. 1 & Table 1), and glucose transport activity reconstituted into lecithin liposomes, usually yield lower responses than the transport activity in intact p.m. preparations (Table 1). This difference could be due to the contamination of p.m. with intracellular microsomes which would increase the reconstituted glucose transport activity and the number of transporters in p.m. from basal cells, but not glucose transport activity in intact vesicles [12].

The degree of cross-contamination of the membrane fractions poses a major problem in interpreting results from experiments involving cell fractionation, and has therefore been analyzed meticulously [15, 18]. As yet there is no reliable marker to assess the contamination of p.m. with low-density microsomes. Comparison between glucose transporters and glucose transport

Fig. 1. Detection of glucose
transporters in adipocyte
membrane fractions [15] photo-
lysed in the presence of IAPS-
forskolin [17] (gifted by
Dr. M. Shanahan, Carbondale,
IL, U.S.A.) for 30 sec.
Labelled membranes were
separated by SDS-PAGE, and
the dried gels were auto-
radiographed for 2 days.
The adipocytes had been incubated with insulin (8 nM;
fractions denoted I) or
without insulin (basal, B).
PM = plasma mebrane (p.m.
in Table 1 and text);
LD = low-density microsomes
(1-d. in Table 1).

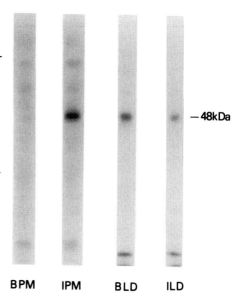

—48kDa

BPM   IPM   BLD   ILD

in basal p.m.   allows an estimate.   Low-density microsomal
glucose transporters do not contribute to the transport activity
in p.m. vesicles [12], probably because of the small volume
of the microsomal vesicles.   Hence cross-contamination of
p.m. with low-density microsomes increases the number of glucose
transporters, the reconstituted transport activity and the
amount of photolabel incorporated in basal p.m. without a
corresponding increase in the transport activity of the intact
vesicles.   On the basis of this discrepancy, we estimate
that our p.m. preparations are ~10% cross-contaminated with
low-density microsomes.

Using adenylate cyclase and 5-nucleotidase as established
markers for p.m., we found a significant cross-contamination
(10-15%) of the low-density microsomes with p.m. (Table 1).
However, a much lower cross-contamination (<5%) was found
with other proteins as markers, e.g. the proto-oncogene K-*ras*
(Table 1).   This discrepancy might indicate that the concept
of using marker enzymes for assessing cross-contamination
has to be re-considered.   Cross-contamination of membrane
fractions is caused primarily by overlaps in density and size
of the different populations of vesicles generated by the
shearing forces of homogenization[*].   Possibly this overlap
may not reflect a random distribution of membrane regions,

---

[*]*Note by Ed.*- Enquiries welcomed about *Subcellular Organelles* and
other early vols. with oft-neglected lore (D.J. Morré and others)
on fraction-purity.   For the adipocyte homogenates, differential
pelletting was followed by isopycnic centrifugation.

but certain domains of the membrane may be especially prone to form smaller or larger vesicles which would contaminate the other membrane fractions. Hence cross-contamination would ultimately have to be assessed with a specific marker closely related to the protein under investigation or to the cell organelle with which the target protein is associated.

DETECTING AND ASSAYING ADIPOSE-TISSUE AND MUSCLE glcT's

The first method for detecting and quantitating glcT's took advantage of the specific glucose-inhibitable binding of cytochalasin B to the transport protein [1, 15]. Another approach uses forskolin having an iodinated photoreactive group [17]: IAPS-forskolin binds specifically to the glcT [19] and, as Fig. 1 shows, labels a 48 kDa protein in adipocyte membranes which is translocated from an intracellular pool to the p.m. in response to insulin.

Recently it has been shown that adipose and muscle tissue express two different species of glcT proteins: the erythrocyte/brain-type, $glcT_1$ [20], and the tissue-specific adipose/muscle-type, $glcT_4$ [21, 22] whose M_r (48 kDa) slightly exceeds that of $glcT_1$ (45 kDa; Fig. 2). Their sequence homology is only 63.5%; hence they can be distinguished by immunoblotting with specific antisera (Fig. 2). Both transporters are translocated to the p.m. in response to insulin. Since they bind cytochalasin B with identical affinities, this binding assay determines the sum of $glcT_1$ and $glcT_4$. Based on the assumption that the photolabel IAPS-forskolin labels both transporters with equal efficiency, a quantitative comparison of labelled, immunoprecipitated glcT's indicated that $glcT_4$ is the predominant species (75-90%) in adipocytes [23].

The discovery of a family of glucose transporters raises the question whether insulin sensitivity of a cell is conferred by the amino acid sequence of its transporter and/or by some unknown ability of the cellular machinery. If, therefore, the primary structure of the transporter confers insulin sensitivity on muscle and adipose tissue, the important structural regions must be homologous with $glcT_1$ and $glcT_4$. However, studies in 3T3-L1 cells suggest that insulin sensitivity of glucose transport also requires the formation of an intracellular pool of glcT's which is formed during differentiation of the cells to an adipocyte-like phenotype [24].

STUDYING INSULIN EFFECTS ON ADIPOCYTE PROTEIN PHOSPHORYLATION

Insulin initiates its action by binding to a membrane receptor whose intracellular domain possesses intrinsic tyrosine kinase (TK) activity [10]. After insulin binds, the receptor triggers a cascade of target-enzyme phosphorylation and, equally

Fig. 2. Immunochemical
detection [24] of the
glcT₁ and glcT₄ species
in membranes from adipo-
cytes incubated with
insulin (8 nM; I) or
without insulin (B).
The membrane fractions
[15] (P, p.m., = PM in
Fig. 1; L , low-density
microsomes, =LD in
Fig. 1) were subjected
to SDS-PAGE, and trans-
ferred to nitrocellulose
membranes for immuno-
assay. The anti-glcT₁
antiserum, α-GT₁ *(left
panel)*, was prepared
with purified red cell
glcT₁; *α-CT3* was raised
against a peptide corres-
ponding to the C-terminus
of glcT₄ [24].

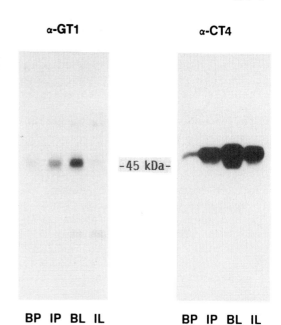

α-GT1   α-CT4

-45 kDa-

BP IP BL IL   BP IP BL IL

important, dephosphorylation. Serine kinases are obviously
involved, since insulin activates acetyl-CoA carboxylase by
serine-residue phosphorylation [25]. Several insulin-dependent
serine kinases have been identified, but their activation
through a Tyr phosphorylation or mediator remains unproved [26].
Interestingly, one such kinase phosphorylates the microtubule-
associated MAP-2 [27], possibly the link between insulin-stimulated
protein phosphorylation and translocation.

The insulin receptor's TK activity and its response
to insulin can be assayed in partially purified receptor
preparations *in vitro* in two ways: by measurement of phosphate
incorporation into the β-subunit of the receptor, or with
the aid of exogenous substrates such as histone or co-polymerized
Glu/Tyr [28]. However, it is preferable to stimulate intact
cells *in vivo* and then perform the kinase assay [29].   More
recently, anti-phosphotyrosine Ab's have become available which
allow phosphate incorporation into Tyr residues of the insulin
receptor to be determined by Western blotting.

The phosphorylation of target proteins in response to
insulin can be studied in adipocytes treated for 60-90 min
with tracer [³²P]phosphate, which with appropriate conditions
[30] equilibrates with the ATP pool.   Phosphoproteins can
be isolated in the presence of phosphatase inhibitors from
membrane fractions or lysates of whole cells.   Whereas no

insulin-dependent protein phosphorylations are manifest in p.m., SDS-gels of crude membrane fractions manifest phosphorylation of several intracellular proteins [31, 32]. These effects can be used as an independent control for the responsiveness of the cells to insulin after the labelling period.

In low- and high-density microsomes, the phosphorylation of ribosomal protein S6 (32 kDa) and of ATP-citrate lyase (115 kDa) can be detected; in our hands, insulin increases the phosphorylation of these proteins by ~5- and 2-fold respectively. In addition, the phosphorylation of several unidentified proteins in the cytosol is stimulated by insulin [32]. Less abundant phosphoproteins, e.g. the insulin receptor or glcT's, have to be isolated by immunoprecipitation with specific antisera. Thereby it was shown that autophosphorylated insulin receptors are internalized into the low-density microsomes, where they are probably dephosphorylated and recycled to the p.m. [31]; it seems plausible that this phosphorylation/dephosphorylation cycle represents the signal for the cellular translocation of the receptor.

Glucose transporters do not appear to be the substrates for an insulin-controlled phosphorylation reaction. Adipocyte $glcT_1$ contains no measurable amounts of phosphate either in the basal state or after exposure to insulin or isoproterenol [32], while $glcT_4$ contains phosphate in the basal state but insulin fails to alter its content (Fig. 3). Isoproterenol, in contrast, produces a moderate rise in $glcT_4$ phosphorylation [33], as Fig. 3 confirms. **In conclusion,** the translocation of glcT's is probably not signalled by the phosphorylation state of the transporter protein itself, but other phosphoproteins localized in the intracellular vesicles might be the target of the phosphorylation cascade initiated by the insulin receptor.

G-PROTEINS AND INSULIN ACTION

Several lines of research have suggested that G-proteins might be involved in the mechanism of signalling by insulin. It has been reported to inhibit catecholamine-induced adenylate cyclase in adipose tissue [34, 35], and it antagonizes cAMP-dependent pathways by stimulating the low-K_m phosphodiesterase which is probably regulated through interaction with a G-protein [36, 37]. More recently, a direct interaction of G-proteins with the insulin-sensitive glcT has been postulated on the basis of evidence for catecholamine-induced, cAMP-dependent changes in the intrinsic activity of the glcT [38, 39]

To test the hypothesis that G-proteins are involved in the signal mechanism of insulin and/or in glucose transport regulation, we have used three approaches. (1) We tested

Fig. 3. Phosphate incorporation, in response to insulin
and isoproterenol, into the
glcT$_4$ of isolated adipocytes,
equilibrated with [^{32}P]-
phosphate (0.2 mCi/ml) for
90 min. After exposure to
agents specified below, the
cells were centrifuged
through silicone oil, rapidly
frozen in methanol/dry ice,
and solubilized in a buffer
containing Triton X-100 and
phosphatase inhibitors [40].
Immunoprecipitates prepared
from the extracts were separated electrophoretically and
autoradiographed for 1 day at
-70° with the aid of an
enhancing screen. PI, pre-
immune serum; α-CT3, immune
serum against the glcT$_4$.
Bs, basal (no agents).
Ins, insulin (8 nM).
Iso, insulin (8 nM) for 20 min,
then isoproterenol (1 μM) plus
adenosine deaminase
(2.5 μg/ml) for 15 min.

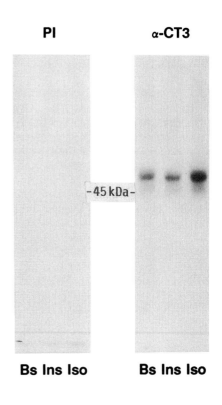

whether G-proteins are substrates for insulin-controlled phosphorylation reactions. Using a peptide-derived Ab against
a highly conserved sequence, we immunoprecipitated the α-
subunits of G$_i$ and G$_s$ from adipocytes equilibrated with [^{32}P]-
phosphate. The data indicated that phosphorylation of neither
G-protein was stimulated by insulin [40]. (2) We have studied
the subcellular distribution of the G-proteins in adipocytes,
postulating that if an association of the G-proteins with
the transporter existed, we would detect intracellular
G-proteins. Low-density microsomes indeed were rich in low-mol.
wt. G-proteins [41], but had less of the G$_i$ and G$_s$ subunits.
(3) We found the glucose transport activity reconstituted
from membrane fractions to be reduced by guanine nucleotides,
provided that the cells were treated with insulin prior to
isolation of membrane fractions [42]. Although this effect
suggests that GTP-dependent mechanisms are involved in insulin
action and the regulation of glucose transport, the GTP-binding
protein responsible for the observed effect has yet to be
identified.

References

1. Cushman, S.W. & Wardzala, L.J. (1980) *J. Biol. Chem.* 255, 4758-4762.
2. Suzuki, K. & Kono, T. (1960) *Proc. Nat. Acad. Sci.* 77, 2542-2545.
3. Wardzala, L.J. & Jeanrenaud, B. (1981) *J. Biol. Chem.* 256, 7090-7093.
4. Wardzala, L.J., Simpson, I.A., Rechler, M.M. & Cushman, S.W. (1984) *J. Biol. Chem.* 259, 8378-8383.
5. Davis, R.J., Corvera, S. & Czech, M.P. (1986) *J. Biol. Chem.* 201, 8708-8711.
6. Simpson, I.A. & Cushman, S.W. (1986) *Annu. Rev. Biochem.* 55, 1059-1089.
7. Joost, H.G. & Weber, T.M. (1989) *Diabetologia 32*, 831-838.
8. Denton, R.M., Brownsey, R.W. & Belsham, G.J. (1981) *Diabetologia 21*, 347-362.
9. Avruch, J., Alexander, M.C., Palmer, J.L., Pierce, M.W., Nemenoff, R.A., Blackshear, P.J., Tipper, J.P. & Witters, L.A. (1982) *Fed. Proc. 41*, 2629-2633.
10. Kasuga, M., Karlsson, F.A. & Kahn, C.R. (1982) *Science 215*, 185-187.
11. Martin, D.B. & Carter, J.R. (1970) *Science 167*, 873-874.
12. Joost, H.G., Weber, T.M. & Cushman, S.W. (1988) *Biochem. J. 249*, 155-161.
13. Rodbell, M. (1964) *J. Biol. Chem. 239*, 375-380.
14. Gliemann, J. & Rees, W.D. (1983) *Curr. Topics Memb. Transp. 18*, 339-379.
15. Weber, T.M., Joost, H.G.. Simpson, I.A. & Cushman, S.W. (1988) in *Receptor Biochemistry and Methodology. Insulin Receptors Part B: Biological Responses and......* (Kahn, C.R. & Harrison, L., eds.), Liss, New York, pp. 171-187.
16. Green, H. & Kehinde, O. (1974) *Cell 1*, 113-116.
17. Wadzinsky, B.E., Shanahan, M.F. & Ruoho, A.E. (1987) *J. Biol. Chem. 262*, 17683-17689.
18. Simpson, I.A.. Yver, D.R., Hissin, P.J., Wardzala, L.J., Karnieli, E., Salans, L.B. & Cushman, S.W. (1983) *Biochim. Biophys. Acta 763*, 393-407.
19. Joost, H.G. & Steinfelder, H.J. (1987) *Mol. Pharmacol. 31*, 279-283.
20. Mueckler, M., Caruso, C., Baldwin, S.A., Panico, M., Blanch, I., Morris, H.R., Allard, W.J., Lienhard, G.E. & Lodish, H.F. (1985) *Science 229*, 941-945.
21. James, D.E., Strube, M. & Mueckler, M. (1989) *Nature 338*, 83-87.
22. Birnbaum, M.J. (1989) *Cell 57*, 305-315.
23. Zorzano, A., Wilkinson, W., Kotliar, N., Thoidis, G., Wadzinsky, B.E., Ruoho, A.E. & Pilch, P.F. (1989) *J. Biol. Chem. 264*, 12358-12363.

24. Weiland, M., Schürmann, A., Schmidt, W.E. & Joost, H.G.
    (1990) *Biochem J. 270*, 331-336.
25. Brownsey, R.W. & Denton, R.M. (1982) *Biochem. J. 202*, 77-86.
26. Yu, K.T., Khalaf, N. & Czech, M.P. (1987) *Proc. Nat. Acad.
    Sci. 84*, 3972-3976.
27. Ray, L.B. & Sturgill, T.W. (1988) *Proc. Nat. Acad. Sci. 85*,
    3753-3757.
28. Joost, H.G., Steinfelder, H.J. & Schmitz-Salue, C.
    (1985) *Biochem. J. 233*, 677-681.
29. Klein, H.H., Freidenberg, G.R., Kladde, M. & Olefsky, J.M.
    (1986) *J. Biol. Chem. 261*, 4691-4697.
30. Hopkirk, T.J. & Denton, R.M. (1986) *Biochim. Biophys. Acta
    885*, 195-205.
31. Weber, T.M., DiPaolo, S., Joost, H.G., Cushman, S.W. &
    Simpson, I.A. (1988) in *Progress in Endocrine Research and
    Therapy, Vol. 4, Insulin Action and Diabetes* (Goren, H.J.,
    Hollenberg, M.D. & Roncari, D.A.K., eds.), Raven Pr., N.York, pp.151-156.
32. Joost, H.G., Weber, T.M., Cushman, S.W. & Simpson, I.A.
    (1987) *J. Biol. Chem. 262*, 11261-11267.
33. James, D.E., Hiken, J. & Lawrence, J.C. (1989) *Proc. Nat.
    Acad. Sci. 86*, 8368-8372.
34. Hepp, K.D. (1972) *Eur. J. Biochem. 31*, 266-276.
35. Heyworth, C.M. & Houslay, M.D. (1983) *Biol. J. 214*, 547-552.
36. Vaughan, M., Danello, M.A., Manganiello, V.C. &
    Strewler, G.J. (1981) *Adv. Cyclic Nucleotide Res. 14*, 263-271.
37. Heyworth, C.M., Wallace, A.V. & Houslay, M.D. (1983)
    Biochem. J. 214, 99-110.
38. Joost, H.G., Weber, T.M., Cushman, S.W. & Simpson, I.A.
    (1986) *J. Biol. Chem. 261*, 10033-10036.
39. Kuroda, M., Honnor, R.C., Cushman, S.W., Londos, C. &
    Simpson, I.A. (1987) *J. Biol. Chem. 262*, 245-253.
40. Joost, H.G., Schmitz-Salus, C., Hinsch, K.D., Schultz, G.
    & Rosenthal, W. (1989) *Eur. J. Pharmacol. 172*, 461-469.
41. Schürmann, A. & Joost, H.G. (1991) *Arch. Pharmacol. 343
    (Suppl.)*, 239.
42. Schürmann, A., Rosenthal, W., Hinsch, K.D. & Joost, H.G.
    (1989) *FEBS Lett. 255*, 259-264.

#C-6

INVESTIGATION OF THE RAPID ACTION OF
GLUCOGENIC HORMONES ON LIVER METABOLISM

Sibylle Soboll, Holger Hummerich, Michael Görlach
and Helmut Sies

Institut für Physiologische Chemie I,
Universität Düsseldorf, Universitätstrasse 1,
D-4000 Düsseldorf, Germany

This article concerns approaches to investigating hormonal action on the liver cell in relation to metabolic output, intracellular signalling and protein modification. The influence of glucogenic hormones such as glucagon, VP* and L-T_3 on Ca^{2+} uptake, glucose production and O_2 consumption was measured in isolated rat-liver perfusates. With fed or fasted rats these parameters were increased by all three hormones, but with differences in time courses.

Both glucagon and L-T_3 increased the mitochondrial/cytosolic pH difference, as studied by pre-equilibrating liver cells in situ with ^{14}C-DMO, then freeze-clamping and freeze-drying the liver and subjecting it to density gradient centrifugation in non-aqueous solvents. Glucagon stimulated the cAMP-dependent phosphorylation of particular mitochondrial proteins present in the outer or inner membrane. Glucogenic hormones (e.g. VP) not dependent on cAMP were ineffective, and evidently exert their effects on liver metabolism through $InsP_3$-mediated changes in $[Ca^{2+}]_c$.

Glucogenic hormones are notable for their rapid action on metabolism. In liver within <1 min a stimulation of O_2 consumption, gluconeogenesis or glycogenolysis and other metabolic pathways is observed [1]. They differ in signalling pathways: β-agonists and glucagon act *via* cAMP, whereas α-agonists and VP act *via* the $InsP_3$ pathway; yet they are similar in provoking an initial rise in cytosolic Ca^{2+} which can function as a third messenger within the cell.

The effects of glucagon, VP and L-T_3 on liver metabolism were studied in an isolated liver perfusion system, where glucose output, O_2 and $[Ca^{2+}]$ can be monitored in the effluent

*Abbreviations.- Subscripts in concentration expressions such as $[Ca^{2+}]_c$: c = cytosolic (usually equivalent to intracellular, i, as in other arts.), e = external, m = mitochondrial. cAMP, cyclic AMP; $InsP_3$, inositol trisphosphate; DMO, dimethyloxazolidine; TPMP, triphenylmethylphosphonium; p.m., plasma membrane; VP, vasopressin; L-T_3, L-triiodothyronine. Fura 2/AM = acetoxymethyl ester of the fluorescent probe fura 2.

perfusate.   However, to get a more detailed picture, one
has to study the intracellular events directly.   Therefore,
additionally, tissue fractionation and protein phosphorylation
studies were performed to investigate the hormonal action
on the liver cell at the level of its metabolic output,
intracellular signalling and protein modification.

METHODS

Haemoglobin-free liver perfusion

Livers of male Wistar rats were perfused as described
in [2] with Krebs-Henseleit bicarbonate buffer in an open
system.   Substrates were added as indicated in Tables and
Figures.   Oxygen was measured in the effluent perfusate
with a Clarke-type electrode, and $[Ca^{2+}]$ recorded with a
Ca^{2+}-sensitive electrode.   Glucose was measured in effluent
perfusate by enzymatic analysis [3].   For determining subcellular pH values, 5 µCi $2-^{14}C$-DMO was added to the perfusate
5 min before termination of the perfusion experiment by freeze-
clamp and application of procedures described below.

Membrane potentials were measured in separate experiments
by perfusing the liver for 15 min as described above and
then switching for 10 min to a closed system containing,
per 200 ml, 15 µCi ^{14}C-TPMP. The liver was then freeze-clamped.

Determination of mitochondrial and cytosolic concentrations of DMO and TPMP in intact liver tissue (Fig. 1)

The freeze-stopped tissue was freeze-dried and fractionated
using density gradient centrifugation in non-aqueous solvents
[4].   The dried tissue (~300 mg) was sonicated in heptane/CCl_4
and put on top of a density gradient consisting of the
same media (density 1.29-1.38 g/ml).   The gradient yielded
8 fractions differing in the proportions of mitochondrial
and cytosolic protein.   By elimination of water from the
fractionation system, metabolic disturbances are avoided after
fixation of liver metabolic status by freeze-clamp.   The
fractions were dried and suspended in buffer.

Mitochondrial and cytosolic proportions in each fraction
were determined by measuring the specific activities of marker
enzymes therein (phosphoglycerate kinase for cytosol, citrate
synthase for mitochondria).   Mitochondrial and cytosolic
contents of DMO and TPMP were obtained from the radioactivities
in the fractions by extrapolation to pure mitochondria and
pure cytosol [4].   Contents were converted into concentrations
using water spaces of 0.8 µl/mg mitochondrial protein and
3.8 µl/mg cytosolic protein respectively.   The mitochondrial/
cytosolic pH difference was calculated from the DMO concentration ratio according to equation (1).

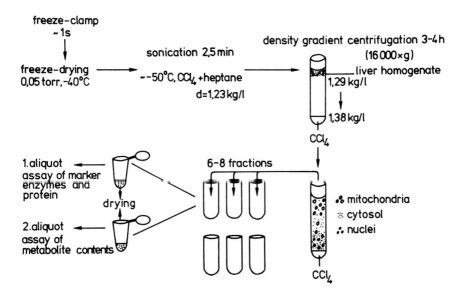

Fig. 1. Non-aqueous fractionation for determining metabolites in mitochondrial and cytosolic fractions.
....

$$\Delta\,pH_m \;=\; \log\,[DMO]_m/[DMO]_c \tag{1}$$

The membrane potentials were calculated from the concentrations of TPMP across the membrane (inside *vs.* outside) according to equation (2) [RT and F have the usual thermodynamic meaning]:

$$\Psi_{in\text{-}out} \;=\; 2.3\;RT/F\,\log\,[TPMP]_{in}/[TPMP]_{out} \tag{2}$$

Binding of TPMP to hepatocytes was ascertained with hepatocyte suspensions to be 25% and was subtracted from the overall TPMP content in the tissue before calculation of subcellular TPMP contents.

Hepatocyte studies, including protein phosphorylation

Hepatocytes were prepared from the livers of fed rats essentially as in ref. [5]. Loading of the cells with the Ca^{2+}-binding fluorescent dye fura 2/AM and measurement of $[Ca^{2+}]_c$ was performed as described in [6]. To investigate protein phosphorylation, hepatocytes were washed twice with low-phosphate (0.2 mM) Krebs-Henseleit buffer and brought to a final concentration of 15 mg cellular protein per ml.

Before preparing subcellular fractions, hepatocytes were incubated for 60 min as described in [6] in the presence of 100 µCi/ml ^{32}P-orthophosphate and then stimulated with the hormone for 10-20 min. The incubation was terminated by centrifugation of the hepatocyte suspension. The cells were resuspended in 250 mM sucrose/10 mM Tris/5 mM EDTA, pH 7.4, and disrupted by Ultraturrax. Cytosol and mitochondria were obtained by differential centrifugation [7]. To obtain the outer and inner membranes, mitochondria were further fractionated essentially as described in [8]. Cytosol and outer membrane and inner membrane fractions, in SDS-buffer medium, were subjected to SDS-polyacrylamide gel electrophoresis [9]. Phosphorylated bands were obtained by autoradiography. ·

RESULTS AND DISCUSSION

Glucagon released Ca^{2+} into the effluent perfusate within the first minute, followed by an uptake within the next 2-3 min which was accompanied by a stimulation of O_2 uptake and gluconeogenesis. Subsequently Ca^{2+} was released for ~5 min and then taken up again for >15 min (Fig. 2A). Ca^{2+} uptake was parallelled by an increase in O_2 uptake and glucose output, and *vice versa* (not shown). Essentially the same time course is observed with VP but with somewhat faster kinetics [6]. The release of Ca^{2+} from the liver by glucagon within one minute is noteworthy in that glucagon does not act *via* $InsP_3$, which is known to release Ca^{2+} from intracellular stores. (Fig. 2A also shows pH changes.)

L-T_3 causes qualitatively the same effects on perfused liver, but with completely different kinetics (Fig. 2B). Firstly, the Ca^{2+} movements are not biphasic: Ca^{2+} is not released within the first minute but is taken up by the liver from the outset of L-T_3 infusion. Secondly, the time course is much slower than with the other hormones. Since no intracellular messenger for the fast actions of L-T_3 is known, it is suggested that L-T_3 exerts its action in the cell only by mediating an influx of Ca^{2+} into the liver cell [6].

In a further set of experiments the liver cell p.m. potential was decreased by infusing 20 mM K^+ (Fig. 3A). Also in this experiment perfusate Ca^{2+} was decreased with a monophasic time course, but gluconeogenesis was not stimulated. Consistently, $InsP_3$ in fura 2/AM-loaded hepatocytes decreased rather than increased in the presence of 20 mM K^+ (Fig. 3B). In contrast, VP, glucagon and L-T_3 caused a rise in $InsP_3$ by ~50 nM [6]. By decreasing the p.m. potential, Ca^{2+} may become bound to the membrane, therefore lowering extracellular and intracellular Ca^{2+}.

Comments on Ca^{2+} involvement.- Since the time course of Ca^{2+} changes in the effluent perfusate is very similar to that of the changes in metabolic rates in liver, it

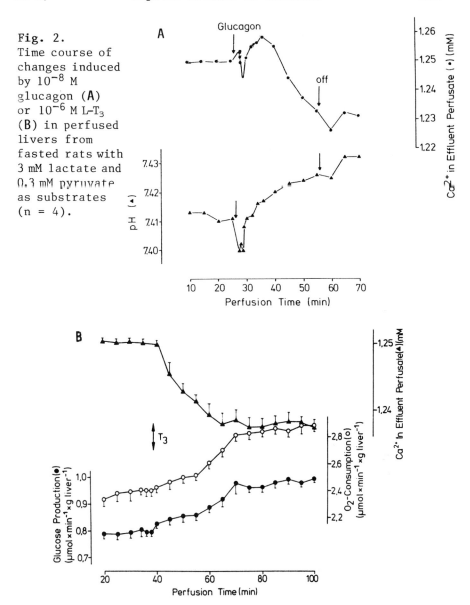

Fig. 2.
Time course of
changes induced
by 10^{-8} M
glucagon (**A**)
or 10^{-6} M L-T_3
(**B**) in perfused
livers from
fasted rats with
3 mM lactate and
0.3 mM pyruvate
as substrates
(n = 4).

is reasonable to conclude that Ca^{2+} is a main mediator
of the metabolic effects of glucogenic hormones. The detailed
mechanism, however, remains unclear, particularly with regard
to the stimulating action on the mitochondria. It is generally
assumed that following hormonal stimulation, Ca^{2+} also enters
the mitochondria, where it stimulates Ca^{2+}-dependent dehydro-
genases [10].   If respiration is to be stimulated, the cell
and the mitochondria may have to be supplied with more substrate. Regulation of cell and mitochondrial membrane transport
could be a site of action.

Fig. 3. Time course
of changes induced
by 20 mM KCl in
metabolism of perfused
livers from fasted
rats (**A**; substrates
as in Fig. 2) and in
the cytosolic Ca^{2+}
of hepatocytes (**B**).
Each value represents
a single experiment.

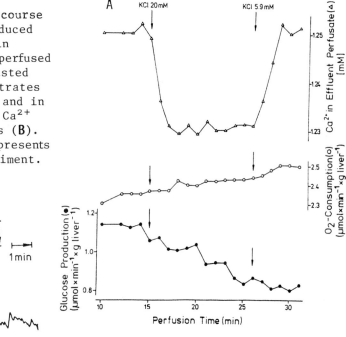

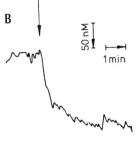

Membrane potential and pH gradient studies

The driving force for several p.m. transport systems
is the p.m. potential, Ψ_p [11], whereas many mitochondrial
transport systems are driven by the proton gradient across
the inner mitochondrial membrane [12]. By means of a special
fractionation technique in non-aqueous solvents developed for
intact freeze-clamped organs, it is possible to differentiate
between cytosolic and mitochondrial metabolites as well as
marker compounds for the pH-gradient and for the electric
potential across cellular membranes, such as DMO and TPMP
respectively. We could show that some glucogenic hormones,
such as glucagon and $L-T_3$, increased not only the p.m.
potential but also the mitochondrial/cytosolic pH difference
(Table 1). The mitochondrial membrane potential underwent
no change, or too small a Ψ_m change to be detected with
our technique. Thus the protonmotive force increased by
~20-40 mV mainly due to an increase in ΔpH_m, reflecting
a higher energization of the mitochondria.

From our data it appears reasonable to conclude that
one way by which glucogenic hormones stimulate liver metabolism
is to increase transport of substrates into the cytosol
and mitochondria of the cell.

Table 1. Effects of glucogenic hormones on the mitochondrial/ cytosolic pH difference and on membrane potentials in perfused rat liver (with 3 mM lactate, 0.3 mM pyruvate). No. of experiments: 4-6 (±S.E.M.) or 2.

	Ψ_m, mV	Ψ_p, mV	ΔpH_m
Fasted rats	144 ± 10	63 ±5	0.22 ±0.04
+ glucagon 10^{-8}M	144 (143, 145)	87 (76, 98)	0.49 ±0.07
+ L-T_3 10^{-6}M	142 ±10	80 ±5	0.61 ±0.07

Protein phosphorylation

It is well established that cAMP as well as Ca^{2+} acts on metabolism *via* phosphorylation-dephosphorylation of cellular proteins. This could also be a mechanism by which hormones change the electrical conductance and/or proton conductance of membranes since, as discussed above, the p.m. potential and the mitochondrial/cytosolic pH-difference were increased by glucagon and L-T_3. Whereas phosphorylation of some cytosolic and mitochondrial matrix proteins has already been demonstrated, especially with glucagon, little is known on phosphorylation of membrane proteins. We found that two mitochondrial membrane proteins were phosphorylated in glucagon-stimulated hepatocytes, one in the outer membrane with M_r 50000 and one in the inner membrane with M_r 20000 (Fig. 4, a-d). Glucogenic hormones not dependent on cAMP had no effect.

From this finding it is suggested that despite their similar effects on liver metabolism, glucogenic hormones mediate these changes through at least two different pathways: *via* cAMP-dependent protein phosphorylation or *via* InsP$_3$-mediated changes in cellular Ca^{2+} concentrations. The identification of the phosphorylated membrane proteins and the role of phosphorylation awaits further investigation. Ca^{2+}-dependent phosphorylation of mitochondrial membrane proteins appears to play a minor role in stimulating liver metabolism, since phosphorylation of mitochondrial proteins could not be detected with VP and other hormones not dependent on cAMP. A direct interaction of Ca^{2+} with cellular proteins cannot be excluded even for glucagon since it also increases the cytosolic Ca^{2+} concentration.

References

1. Soboll, S. & Sies, H. (1989) *Meths. Enzymol. 174*, 118-130.
2. Sies, H. (1978) *Meths. Enzymol. 52*, 48-59.
3. Bergmeyer, H.U., ed. (1970) *Methoden der Enzymatischen Analyse*, 2nd edn., Verlag Chemie, Weinheim.
4. Soboll, S., Akerboom, T.P.M., Schwenke, W.D., Haase, R. & Sies, H. (1981) *Biochem. J. 200*, 405-408.
5. Berry, M.N. & Friend, D.S. (1969) *J. Cell Biol. 43*, 506-520.
6. Hummerich, H. & Soboll, S. (1989) *Biochem. J. 258*, 363-367.
7. Klingenberg, M. & Sienczka, W. (1959) *Biochem. Z. 331*, 486-517.
8. Parsons, D.F., Williams, D.R. & Chance, B. (1966) *Ann. N.Y. Acad. Sci. 137*, 643-666.

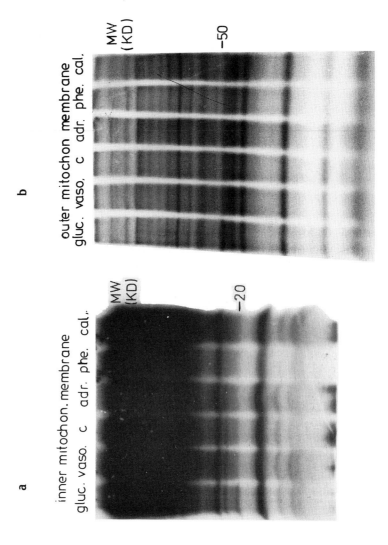

9. Schägger, H. & v. Jagow, G. (1987) *Anal. Biochem. 166*, 368–379.

10. Denton, R.M. & McCormack, J.G. (1985) *Am. J. Physiol. 249*, E543–E554 [& art. in Vol. 19, this series (1989) – *Ed.*].

11. Christensen, H.N. (1984) *Biochim. Biophys. Acta 779*, 255–269.

12. La Noue, K.F. & Schoolwerth, A. (1979) *Ann. Rev. Biochem. 48*, 871–922.

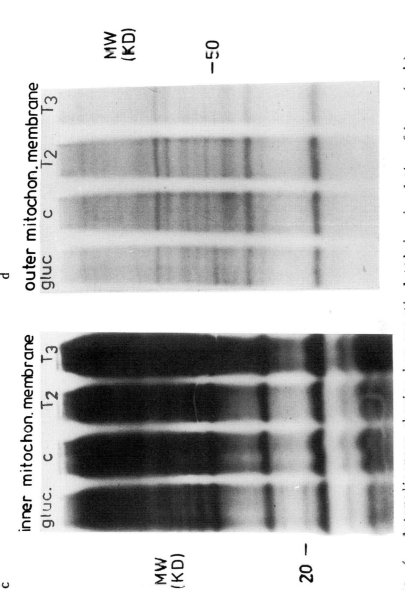

Fig. 4. Autoradiograms showing hormone-stimulated phosphorylation of inner (**a, b**) and outer (**c, d**) mitochondrial membrane proteins: **gluc**, glucagon; **c**, control; **adr**, adrenaline; **phe**, phenylephrine; **cal**, calcimycin; T₂, L-diiodothyronine; T₃, L-triiodothyronine.

#ncC

NOTES and COMMENTS relating to

HORMONE ORIGINATION AND ACTIONS, ESPECIALLY INSULIN AND GLUCAGON

Comments on #C-1: V. Marks – IMMUNONEUTRALIZATION: ISLET STUDIES
#C-3: S.L. Howell – PKC AND B-CELL RESPONSES
#C-4: P-O. Berggren – INSULIN-RELEASE STUDIES

Berggren, to Marks.– We have demonstrated that GABA which is released from the B-cell upon glucose stimulation activates GABA$_A$-receptor Cl$^-$ channels, thereby hyperpolarizing the A-cell and largely inhibiting glucagon release. So this may well account for the inhibition of glucagon release by glucose stimulation. As regards glucagon stimulation of insulin release one wonders whether this occurs only through the cAMP system or whether it may occur through the PL-C system also. **Question from R. Heath.**– Does the system you described also pose the possibility of signal augmentation *via* rate of capillary flow and hence concentrations of hormones impinging on the cells? **Reply.**– Yes; thus, gut hormones influence blood flow. Conceivably, too, there may be substances that influence the direction of blood flow.

G. Milligan asked Howell to comment on the apparent low sensitivity to PMA of PKC-mediated insulin secretion from the islet, and wondered if this reflects the isozyme profile of PKC in these cells. **Reply.**– Indeed there is low sensitivity to PMA; little is known about PKC except that both α- and β-subtypes are present. **Reply to U. Lang:** there was no loss of PKC after permeabilization. **Comment by Lang.**– When we electropermeabilized aortic smooth muscle cells we lost ~40% of membranous PKC activity and ~60% of ANG II binding sites. However, these cells responded to Ca^{2+} addition in a concentration-dependent manner by increased prostacyclin production. **Døskeland asked** (1) for how long do the islet cells stay intact and permeable after electropermeabilization? (2) what is the half-maximally effective concentration of extracellularly added cAMP for stimulation of insulin release from permeabilized cells? **Replies:** (1) 1-2 h for A- and up to 24 h for B-cells; (2) 10-20 μM cAMP.

R. Bruzzone asked Berggren whether he thought that ATP regulates the K$^+$ channels under physiological conditions – for example, an increase in glucose from 5 mM to 8-10 mM, such that neither ATP concentrations nor ATP/ADP ratios should vary significantly. **Answer.**– We need to measure ATP concentrations in single cells (localized changes?) to give a definitive answer.

Question from D. Scheller.- Under physiological conditions (5-10 mM glucose, high ATP) does the regulation of the K^+-channel play a role? **Berggren's reply.**- Yes, it is critical - but it is not related to total ATP; it depends on the ATP/ADP ratio, and other nucleotides may also play a role. **Reply to a further question.**- Tolbutamide and diazoxide were indeed used at rather high concentrations, but agonists/antagonists better in this respect are available; thus 2 µM glibenclamide has similar effects. **S.O. Døskeland asked** (1) whether analogues, maybe non-hydrolyzable, of ATP or ADP had been tested for ability to modulate the K^+ channels, and (2) what effect does ADP alone have on the channels? **Replies:** (1) ATP hydrolysis is not essential; (2) ADP has complex effects, both inhibitory and stimulatory. **Answer to D. Wermelskirchen,** who remarked that pH 6.6 seemed a relatively low intracellular pH.- Introducing fluorescence dyes always changes the intracellular situation, e.g. H^+ or Ca^{2+} changes that affect buffering capacity. So the true values may be higher. Also, the experiments were carried out in Hepes, which changes intracellular pH; but experiments in bicarbonate buffer gave results comparable to those with Hepes. We therefore suggest that true alterations in intracellular pH are being manifested. **Wermelskirchen asked** whether the membrane-potential measurements with bisoxonol, insofar as fluorescent dyes tend to distribute into all lipophilic compartments, reflected mitochondrial membrane potential at least to some extent? **Reply.**- Our measurements are relative, aimed at p.m. potential, and we don't know whether there is some mitochondrial membrane-potential contribution.

Comments on **#C-5:** H.G. Joost - INSULIN ACTIONS ON PROTEINS
        #C-6: S. Soboll - HEPATIC RESPONSES TO HORMONES

Joost, answering G.J. Barritt.- The purity of our p.m. fraction is hard to judge, as some marker enzymes show distribution anomalies. **Answer to G. Milligan:** as yet we have no firm evidence that $glcT_4$ is co-immunoprecipitable with G_S or any other G-protein. **Answer to Berggren:** insulin may have long-term but not acute effects on the glucose transporter in the B-cell. **I.H. Madshus.**- Vesicles containing glucose transporters brought by insulin action to the p.m. probably fuse with the p.m. How is the removal of the transporters from the p.m. induced? Can glucagon induce an increase in the endocytosis of the transporters? **Reply.**- There is a continuous recycling of glucose transporters between the p.m. and the interior of the cell. The effect of glucagon is to inactivate the transporters present in the p.m. (Cf. #F-1.-*Ed.*)

Soboll, in reply to Barritt.- In the fura 2 experiments a physiological level of external Ca^{2+} was used. Where the Ca^{2+} response seemed to be a transient rather than a prolonged increase, this may have reflected the finding that older but not younger cells showed a prolonged increase. **Milligan asked** which of the glucogenic actions of glucagon that you measure can be mimicked by analogues such as TH-glucagon which provide a DAG signal but not a stimulation of cAMP? **Reply:** such analogues, not yet tried, would indeed be worth studying.

Editor's note concerning hepatic 'non-aqueous nuclei'
(cf. Soboll's 'non-aqueous mitochondria', #C-6)
This type of nuclear preparation [1] serves to check possible leakages from 'aqueous' nuclei. A micro version has been used for L-cells ([2]; a column-operation variant is also described, for 'aqueous' nuclei isolated from freeze-dried liver). Verification of survival or, where applicable, of true occurrence in the nucleus has been obtained for NAD pyrophosphorylase, LDH and aldolase [2], for DNA polymerase [2], and for enzymes of RNA and ribonucleotide metabolism including RNases and UMP and UDP kinases [3].

1. Siebert, G. (1961) *Biochem. Z. 334*, 369-387.
2. Siebert, G., *et al.* (1974) in *Subcellular Studies* [Vol. 4, this series] (Reid, E., ed.), Longman, London, pp. 13-29. *[Sole stockist now: Guildford Acad. Ass., 72 The Chase, Guildford, U.K.]*
3. Reid, E., El-Aaser, A.A., Turner, M.K. & Siebert, G. (1964) *Hoppe-Seyl. Z. Physiol. Chemie 339*, 135-149.

SOME LITERATURE PERTINENT TO 'C' THEMES, noted by Senior Editor
(besides some citations near end of #ncE & #ncF)

Lynch, A. & Best, L. (1990) *Biochem. Pharmacol. 40*, 411-416.-
Commentary: 'Cytosolic **pH** and pancreatic β-cell function'; the focus is on glucose effects, taking into account that the main systems responsible for $[pH]_i$ homeostasis are Na^+/H^+ and HCO_3^-/Cl^-.
Rustenbeck, I., *et al.* (1989) *J. Planar Chromatog. 2*, 207-210.-
Phospholipids in subcellular fractions of insulin-secreting cells.
Zawalich, W.S. (1991) *Biochem. Pharmacol. 41*, 807-813.-
'**Glycuride** priming of beta cells. Possible involvement of phosphoinositide hydrolysis' (increase may mediate effects).
Metz, S.A. & Dunlop, M., *ibid.*, R1-R4.- **Propranolol** inhibition, in intact islets, of Ptd-ethanol and PA metabolism, and stimulation of insulin release that may involve lysophospholipids.

Espinal, J. (1989) *Understanding Insulin Action*, Horwood, Chichester, 130 pp.
Vicario, P.P. & Bennun, A. (1989) *Biochem. Soc. Trans. 17*, 1110-1111.- **Insulin-receptor TK** and **adenyl cyclase**: metabolic control by glucose metabolites; a decreased concentration of divalent metal activators (glycolysis produces chelating metabolites) could regulate these activities. The kinetics of the TK in relation to Mg^{2+}, Mg-ATP and ATP have been studied by the same authors [(1990) *Arch. Biochem. Biophys. 2 8*, 99-105].

Gulati, P. & Skett, P. (1989) *Biochem. Pharmacol. 38*, 4415-4418.- An **insulin mediator** was extracted from a hepatic memb-rane fraction after insulin treatment, and (as for insulin) was found to stimulate the metabolism of an androgenic steroid.

Glucagon and **VP** hepatocyte effects: Combettes, L., *et al.* (1986) *Biochem. J. 237*, 675-683 (same lab. as for #B-5; Ca^{2+} movements).

Conelly, P.A., *et al.* (1987) *J. Biol. Chem. 262*, 10154-10163.- Activation of the multifunctional Ca^{2+}/calmodulin **protein kinase** in response to hormones (glucagon and VP) that increase $[Ca^{2+}]_i$.

Cherqui, G., *et al.* (1987) *Endocrinology 120*, 2192-2194.- Decreased insulin responsiveness in **adipocytes** rendered PKC-deficient by phorbol ester treatment.

Gross, D.J., Villa-Komaroff, L., Kahn, C.R., Weir, G.C. & Halban, P.A. (1989) *J. Biol. Chem. 264*, 21486-21490.- 'Deletion of a highly conserved tetrapeptide sequence inhibits **proinsulin to insulin** conversion by transfected pituitary corticotroph (ALT 20) cells'.

Tanner, L.I. & Leinhard, G.E. (1987) *J. Biol. Chem. 262*, 8975-8980.- 'Insulin elicits a redistribution of **transferrin receptors** in 3T3-L1 adipocytes through an increase in the rate constant for receptor externalization'.

Reyl-Desmars, F., Laboisse, C. & Lewin, M.J.M. (1986) *Regul. Pept. 16*, 207-216.- 'A **somatostatin receptor** negatively coupled to adenylate cyclase in the human gastric cell line HGT-1'.

Childs, G.V., Marchetti, C. & Brown, A.M. (1987) *Endocrinology 120*, 2059-2069.- **ACTH release**: regulation involves Na^+ channels and 2 types of Ca^{2+} channels.

#D

INDIVIDUAL-CELL STUDIES, ESPECIALLY ON Ca^{2+}

#D-1

SPATIAL AND TEMPORAL ANALYSIS OF Ca²⁺ SIGNALS IN NEUTROPHILS USING FLUORESCENCE IMAGING OF FURA 2

M.B. Hallett, E.V. Davies and +A.K. Campbell

Departments of Surgery and +Medical Biochemistry,
University of Wales College of Medicine,
Heath Park, Cardiff CF4 4XN, U.K.

To fully understand the role of $[Ca^{2+}]_i$* in intracellular signalling, both the time course of the Ca^{2+} signal and its distribution between individual cells in the cell population and within individual cells has to be established. These temporal and spatial aspects of Ca^{2+} signalling have been investigated by visualizing $[Ca^{2+}]_i$ using ratio imaging of fura 2 fluorescence. Details and problems of this approach are described for neutrophils, which showed heterogeneity in signal timing and size that underlies the smooth Ca^{2+} transient seen in populations. With extracellular Ca^{2+} present, $[Ca^{2+}]_i$ rises uniformly throughout the cytoplasm whereas, in its absence, Ca^{2+} release from an intracellular store is seen, manifesting a localized 'cloud' of elevated $[Ca^{2+}]_i$. Imaging of $[Ca^{2+}]_i$ has thus given important insights into the mechanisms and significance of Ca^{2+} signalling and, for neutrophil responses, has raised new questions about it.

The central role of changes in $[Ca^{2+}]_i$* in cell signalling is well recognized [1].[@] In neutrophils, elevation of $[Ca^{2+}]_i$ induced by some receptor agonists including fMLP triggers some cellular end-responses, including oxidase activation [2, 3]. Much of the pioneering work on signalling by $[Ca^{2+}]_i$ involved its measurement in large non-mammalian cells micro-injected with Ca^{2+}-sensitive photoprotein [1, 4, 5]. Ca^{2+} binding results in the discharge of (at most) a single photon, the rate of photon emission depending on this binding [4, 5]. In giant cells, where sufficient photoprotein can be micro-injected, the spatial distribution of the Ca^{2+} signal can be observed by image intensification techniques [1, 6, 7]. To overcome the difficulty of introducing photoproteins into the cytoplasm of smaller mammalian cells, some strategies have been devised [4, 5]. In neutrophils the photoprotein obelin has been introduced into the cytosol by fusion with erythrocyte ghosts [8, 9] or liposomes [10] containing the protein. However, the amount thus introduced is small, and the total photon emission too low to allow single-cell analysis of the timing and subcellular distribution.

---

*$[Ca^{2+}]_i$ *denotes* cytosolic free Ca^{2+}; fMLP, fMet-Leu-Phe; **ex**, excitation; AM, acetoxymethyl ester (of the probe fura 2).
[@]Vol. 19 [see refs. list, 1.] has arts., e.g. by Campbell, pertinent to the present art.- *Ed.*

The development of synthetic fluorescent Ca^{2+} chelators has revolutionized the study of $[Ca^{2+}]_i$ in small mammalian cells [11, 12]: the AM esters, which cytosolic esterases convert to the hydrophilic indicator, are membrane-permeant, overcoming the problem of introduction into small cells. Moreover, the spectral shift with Ca^{2+} binding allows ratio imaging techniques to be applied, enabling $[Ca^{2+}]_i$ signals in small cells to be visualized.

Principles of $[Ca^{2+}]_i$ measurement using fura 2

Unlike quin 2 and some other Ca^{2+}-sensitive fluorescent indicators, fura 2 is significantly fluorescent in both the Ca^{2+}-bound and Ca^{2+}-free forms; but the **ex** spectrum depends on its Ca^{2+} saturation [13]. When Ca^{2+}-saturated, its **ex** spectrum peaks at 340 nm, while a decrease in saturation causes a shift towards longer wavelengths. The spectral shift on binding Ca^{2+} shows a characteristic 'cross-over' point near 360 nm where the emission intensity is independent of the Ca^{2+} concentration (Fig. 1). These features allow dual wavelength measurements to be performed, a feature essential for single-cell imaging and overcoming difficulties (see below) in single-wavelength measurement.

A common strategy is to determine emission intensities at 2 **ex** wavelengths – giving a 340/380 ratio (**R**) – which give maximal deviations as the indicator shifts from Ca^{2+}-bound to Ca^{2+}-free form. From mass-action considerations,

$$[Ca^{2+}]_i = Kd.\beta.(R - R_{min})/(R_{max} - R)$$

where Kd = dissociation constant, β = ratio of 380 nm intensities in the absence of Ca^{2+} and with maximum Ca^{2+} binding, and $_{min, max}$ are ratio values from fura 2 with minimum and maximum Ca^{2+} binding [13]. Since the Kd for Ca^{2+} of fura 2 is ~200 nM, the indicator will be between 9% and 90% saturated with Ca^{2+} at ~20-2000 nM free Ca^{2+}. The indicator is thus ideally suitable for measuring changes in $[Ca^{2+}]_i$ which are often within this range.

Besides Kd the key parameters are the values of R_{max} and R_{min} – the extreme achievable ratio values obtained from the indicator in the totally Ca^{2+}-saturated and totally Ca^{2+}-free forms. One approach is to measure these values in solutions thought to mimic the intracellular environment. This approach, however, may not take into account all intracellular components which influence the ratio values, e.g. the local microviscosity or spectral absorption of the cytoplasm. We routinely find R_{max} values that are observed within individual neutrophils to be lower than from external calibration procedures. This may be due to the influence of the intracellular environment, perhaps reflecting local viscosity where

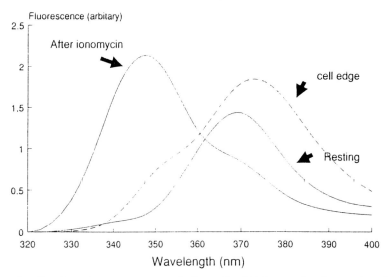

Fig. 1. Excitation spectra of fura 2 within a single neutrophil at rest and after ionomycin treatment to elevate $[Ca^{2+}]_i$. The characteristic wavelength shift is seen with elevated $[Ca^{2+}]_i$, although the spectra are distorted at shorter wavelengths by the glass optics of the microscope. The *interrupted line* shows the ex spectrum for a region near the cell edge: the ratio values (resting) are artificially low. After elevation of $[Ca^{2+}]_i$ the spectrum of this region shifts to that seen throughout the cell ('After ionomycin').

the cytosol contains polymerized actin [14] or due to 340 nm absorbtion by endogenous NADH and flavin nucleotides at mM levels. Since with $[Ca^{2+}]_i >20$ μM the indicator will be >99% saturated, this value can be recorded by elevating $[Ca^{2+}]_i$ using a non-fluorescent ionophore such as ionomycin or Br-A23187. The value for R_{min} can be obtained subsequently by adding EGTA to lower $[Ca^{2+}]_i$ to <2 nM, when 99% of the fura 2 will be in the Ca^{2+}-free form. Agents such as digitonin which totally disrupt the plasma membrane are inappropriate for imaging because the influence of the intracellular environment may be lost and the indicator will be lost into the medium and hence cell imaging prevented.

Necessity for ratio measurements in the determination of the Ca^{2+} distribution in single neutrophils

Single-wavelength measurements may serve to determine the 'representative' $[Ca^{2+}]_i$ signal in a cell population, e.g. using quin 2 and fluo 3 [11, 12, 15]. However, single-cell imaging is precluded because of several problems, notably because the fluorescence intensity of the cellular indicator depends not only on its Ca^{2+} saturation but also on its

amount.   When loading is achieved with the AM ester, the
amount of indicator within each cell may be different (Fig. 2).
Also, the 3-D shape of an individual cell may produce regions
of cytoplasm of differing thickness and hence the pathlength
of the **ex** light will depend on the cellular location within
an individual cell. This is a particular problem with neutrophils which 'spread out', forming pseudopodia and presenting
various 2-D shapes with, for example, the cytoplasmic thickness
in the extended pseudopod being markedly less than that in
the cell body.   Thus it would not be possible to distinguish
cells or areas within a cell in which $[Ca^{2+}]_i$ was high
from cells or areas within cells in which the amount of
indicator was high.

The use of two **ex** wavelengths, however, overcomes these
problems.   One wavelength could be set to the Ca^{2+}-insensitive
point of the spectrum (~360 nm) to provide a monitor of
the amount of fura 2 in an individual cell or part of a
cell.   By taking the ratio of the intensity emitted by
ex at this wavelength to that with a Ca^{2+}-sensitive wavelength,
the signal will be independent of the amount of fura 2.
In practice it is more useful to take two wavelengths which
are Ca^{2+}-sensitive but which change in opposite directions.
The ratio of these signals will be independent of fura 2
amount, but the magnitude of the ratio change will be larger.
This may be particularly important with microscopic imaging
where transmission losses at the shorter wavelengths (Fig. 1)
may reduce the dynamic range of the ratio.

Two-wavelength signal recording also enables fluorescence
changes not caused by $[Ca^{2+}]_i$ changes to be excluded.   Since
the Ca^{2+}-dependent changes in the signals at each wavelength
are mathematically related, it is possible to exclude fluorescent
changes which result from 'baseline shifts' due to changes
in autofluorescence, cellular absorbance and other mechanisms
unrelated to Ca^{2+} ([15, 16] & see below).

Loading fura 2 into neutrophils

Optimum uptake of fura 2/AM (which then yields the free
acid) is achieved after 30 min by adding stock fura 2/AM
(in dimethyl sulphoxide with Pluronic present [15]) to
neutrophil suspensions containing $>10^6$ cells/ml at room
temperature.   Fura 2 is produced within the cytosol [16]
with uniform fluorescence throughout (Fig. 3, a & b).   Under
some circumstances and with some cell types, fura 2 may be
generated in organelles within the cell [17-19].   This
possibility may be excluded by microscopic examination and
subcellular fractionation [16]. Fluorescence anisotropy measurements suggest that fura 2 within neutrophils has restricted

Fig. 2. Population distribution of fura 2 loaded as fura 2/AM. The 360_{ex} intensity served as a $[Ca^{2+}]_i$-independent measure of $[fura\ 2]_i$. The distribution is unimodal and approximately normal. *Inset:* the distribution of fluoresceinated-bovine serum albumin loaded into neutrophils by liposomal fusion, showing preferential uptake by a sub-population of cells.

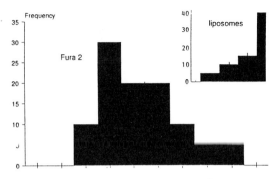

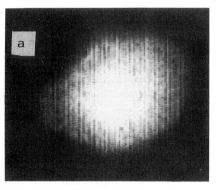

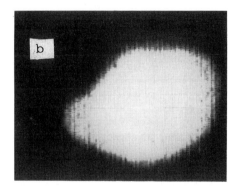

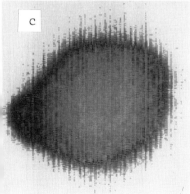

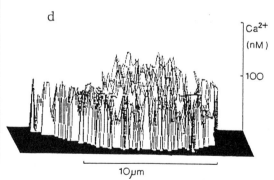

Fig. 3. Measurement and manifestation of $[Ca^{2+}]_i$ in an individual neutrophil using fura 2: fluorescence with 350 (**a**) or 380 (**b**) nm ex, ratio of images a/b (**c**), and $[Ca^{2+}]_i$ intracellular distribution as a '3-D map' (**d**). *Note* uniformity of fura 2 loading without obvious accumulation in organelles (**a, b**), ratio-image uniformity (no 'bull's-eye' pattern typical of uncorrected intensity-dependent ratio error), and artefactual ring of low ratio near the cell periphery.

rotational mobility, explicable if the intracellular environment has a 'micro-viscosity' greater than water but less
than specific granules [16]; so conceivably fura 2 may become
associated with cytosolic proteins within the neutrophil.

It has been reported that some batches of fura 2/AM
are incompletely deesterified, generating a fluorescent but
Ca^{2+}-insensitive product which is sufficiently hydrophilic
to become entrapped within the cell [20]. If the proportion
of partially esterified fura 2 is high, this can be readily
checked from the **ex** spectra of the indicator in the presence
and absence of Ca^{2+}; if the proportion is low, its amount
can be calculated from its Ca^{2+}-insensitivity [15]. This
can then be treated as 'background' autofluorescence and
subtracted from the signals at each wavelength before calculating the ratio.

A possible problem is leakage of fura 2 from the cells,
reducible in some cells by using anionic blockers [21]. This
problem does not arise in neutrophils, particularly if stored
at 0° after loading [16]. Since leaked fura 2 will be totally
saturated with Ca^{2+}, $[Ca^{2+}]_i$ in a cell population containing
leaked fura 2 will tend to be overestimated. With imaging
techniques the spatial resolution provides a 2-D 'Ca^{2+} map'
which thereby excludes extracellular fura 2 and thus may
provide an estimate of $[Ca^{2+}]_i$ lower than those obtained
in cell suspensions. In the cell population, however, the
contribution to the total signal of 10% of the fura 2 in
an extracellular space will result in only a 36% overestimation
of $[Ca^{2+}]_i$ [15].

Imaging of fura 2 ratio signals from neutrophils

Practical aspects of ratio imaging have been excellently
reviewed [e.g. 22, 23]. There are essentially 3 elements of
an imaging system for fura 2 fluorescent signals: (1) the
ex system, which provides alternately the illumination at
the appropriate wavelengths, (2) the microscope and camera
to acquire the 'raw' images, and (3) the digitising and processing system, which allows computation of the 'raw' data after
conversion of the video images into digital pixels.

(a) Excitation system

Amongst possible **ex** systems, to provide light at the
two **ex** wavelengths, two are currently used: (1) a rotating
filter placed in front of the fluorescent **ex** lamp, and (2) a
rotating optical chopper alternating the outputs from two
monochromators. While (1) is cheaper and may provide more
illumination, (2) has the advantage of allowing flexibility
in wavelength choice and spectral analysis. With two lamps

and two monochromators it is also possible to balance the illumination intensities at either wavelength to provide equal illumination at each wavelength and give the maximum range of ratio values. With either system it is obviously important for maximum ratio shifts that there be no significant spectral overlap of the **ex** sources. The filter-wheel or optical-chopper rotation is usually controlled by software which allows complete video frames to be taken at each wavelength without inclusion of frames acquired during the changing between wavelengths. The rate of change between wavelengths thus cannot exceed video rate (30 or 25/sec, U.S.A. or European standards respectively), which thus limits the maximal temporal resolution of the system to ~10 ratio images/sec. This rate is rarely achievable, 1 ratio image/sec being more realistic. With neutrophils the $[Ca^{2+}]_i$ signal is slow, with time courses up to 2 min; thus a slower rate is acceptable.

(b) Microscope and video camera

Since the microscope provides the 'raw' image data, it is arguably the most important single element of the system. Both conventional and epi-fluorescent microscopes have been used. Inverted microscopes give easier access to the cells under observation and thus allow other manipulations to be performed. However, there is the penalty of capture of stray ambient light that would contaminate the acquired images (particularly if used with the highly light-sensitive camera; see below). High-resolution short working distance objectives may also be more difficult to use with inverted microscopes. Ideally the microscope optics should transmit and the dichroic mirror reflect both wavelengths equally with high efficiency. Quartz optics are required to transmit 340 nm light, although glass optics transmit sufficient light to allow use of 350 nm instead (cf. Fig. 1).

The video camera, probably the next most important element of the imaging system, has 3 important desirable features: low image persistence, high light sensitivity and light-response linearity. Image persistence after switching to the second wavelength will obviously invalidate ratio-value calculations. Most conventional video cameras have significant 'lag' or after-image, whereas CCD cameras exhibit low image persistence. With high contrast images, e.g. a 'bright' fura 2-loaded cell against a black background, CCD may also surpass conventional cameras in which the bright object tends to 'flare' into the surroundings. High light-sensitivity is desirable for several reasons.- (i) $[fura 2]_i$ must be minimized to avoid excessive Ca^{2+}-buffering effects [15, 18]. (ii) High illumination intensities, best obviated, may cause photobleaching of the indicator and lead to production of toxic radicals. (iii) High sensitivity allows fast acquisition of usable ratio images, for which the number of frames to be averaged constrains data-acquisition speed.

Since the strength of the video signal from the camera is used quantitatively in calculating $[Ca^{2+}]_i$, it is essential that the signal increase linearly with light intensity. In resting near-spherical neutrophils the intensity of the fura 2 signal will be greatest in the centre and diminish towards the cell edge as cell thickness decreases. Video non-linearity will result in an intensity-dependent error. The resultant pattern, often described as a bulls-eye [22], cannot be taken as evidence for a $[Ca^{2+}]_i$ gradient. The absence of such a pattern, however, suggests camera-response linearity (Fig. 3c).

(c) Digitising and processing system

Technical aspects of assembling such systems from commercially available components have been discussed [e.g. 22, 23]. Packages are also now available (e.g. Spex IM system, Joyce-Loebl Magical system) for acquiring and digitising the images and processing the data. Essentially, the digitising system converts the analogue video signal into a digital signal, the image being divided into individual picture elements (pixels) each with a numerical value. The processing system is essentially software which enables digitised images to be analyzed and mathematically manipulated.

Presenting and recording image data

From the original 'raw' digitised fluorescent images, ratio or Ca^{2+} data can be obtained by mathematical manipulation of each pixel value to produce a new pixel array, each pixel having a numerical value (usually 0-255). These data are essentially a map of the Ca^{2+} concentrations in the field. They can be presented visually by assigning a colour or shade to each numerical value, using a 'look-up table'. A pseudo-colour image is thus generated where regions of similar colour represent regions of similar $[Ca^{2+}]_i$. Since the choice of colours used is arbitrary (and varies between different workers), a colour stripe must also be provided to enable the viewer to know whether, for example, 'green' represents a higher or lower $[Ca^{2+}]_i$ value than 'yellow'.

An alternative is to produce a pseudo-grey image, where black indicates Ca^{2+} absence and white the presence of saturating Ca^{2+} (>500 nM). This produces an immediately intelligible image with no need to refer to a colour scale, although a stripe will be required to read off the approximate $[Ca^{2+}]_i$ value from the grey level. This approach has the advantage that the image may be reproduced cheaply in journals, but has the disadvantage that, since the eye distinguishes different levels of grey less readily than different colours, it becomes difficult to show small changes dramatically. A further problem with pseudo-grey images is choosing a level of grey for the extracellular areas, since as the cell approaches this grey level it merges with the background.

The numerical array can also be represented as a 3-D plot, with the x- and y-axes in the horizontal planes being lengths and the z-axis in the vertical plane being $[Ca^{2+}]_i$ (Fig. 3d). This is achievable using commercially available software or by reading off lines of the image data into a plotting program.

The amount of information present in a ratio image, ~0.25 Mbytes, can cause problems with storage. Images can be recorded in video buffer memory before loading onto hard disc or tape streamer for analysis and processing. An alternative approach may be to store the video signals in analogue form on video tape for subsequent digitisation and analysis.

Interpretation of fura 2 ratio images

Potential problems of two kinds may arise when interpreting fura 2 ratio images: those associated with the indicator and those caused by the imaging. The former, discussed elsewhere [e.g. 15], include distortion of the $[Ca^{2+}]_i$ signal caused by the buffering capacity of the fura 2, interference from other ions, and the effect of local viscosity [14, 15, 24].

Confidence in fura 2 as a $[Ca^{2+}]_i$ indicator in neutrophils comes from the accord with values obtained with a Ca^{2+}-sensitive photoprotein [9, 16]. In neutrophil populations loaded with both fura 2 and obelin, there was good agreement in time-course and $[Ca^{2+}]_i$-signal values (Fig. 4). Chloroadenosine, however, had little effect on the fura 2 signal but totally abolished the obelin signal (Fig. 4). This discrepancy may have resulted from the liposomal method used to load the obelin, uptake with this method being restricted to a sub-population of neutrophils (Fig. 2). This emphasizes the problem of inter-preting Ca^{2+} signals measured from cell populations, and illustrates the need for measurements to be performed in individual cells.

Perhaps the biggest problem in using fura 2 for imaging $[Ca^{2+}]_i$ signals is posed by the determination of the non-fura 2 fluorescent signals at each wavelength from the cell and as a background. These signals must be subtracted from the total fluorescent signals at each wavelength before producing a ratio. If unsubtracted background signals are large compared with the fura 2 signals, then as the cell thickness decreases towards the edge and the intensity of the cellular fura 2

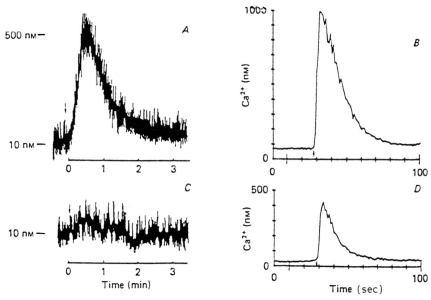

Fig. 4. Comparison of $[Ca^{2+}]_i$ signals in a neutrophil population loaded with the fura 2 (as AM ester) and obelin (liposomal
uptake) jointly. A,C: obelin signals. B,D: fura 2 signals
after stimulation with 1 μM fMLP. In C & D 10 μM 2-chloroadeno-
sine was present.

signal diminishes, the contribution from the background escalates.
This would produce a systematic error which increases as
the cell edge is approached, giving rise to the characteristic
bull's-eye pattern.

Obtaining suitable quantitatively and spatially correct
autofluorescent images poses a problem. Images of the cell
could be taken at each wavelength before fura 2 loading and
subsequently subtracted from the fura 2-loaded images before
ratio calculation. This, however, is not feasible for neutrophils since changes in the cell shape which may occur during
loading would invalidate the exercise. Moreover, it must
be assumed that the autofluorescence from the neutrophils,
resulting mainly from NADPH and flavin nucleotides, does
not change during loading or during an experiment in which
the cell is stimulated. It is unlikely that either assumption
will hold for neutrophils, since fura 2 loading can reduce
their ATP level by 50% [16] and activation of the neutrophil
oxidase which accompanies stimulation can cause large changes
in the NADPH concentration within the cell [25].

The autofluorescent level from the cell can be calculated
at the end of the experiment with fura 2-loaded cells, by
ionomycin treatment followed by EGTA [15], and it could be
subtracted from all preceding images before calculating

ratios. This overcomes the problem of shape change and possible autofluorescence change during loading, but again gives data for only a single time-point. The solution we have adopted is to subtract non-changing uniform background signals and use sufficient fura 2 to make the contribution from the changing cellular autofluorescence negligible. Hence we use 50-100 µM fura 2. The assumption that the changes in autofluorescence are negligible can be tested by summation of the signals at each wavelength (after multiplication by the appropriate proportionality constant) to furnish a signal proportional to the total fluorescence being recorded [15, 16]:- $A.F_{s1} + F_{s2}$ is constant, the term A being $(F_{f2} - F_{b2})/F_{b1} - F_{f1})$, where F is the fluorescent signal under the condition of the subscripts s, f and b referring to the experiment, the Ca^{2+}-free and the Ca^{2+}-bound, 1 and 2 referring to ex wavelengths 350 and 380 nm respectively. Any significant change in autofluorescence would cause this parameter to deviate from its initial value (amplification in [15]).

Accurate conversion of the ratio signals into $[Ca^{2+}]_i$ values requires a number of parameters, especially actual determination of R_{max} and R_{min}; for neutrophils, as discussed earlier, mere use of a 'look-up table' may be unsafe.

Light-microscopic resolution, ~0.2 µm (at 510 nm with aperture 1.6), limits the precision with which ratio images can be resolved. However, since this image is a 2-D representation of the 3-D cell, interpretation of the spatial relationships between features in the image needs care. With non-focal conventional fluorescence microscopy, it must be remembered that the signals from an 'area' of the cell include signals from the whole volume of cytoplasm above and beneath that area. Also, non-confocal fluorescence is by its nature 'out-of-focus', even if the camera introduces no additional 'blur' to the image. This leads to imprecision in defining the spatial nature of some Ca^{2+} signals. Thus, the image of a $[Ca^{2+}]_i$ 'cloud' which is seen in ~30% of neutrophils stimulated by fMLP, in the absence of extracellular Ca^{2+} and presumably due to release from an intracellular Ca^{2+} store, occupies an area of ~2 µm. Yet precision about the cloud's true dimensions is precluded, except that it is probably <2 µm. Nor can its 3-D shape be determined, or the volume of cytosol with elevated $[Ca^{2+}]_i$. Also, possibly only 30% of neutrophils examined were in a focus or configuration optimal for visualising the cloud.

An artefact we have observed in neutrophils examined under high magnification is a ring of cytosol with low ratio values near the cell edge, but separated from it by a region with normal resting $[Ca^{2+}]_i$ (Fig. 3d). This artefact is transient, disappearing as $[Ca^{2+}]_i$ rises either locally or

uniformly throughout the cell.   The **ex** spectra of the fura 2
in this region *vs*. the rest of the cell are not consistent
with interpretation as a region of lowered $[Ca^{2+}]_i$, since
the characteristic   spectral   shift with 'cross-over' and
decreased peak intensity (Fig. 1) is not seen.   Instead,
the decreased ratio in this region is seen to result from
an enhancement of fluorescence at the longer *vs*. the shorter
wavelengths.   A similar spectral distortion has been observed
with increasing viscosity [14, 23].   It is thus possible
that this cortical region of the cytoplasm has elevated micro-
viscosity, perhaps due to the cortical actin network which
exists in these cells. This would also explain its disappearance
with elevated $[Ca^{2+}]_i$, since actin disassembly may be triggered
under these conditions.   Care must therefore be taken in
interpreting changes in the fura 2 ratio, particularly in
neutrophils where gel-sol transitions occur during chemotaxis.
For example, does the low ratio seen in the forming pseudopod
of the neutrophil shown in Fig. 3 reflect low $[Ca^{2+}]_i$ or
increased micro-viscosity?   Spectral analysis of this region
would be required before attempting to answer such a question.

Some discoveries made by fura 2 imaging of individual neutrophils

Several such discoveries have implications for our understanding of signalling processes in neutrophils. Thus, imaging
has established that the asynchrony and heterogeneity of
the exocytosis and activation of the NADPH-oxidase [26, 27]
arise from variability in the generation of the $[Ca^{2+}]_i$ signal.
$[Ca^{2+}]_i$ rose throughout the neutrophil cytoplasm in one of
three ways: (1) a fast transient rise occurring within 6 sec;
(2) a sustained oscillating rise; or (3) a rise occurring
after a delay period of up to 56 sec [28]. These differences
were not governed by fura 2-loading discontinuities or by
when the stimulus arrived at the cell.   The changes in
$[Ca^{2+}]_i$ were correlated with oxidase activation, assessed by
detection of formazan deposited through reduction of Nitro
blue tetrazolium.   Whatever the timing or pattern (transient
or oscillating) of the $[Ca^{2+}]_i$ rise, the oxidase always became
activated in those neutrophils in which $[Ca^{2+}]_i$ rose to >250 nM.

In the absence of extracellular Ca^{2+}, a localized Ca^{2+}
'cloud' was observed in ~30% of cells (having $[Ca^{2+}]_i$ up
to ~500 nM); it was confined to a single region of the
cytosol.   By also using fluorescent stains for the nucleus,
mitochondria and endoplasmic reticulum (e.r.), the location
of this store was shown to be between the nuclear folds
and the plasma membrane (p.m.) and at a site which also
stained with the e.r. marker [29]. The location of this store
did not accord with the distribution of 'calciosomes'.   As
observed in the presence of extracellular Ca^{2+}, the release
from the store by fMLP occurred after variable delays, up

to 120 sec, and was transient: $[Ca^{2+}]_i$ declined to the resting level without oscillation. When $[Ca^{2+}]_i$ near the p.m. exceeded 250 nM, a localized activation of the NADPH-oxidase occurred, as demonstrated by reduction of Nitro blue tetrazolium [29].

An intriguing feature observed in the absence of extra-cellular Ca^{2+} was a cytoplasmic region, between the Ca^{2+} cloud and the site of localized oxidase activation, which remained at the resting $[Ca^{2+}]_i$ level. This suggested that there was an intermediary between the region of elevated $[Ca^{2+}]_i$ and the site of oxidase activation at the p.m., possibly a Ca^{2+}-binding protein, a kinase or a component of the oxidase.

It is unlikely that any of these phenomena would have been discovered without the ability to image the $[Ca^{2+}]_i$ signal in individual neutrophils. This approach is likely to lead to further discoveries of importance to the cell signalling process.

Acknowledgements

We are grateful to the Arthritis and Rheumatism Council (U.K.) for supporting the work presented.

References

1. Campbell, A.K. (1983) *Intracellular Calcium: its Universal Role as Regulator*, Wiley, Chichester, 556 pp.
   Also (added by Ed.): (1989) in *Biochemical Approaches to Cellular Calcium* [Vol. 19, this series] (Reid, E., *et al.*, eds.), Roy. Soc. of Chem., Cambridge, pp. 1-14. *Other arts. also relevant.*
2. Hallett, M.B. & Campbell, A.K. (1984) *Cell Calcium* 5, 1-19.
3. DiVirgilio, F., Stendahl, O., Pittet, D., Lew, D.P. & Pozzan, T. (1990) in *Current Topics in Membranes and Transport*, Vol.15, Academic Press, New York, pp. 179-205.
4. Hallett, M.B. & Campbell, A.K. (1982) in *Chemical and Biochemical Luminescence* (Kricka, L.J. & Carter, T.J.N., eds.), Marcel Dekker, New York, pp. 89-152.
5. Campbell, A.K., Dormer, R.L. & Hallett, M.B. (1985) *Cell Calcium* 6, 69-82.
6. Gilkey, J.C., Jaffe, L.F., Ridgway, E.B. & Reynolds, G.T. (1978) *J. Cell Biol.* 76, 448-466.
7. Rose, B. & Lowenstein, W.R. (1975) *Nature* 254, 250-252.
8. Hallett, M.B. & Campbell, A.K. (1982) *Nature* 295, 155-158.
9. Campbell, A.K. & Hallett, M.B. (1983) *J. Physiol.* 338, 537-550.
10. Campbell, A.K., Patel, A.K., Razavi, Z.S. & McCapra, F. (1988) *Biochem. J.* 252, 143-147.
11. Tsien, R.Y., Rink, T.J. & Poenie, M. (1985) *Cell Calcium* 6, 145-157.

12. Tsien, R.Y. (1989) *Annu. Rev. Neurosci. 12*, 227-255.
13. Gryniewicz, G., Poenie, M. & Tsien, R.Y. (1985) *J. Biol. Chem. 260,* 3440-3450.
14. Poenie, M. (1990) *Cell Calcium 11*, 85-92.
15. Hallett, M.B., Dormer, R.L. & Campbell, A.K. (1990) in *Peptide Hormones: a Practical Approach* (Siddle, K. & Hutton, J.C., eds.), IRL Press, Oxford, pp. 113-148.
16. Al-Mohanna, F.A. & Hallett, M.B. (1988) *Cell Calcium 9*, 17-26.
17. Taniguchi, S., Marchetti, J. & Morel, F. (1989) *Pflügers Arch. 414*, 125-133.
18. Almers, W. & Neher, E. (1985) *FEBS Lett. 192*, 13-18.
19. DiVirgilio, F., Steinberg, T.H. & Silverstein, S.C. (1990) *Cell Calcium 11*, 57-62.
20. Scanlon, M., Williams, P.A. & Fay, F.S. (1987) *J. Biol. Chem. 262*, 6308-6312.
21. DiVirgilio, F., Steinberg, T.H., Swanson, J.A. & Silverstein, S.C. (1987) *J. Immunol. 140*, 915-920.
22. Tsien, R.Y. & Harootunian, A.T. (1990) *Cell Calcium 11*, 93-110.
23. Moore, E.D.W., Becker, P.L., Fogarty, K.E., Williams, D.A. & Fay, F.S. (1990) *Cell Calcium 11*, 157-180.
24. Roe, M.W., Lemasters, J.J. & Herman, B. (1990) *Cell Calcium 11*, 63-74.
25. Selvaraj, R.T. & Sbarrra, A.J. (1967) *Nature 211*, 1272-1275.
26. Hallett, M.B. (1985) *Biochim. Biophys. Acta 847*, 15-19.
27. Patel, A.K., Hallett, M.B. & Campbell, A.K. (1987) *Biochem. J. 248*, 173-180.
28. Hallett, M.B., Davies, E.V. & Campbell, A.K. (1990) *Cell Calcium 11*, 655-663.
29. Davies, E.V., Hallett, M.B. & Campbell, A.K. (1991) *Immunology*, in press.

#D-2

SPATIAL ASPECTS OF THE CALCIUM SIGNAL THAT TRIGGERS EXOCYTOSIS REVEALED BY IMAGING TECHNIQUES

[1]Timothy R. Cheek[†], [2]Robert D. Burgoyne
and [1]Michael J. Berridge

[1]AFRC Laboratory of Molecular Signalling,
Department of Zoology, Downing Street, Cambridge CB2 3EJ, U.K.

[2]Department of Physiology, University of Liverpool,
Brownlow Hill, P.O. Box 147, Liverpool L69 3BX, U.K.

New insights into the role of Ca^{2+} ions in triggering the exocytotic secretion of catecholamines from the chromaffin cells of the mammalian adrenal gland are beginning to emerge from experiments in which the rise in $[Ca^{2+}]_i$ can be visualized and secretion monitored simultaneously from the same cell. These results show that different classes of stimuli give rise to different spatial localizations of intracellular Ca^{2+}, and that only one specific pattern of internal Ca^{2+} is associated with a full secretory response.- Potent (depolarizing) secretagogues cause Ca^{2+} to be localized initially to the subplasmalemmal region of the cell and exocytosis to be initiated from this same cell region. In response to weaker ($InsP_3$-mobilizing) secretagogues, the rise in $[Ca^{2+}]_i$ develops from only one specific area within the cell and then exists as a continuous gradient at the peak of the response. This triggers either no secretion or a secretory response that is polarized to the cell area in which the $[Ca^{2+}]_i$ rise is largest. Imaging studies on Fura 2 quenching by Mn^{2+} entry suggests that the polarized secretion may be due to a localized influx of external Ca^{2+} that accompanies and/or follows Ca^{2+} mobilization. These results strongly suggest that exocytosis from adrenal chromaffin cells is triggered by Ca^{2+} activation of the subplasmalemmal region of the cell, and that this is only achieved by the promotion of Ca^{2+} influx.*

Exocytosis, the process by which intracellular vesicles fuse with the inner surface of the p.m.* and release their contents into the surrounding medium, is the mechanism underlying the secretion of many physiologically important mediators such as hormones, enzymes and neurotransmitters. The process is often regulated by an external signal which stimulates release by altering the level of an intracellular second messenger.

[†]addressee for any correspondence

*Abbreviations.- ACh, acetylcholine; ANG II, angiotensin II; $[Ca^{2+}]_i$, intracellular free Ca^{2+} concentration; DBH, dopamine β-hydroxylase (EC 1.14.17.1); InsP, an inositol phosphate, $InsP_3$ usually signifying $Ins(1,4,5)P_3$; PBS, phosphate-buffered saline; p.m., plasma membrane. Fura 2 is a fluorescent probe.

In the chromaffin cells of the mammalian adrenal gland exocytosis is triggered by ACh that is released from the splanchnic nerve terminal.   Activation of the cell surface nicotinic receptors results in depolarization, the opening of voltage-dependent Ca^{2+} channels, and a rise in cytosolic Ca^{2+} that is due to an influx of extracellular Ca^{2+} [1]. Although secretion from these cells is also controlled by protein kinase C [2], a guanine nucleotide-binding protein [3] and an unknown second messenger that mediates disassembly of the cortical cytoskeleton [4], Ca^{2+} is an absolute requirement for exocytotic fusion to occur as micromolar Ca^{2+} is sufficient to stimulate secretion from permeabilized chromaffin cells [5, 6].

In addition to nicotinic receptors, chromaffin cells also possess receptors for muscarinic agents and the neuro-peptides ANG II and bradykinin [7, 8]. These receptors activate phospholipase C, release $InsP_3$ and then mobilize Ca^{2+} from internal stores that probably reside somewhere in the endoplasmic reticulum (see [9] and arts., e.g. by A.P. Dawson, in a previous vol. [10]).   In various secretory cells this is a major pathway leading to secretion [11]. In bovine chromaffin cells, however, these agonists trigger little if any secretion from cell populations despite raising $[Ca^{2+}]_i$ [7, 8, 12]. Since ACh and nicotine trigger secretion by raising $[Ca^{2+}]_i$ to ~200 nM above basal, it has been assumed that the modest rise in cell-population $[Ca^{2+}]_i$ seen in response to $InsP_3$-mobilizing stimuli (15-98 nM above basal) is insufficient to effectively trigger exocytosis [12-14].   However, recent reports suggest that this assumption is wrong.   Firstly, monitoring of $[Ca^{2+}]_i$ in single cells has shown that some chromaffin cells can respond to $InsP_3$-mobilizing stimuli with rises in $[Ca^{2+}]_i$ that approach the magnitude of those seen in response to nicotine, whereas others show no rise in $[Ca^{2+}]_i$ [15]; cell population measurements do not therefore accurately reflect the Ca^{2+} changes that are occurring at the single-cell level. Secondly, direct activation of phospho-lipase C by intracellularly applied GTP-γS resulted in large Ca^{2+}-transients (>400 nM) but negligible secretion from a chromaffin whole-cell patch [16].

It is, then, unknown why $InsP_3$-mobilizing stimuli are ineffective in triggering secretion from these cells.   It is unlikely that these stimuli result in activation of some pathway that is inhibitory to exocytosis at a step distal to the rise in $[Ca^{2+}]_i$ since muscarinic receptor activation does not inhibit nicotine-induced secretion [12, 17].

In order to gain further insight into why Ca^{2+} influx, but not mobilization of internal Ca^{2+}, appears to trigger

full secretion from these cells, fluorescence imaging techniques have been used with cells loaded with the Ca^{2+}-sensitive
fluorescent dye fura 2 in order to compare the spatial distribution of the Ca^{2+} signal induced by nicotinic and other
depolarizing stimuli with that induced by $InsP_3$-mobilizing
stimuli. The present article amplifies and updates the
account we gave in a previous volume [10].

The results show that the distribution of intracellular
Ca^{2+} induced by depolarizing stimuli is markedly different
from that induced by $InsP_3$-mobilizing stimuli. Potent (depolarizing) otimuli rcoultcd in an immodiato olovation of Ca^{2+}
throughout the entire subplasmalemmal region, whereas $InsP_3$-
mobilizing stimuli resulted in Ca^{2+} being largely confined
to one pole of the cell. In most cells this restricted
localization of Ca^{2+} was not capable of triggering or maintaining
exocytosis as assessed by a novel co-culture technique (see
METHODS). A minority of chromaffin cells, however, responded
with a secretory response that was polarized to the area
of the p.m. that was activated by the Ca^{2+}. Studies on
fura 2-quenching by Mn^{2+} indicate that this may have been
due to a localized entry of external Ca^{2+}.

METHODS

Chromaffin cells were isolated from bovine adrenal medullas
by enzymatic digestion in Krebs-Ringer buffer as previously
described [3, 12, 18].

Measurement of $[Ca^{2+}]_i$ in cell populations (freshly isolated)
loaded at room temperature with 1 μM fura 2/AM (acetoxymethyl ester;
30 min and, after washing, a further 20 min) was done by
continuous monitoring at 37° with magnetic stirring, in the
cuvette of a Perkin-Elmer LS-5 Spectrophotometer (ex.: 340 nm;
em.: 510 nm). The suspending medium for the loading, and for
re-suspending the cells after centrifuging (1000 rpm, 5 min) after
loading, was Krebs-Ringer buffer containing 3 mM $CaCl_2$ [18].

Measurement of catecholamine secretion from cell populations, freshly isolated, was performed as described elsewhere [12].

Measurement of $[Ca^{2+}]_i$ in single cells entailed loading
primary cultures with 2 μM fura 2/AM for 30 min at room temperature, washing, equilibration to 37° for 3 min, and then
immediate imaging as described [15, 18]. Fluorescent images
were obtained by excitation at 340 nm and 380 nm alternately
(40 m-sec at each wavelength) using an image-processing system
(Imagine; Synoptics, Cambridge, U.K.) interfaced to a DEC
Microvax II microcomputer. The ratio image was obtained
at constant video rate and filtered with a time constant

of 200 m-sec (i.e. 5 ratio images/sec). The 3-D plots were generated by Imagine from the ratio image, and depict either the resting distribution of $[Ca^{2+}]_i$ or the distribution and qualitative rise in $[Ca^{2+}]_i$ elicited by the stimulus.

Measurement of secretion from single cells.- In order to simultaneously measure secretion and $[Ca^{2+}]$ the chromaffin cell was co-cultured with fura 2-loaded NIH-3T3^t fibroblasts [19]. These cells do not respond directly to certain chromaffin cell secretagogues but give a strong rise in $[Ca^{2+}]_i$ in response to ATP which is co-released with catecholamine from the chromaffin cell. A rise in $[Ca^{2+}]_i$ in the NIH-3T3^t cells therefore indicates a secretory response from the chromaffin cell.

Immunofluorescence localization of exocytotic sites was carried out as described [19]. Cells in primary culture were incubated with the agonist for 15 min at 37° in the presence of an antibody to the chromaffin granule-membrane protein (DBH; 1/800). Cells were washed and then fixed in 4% (w/v) formaldehyde in PBS overnight, and washed in PBS. Sequential incubations were performed: with 0.3% bovine serum albumin in PBS for 30 min; with anti-rabbit biotin (1/100) for 60 min; with Texas Red-linked streptavidin (1/50) for 30 min. Cells were then mounted and photographed.

RESULTS

$[Ca^{2+}]_i$ and secretion in cell populations

Addition of nicotine to 10 μM (maximal dose [7, 12]) to a population of fura 2-loaded chromaffin cells resulted in a rise in $[Ca^{2+}]_i$ to ~250 nM above basal (Fig. 1a) and a strong secretory response (Fig. 1b), whereas addition of methacholine to 0.3 mM (maximal dose [12, 13]) gave a more modest rise in $[Ca^{2+}]_i$ to ~50 nM above basal (Fig. 1a) and no detectable secretory response above basal (Fig. 1b).

To test whether the non-stimulation of secretion by methacholine was due to the modest $[Ca^{2+}]_i$ rise being insufficient to trigger the exocytotic machinery [13, 14], the concentration of nicotine was titrated down to give a comparably modest elevation in $[Ca^{2+}]_i$. Despite raising $[Ca^{2+}]_i$ by only 58 nM, 1 μM nicotine was able to trigger a significant, albeit reduced, secretory response above basal (Fig. 1c).

$[Ca^{2+}]_i$ and secretion in single cells after depolarizing stimuli

Fura 2 imaging on single cells showed that the Ca^{2+} response to the depolarizing stimulus nicotine (and high $[K^+]$) is generated in two phases (Fig. 2). Initially Ca^{2+}

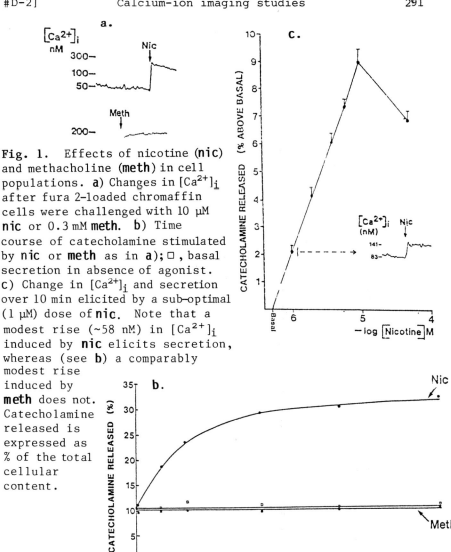

Fig. 1. Effects of nicotine (**nic**) and methacholine (**meth**) in cell populations. **a)** Changes in $[Ca^{2+}]_i$ after fura 2-loaded chromaffin cells were challenged with 10 μM **nic** or 0.3 mM **meth**. **b)** Time course of catecholamine stimulated by **nic** or **meth** as in **a**); □, basal secretion in absence of agonist. **c)** Change in $[Ca^{2+}]_i$ and secretion over 10 min elicited by a sub-optimal (1 μM) dose of **nic**. Note that a modest rise (~58 nM) in $[Ca^{2+}]_i$ induced by **nic** elicits secretion, whereas (see **b**) a comparably modest rise induced by **meth** does not. Catecholamine released is expressed as % of the total cellular content.

is localized exclusively to the entire subplasmalemmal area of the cell. Ca^{2+} then rapidly infills until it is evenly distributed throughout the cell. Using the co-culture technique to monitor secretion, it was found that this Ca^{2+} signal triggered a full secretory response that occurred over the entire surface of the chromaffin cell (Fig. 3). The time course clearly shows that only cell #1, the chromaffin cell, responded initially to the 6-sec application of nicotine. At the peak of the response, the Ca^{2+} was uniformly distributed throughout the cell (Fig. 3; 3-D plot, and cf. Fig. 2).

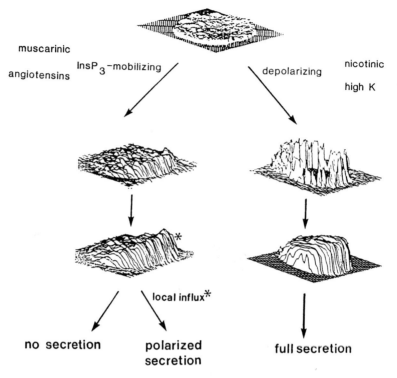

muscarinic

angiotensins   InsP$_3$-mobilizing   depolarizing   nicotinic

high K

local influx*

no secretion   polarized   full secretion
                     secretion

Fig. 2.   The spatial localization of resting and stimulated
[Ca^{2+}]$_i$ in single bovine adrenal chromaffin cells shown in the
form of 3-D plots.   A fura 2-loaded cell of diam. 12 μm is shown
at rest, and at 2.5 and 20 sec after depolarization with 10 μM
nicotine.   This Ca^{2+} signal triggers a secretory response from
the entire surface of the cell *(right)*. *On left:* a cell at 2 and
12 sec after a challenge with muscarine; 65% of cells that
responded with this same Ca^{2+} pattern to ANG II did not secrete,
and the remaining 35% secreted in a polarized manner from the
area of the cell indicated *. This may have been due to a
localized influx of external Ca^{2+} (see text).

Once the application of the nicotine had ceased, there was
no perfusion of medium over the cells.   Subsequently, the
NIH-3T3^t cells (cells #2-#10) responded with a rise in [Ca^{2+}]$_i$
which had a delayed onset that was related to the distance
of the NIH-3T3^t cell from the chromaffin cell, such that
the delay was greatest in those NIH-3T3^t cells that were
furthest from the chromaffin cell.   This is consistent with
ATP being released from the chromaffin cell and then diffusing
to surrounding NIH-3T3^t cells. The latency in the Ca^{2+} responses
of the NIH-3T3^t cells was genuine since subsequent application
of exogenous ATP (100 μM) produced a virtually immediate
(<1 sec) and simultaneous rise in [Ca^{2+}]$_i$ in all the NIH-3T3^t
cells.   This result was seen in 19 out of 20 chromaffin
cells.

Fig. 3. Simultaneous measurements of $[Ca^{2+}]_i$ and secretion in a single chromaffin cell in response to nicotine.
a) Time courses show responses of chromaffin/NIH-3T3[t] co-cultured cells challenged with 10 μM nicotine followed by 100 μM ATP. Event markers show duration of agonist perfusion. Chromaffin cell: #1; NIH-3T3[t] cells: #2-10, with time-course arrangement such that #5 (top trace) was the cell nearest to the chromaffin cell whereas #10 (bottom trace) was the furthest away. The 3-D plot shows the $[Ca^{2+}]_i$ distribution in the chromaffin cell at the peak of the response to nicotine.
b) Cell map indicating position of cells from which data in **a**) were collected. Note that the Ca^{2+} signal in the NIH-3T3[t] cells begins with little delay after cessation of nicotine perfusion in those cells nearest to the chromaffin cell, but after an increasing latency in those cells furthest from the chromaffin cell.

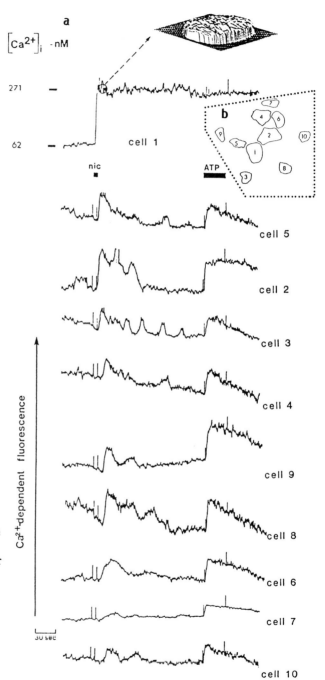

That all the surrounding NIH-3T3 cells detected a secretory response suggested that the chromaffin cell secreted from its entire surface. This was confirmed by an independent immunofluorescence technique in which the sites of exocytosis were labelled using an antibody to the enzyme DBH and subsequently detected by Texas Red fluorescence (Fig. 4). DBH is a major constituent of the chromaffin granule membrane, and its presence in the p.m. after stimulation of the cell therefore indicates where exocytosis has occurred. After a challenge with nicotine, the p.m. of chromaffin cells is completely illuminated by a ring of Texas Red fluorescence (Fig. 4d).

$[Ca^{2+}]_i$ and secretion in single cells after InsP$_3$-mobilizing stimuli

Fura 2 imaging showed that muscarinic stimuli and ANG II result in a highly localized release of internal Ca^{2+} which, in the main, remains confined to the area of the cell from which it originated (Fig. 2, showing the response to muscarine). This internal Ca^{2+} pattern is in marked contrast to the infilling effect seen in response to depolarizing stimuli, and was shown using the co-culture technique not to trigger secretion from 65% of chromaffin cells that gave a large (~160 nM) rise in $[Ca^{2+}]_i$ due to ANG II [19].

These data suggest that full exocytosis is triggered by initial Ca^{2+} activation of the entire subplasmalemmal area of the cell and not by Ca^{2+} that is spatially restricted. However, the restricted Ca^{2+} localization did result, in the remaining 35% of cells, in a release of catecholamine that was polarized to the area of p.m. that was activated by Ca^{2+}, as assessed by the co-culture technique (Fig. 5). In these cells there was also a transient elevation of $[Ca^{2+}]_i$ in the chromaffin cell (Fig. 5a). In contrast to the situation with nicotine, at the peak of the response the Ca^{2+} existed as a continuous gradient with the maximal $[Ca^{2+}]_i$ (252 nM) being recorded only in one pole of the cell (Fig. 5a, 3-D plot area A). Although ~80% of the surrounding cells showed a subsequent rise in $[Ca^{2+}]_i$, indicating secretion from the chromaffin cell, the correlation between delay in ANG II response and distance from the chromaffin cell, seen in response to nicotine, was lost. Furthermore, the NIH-3T3[t] cells which detected the least amount of secretion (i.e. those whose Ca^{2+} response was <50% of the control ATP response) were cell 6 (no response) and cell 3 (38% of the ATP response). The fact that these two cells lie in close proximity to one another (Fig. 5b) indicated that the secretory response could have been polarized with a bias towards area A of the cell (Fig. 5, a & b, 3-D plot).

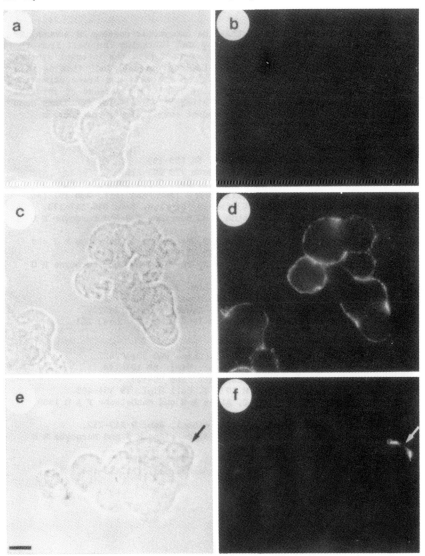

Fig. 4. Anti-DBH staining in chromaffin cells to reveal sites of exocytosis in response to nicotine or ANG II. The 2-day old cells were challenged in the presence of anti-DBH with no agonist (**a**), 10 μM nicotine (**c**), or 0.3 μM ANG II (**e**), for 15 min. The anti-DBH was localized by subsequent staining with Texas Red (see text). This revealed no fluorescence in the absence of agonist (**b**), a continuous ring of fluorescence around the p.m. of most cells after stimulation with nicotine (**d**), and a highly localized fluorescence in a minority of cells in response to ANG II (*arrow,* **f**). It is concluded that secretion originates from the entire surface of the cell in response to nicotine but is polarized in response to ANG II. *Bar represents 10 μm.*

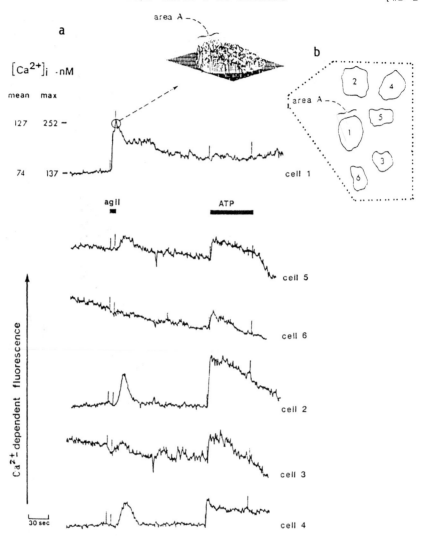

Fig. 5.   Simultaneous measurements of $[Ca^{2+}]_i$ and secretion in a
single chromaffin cell in response to ANG II.   a) Time courses
show response of chromaffin/NIH-3T3^t co-cultured cells (#1 and
#2-6 respectively) challenged with 0.3 μM ANG II followed by
100 μM ATP. Event markers show duration of agonist perfusion.
The time course arrangement is such that NIH-3T3^t cell #5 (top
trace) was the one nearest to the chromaffin cell whereas cell #4
(bottom trace) was the furthest away. The 3-D plot shows the
distribution of $[Ca^{2+}]_i$ in the chromaffin cell at the peak of the
response to ANG II: maximal $[Ca^{2+}]_i$ (252 nM) was achieved only in
area A of the cell.   b) Cell map indicating position of cells
from which data in a were collected; area A of cell #1 (chromaffin cell) corresponds to area A on the 3-D plot. Note that only
in #3 and #6 was the Ca^{2+} signal <50% of the control ATP signal.
                                              [continued opposite

That secretion in a minority of cells responding to ANG II may be polarized was confirmed by the immunofluorescent localization of the exocytotic sites (Fig. 4). In response to 0.3 µM ANG II (Fig. 4f) only a minority of cells secreted, and in those that displayed fluorescence it was quite clearly restricted to one area of the p.m. (arrowed).

Visualization in a single cell of Ca^{2+} entry in response to ANG II

Catecholamine secretion from cell populations in response to ANG II is dependent upon the presence of external Ca^{2+} [7, 8]. We therefore examined the possibility that the observed polarized release was initiated by a localized influx of external Ca^{2+}. A fura 2-loaded cell was challenged with 0.3 µM ANG II in the absence of external Ca^{2+} but with external Mn^{2+} present. Mn^{2+} acts as a surrogate for Ca^{2+} but quenches the fura 2 fluorescence when it enters the loaded cell [20-22]. By visualizing this reaction using imaging techniques and fura 2 excitation wavelengths of 340 and 380 nm, the spatial organization of the quenching reaction can be followed. These experimental conditions do not, however, allow the kinetics of the Mn^{2+} entry to be studied because the fura 2 fluorescence signal is contaminated by that from Ca^{2+} inside the cell.

At the peak of the Ca^{2+} response to ANG II, $[Ca^{2+}]_i$ is maximal in the cell's area A (Fig. 6b, 'max'). The fluorescence in this area then progressively decays until it is virtually abolished ('min'). This decay could be either the normal decay of the peak Ca^{2+} signal, or the decay due to Mn^{2+} entering the cell and quenching the fluorescence. The results suggest that the decay was indeed due to Mn^{2+} entering the cell, because the decay of the control Ca^{2+} signal, with no external Mn^{2+}, displays a completely different pattern. In the control (Fig. 6a), the area of highest $[Ca^{2+}]_i$ at the peak of the response (area A, 'max') is the last region of the cell to decay back to its resting level ('min'). With Mn^{2+}, however, the opposite is the case: the cell region with highest $[Ca^{2+}]_i$ (area A) is the first to decay. This result suggests that not only is the decay in external Mn^{2+} due to Mn^{2+} entry, but also that there is some directionality of the entry such that the Mn^{2+} enters the cell at the pole in which the ANG II-induced rise in $[Ca^{2+}]_i$ is largest.

....

Fig. 5 *legend continued from opposite*

Secretion may therefore have been polarized to area A in these cells because area A is the pole of the cell that is furthest from NIH-3T3^t cells #3 and #6 and is the area of the cell in which the rise in $[Ca^{2+}]_i$ is greatest.

Fig. 6. Visualization of Mn²⁺ entry in a single chromaffin cell in response to ANG II. **b**) One fura 2-loaded cell was challenged with 0.3 µM ANG II in the presence of 1 mM external Mn²⁺. At the peak of the Ca²⁺ signal (max), [Ca²⁺]ᵢ was highest in area A of the cell. Fluorescence in this same cell region is then the first to progressively decrease with time, until eventually fluorescence in the entire cell has decayed (min).

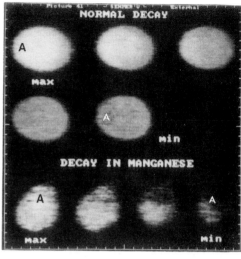

a) A control fura 2-loaded cell in which the Ca²⁺ signal induced by 0.3 µM ANG II is decaying in the absence of external MnCl₂. In this case, the area of highest [Ca²⁺]ᵢ at the peak of the response (area A, max) is the last region of the cell to decay back to its resting level (min). This suggests that not only is the decay of the Ca²⁺ signal with external Mn²⁺ due to Mn²⁺ entry, but also that the Mn²⁺ is entering predominantly into area A of the cell. This is the pole of the cell responsible for the polarized secretion (see Fig. 5 and text).

DISCUSSION

Fura 2 imaging techniques have been used in combination with a novel co-culture technique in order to investigate the relationship between the stimulus-induced rise in $[Ca^{2+}]_i$ and secretion in single bovine adrenal chromaffin cells. Previous studies in which this relationship has been examined in cell populations [e.g. 7, 8, 12, 13] led to the hypothesis that, explaining the failure of muscarinic receptor activation to trigger secretion, the induced rise in $[Ca^{2+}]_i$ (~50 nM; Fig. 1) was too small to trigger the exocytotic machinery [12-14]. The present results argue against this hypothesis since 1 µM nicotine could raise $[Ca^{2+}]_i$ by a comparably modest amount (58 nM) and yet also trigger a detectable secretory response (Fig. 1c). The non-triggering of secretion by muscarinic receptor activation cannot therefore be ascribed to the relatively small $[Ca^{2+}]_i$ rise detected in cell populations. So as to gain further insight into the relationship between the $[Ca^{2+}]_i$ rise and secretion in these cultured cells, we have performed these measurements at the level of the single cell.

In response to the potent secretagogue nicotine, 95% of single cells examined gave a strong (>200 nM) rise in $[Ca^{2+}]_i$ which originated at the cell periphery and then infilled such that the peak $[Ca^{2+}]_i$ was recorded uniformly throughout the cell (Fig. 2), as previously described [15, 18]. This increase in $[Ca^{2+}]_i$ was always followed by a strong secretory response, as indicated by the Ca^{2+} responses elicited in the NIH-3T3[t] cells adjacent to the central chromaffin cell (Fig. 3). That secretion in response to this stimulus originated from the entire surface of the cell was confirmed by an independent immunofluorescence technique in which the exocytotic sites were revealed using an antibody to DBII (Fig. 4). This protein is a component of the chromaffin granule membrane and becomes incorporated into the p.m. during the fusion process, thereby highlighting the sites of exocytosis. This result is consistent with electron-microscopic evidence which also showed that these cells are capable of supporting exocytosis over their entire surface [23].

The responses to the $InsP_3$-mobilizing stimulus ANG II were more variable: (i) ~30% of cells showed no Ca^{2+} response and did not secrete ATP; (ii) ~45% responded with a large Ca^{2+} transient - the $[Ca^{2+}]_i$ peak being observed in only one pole of the cell - and no detectable secretory response; (iii) ~25% of cells responded with a similar Ca^{2+} transient and a secretory response that was localized to the pole of the cell in which the rise in $[Ca^{2+}]_i$ was largest, as assessed by the co-culture technique (Fig. 5) and the immunofluorescent localization of exocytotic sites (Fig. 4f).

The reasons underlying the lack of a $[Ca^{2+}]_i$ response of ~30% of the cells are unknown. One possibility is that the $InsP_3$-sensitive Ca^{2+} store was empty at the time of the challenge [24]. Alternatively, ~30% of adrenal medullary cells have recently been shown to be devoid of ANG II receptors [25]. By directly demonstrating that 45% of these cells are capable of undergoing a large, transient rise in $[Ca^{2+}]_i$ without being stimulated to secrete, these results reiterate the paradoxical observations that although chromaffin cell populations [7, 21] and a high proportion of single cells [18] showed large elevations in $[Ca^{2+}]_i$ due to ANG II, the agent stimulated very little secretion from cells in culture [7, 8]. Muscarinic compounds, which are likewise poor secretagogues, in these cells (Fig. 1) [12, 13, 15], are also capable of eliciting large Ca^{2+} transients in some chromaffin cells (Fig. 2) [15]. The unifying factor linking these results is their strong indication that Ca^{2+} influx, and not mobilization of internal Ca^{2+}, is the most effective trigger for secretion from these cells. This could be because only Ca^{2+} influx results in sufficient Ca^{2+} activation

of the subplasmalemmal exocytotic sites that is needed for
fusion (Fig. 2) [15, 18]. The function of the release
of internal Ca^{2+} may be to trigger alternative cellular
functions such as the synthesis of hormone or peptide precursors
[24] or to generate Ca^{2+} signals in the vicinity of the
nucleus [26]. A recent study done elsewhere [27] in which
$[Ca^{2+}]_i$ and secretion were monitored from a small number
of chromaffin cells also suggested that Ca^{2+} sources are
functionally different in these cells.

The polarized secretion observed in ~25% of cells responding to ANG II could be of physiological importance, as
angiotensins are produced locally within the adrenal gland
([25]; cf. U. Lang, art. #B-2, this vol.- *Ed.*) and are known
to play a key role in the mammalian stress response. However,
it is not known whether the area of the cell responsible
for the polarized release corresponds to the area exposed
to the bloodstream in the intact gland. Similar polarized
exocytotic secretion has been observed in other small secretory
cell types, e.g. mast cells [28] and anterior pituitary
cells (W.T. Mason, pers. comm.). The mechanism underlying
the polarized release is unknown. In the chromaffin cell,
a possibility is that a localized influx of external Ca^{2+}
accompanies and/or follows internal Ca^{2+} release because a
recent study on Ca^{2+} entry reported a small Mn^{2+} influx
component induced by ANG II [25] and, moreover, secretion in
response to ANG II is dependent upon the presence of external
Ca^{2+} [7, 8]. The present study, in which the spatial organization of the fura 2 quenching following Mn^{2+} entry induced
by ANG II was monitored, clearly showed a preferential quenching
of fura 2 fluorescence in the area of the cell responsible
for the polarized release after exposure to ANG II. Since
Mn^{2+} entry mirrors Ca^{2+} entry, it is highly likely that
polarized Ca^{2+} influx plays a key role in initiating the
polarized release. The mechanisms underlying this Ca^{2+} influx
are a mystery, although it is known that entry is not automatic-
ally triggered by release of Ca^{2+} from either the $InsP_3$-sensitive
store [21] or the caffeine-sensitive store [24] in these
cells. As pointed out by Stauderman & Pruss [21], entry
may be through a second-messenger–operated channel involving
$Ins(1,4,5)P_3$, $Ins(1,3,4,5)P_4$, GTP or Ca^{2+}. Alternatively,
the peptide may directly open a receptor-operated channel
in the p.m., as does ADP in platelets [22].

In conclusion, we have used a combination of fura 2
imaging techniques and a novel co-culture technique to directly
demonstrate that influx of external Ca^{2+}, and not release
of internally stored Ca^{2+}, is a vital requirement for the
triggering of exocytosis from these cells. This is probably

because only Ca^{2+} influx results in sufficient Ca^{2+} activation of the subplasmalemmal exocytotic sites to prime these sites. In addition, the results show that the location of the sites of exocytosis in these cells depends on the nature of the stimulus.

Acknowledgements

Grateful thanks are due to Drs. T.R. Jackson, A.J. O'Sullivan and R.B. Moreton for their contributions to these studies and to V.I. Glen for typing the manuscript. This work was supported by the AFRC, by a project grant from the MRC (to R.D.B.) and by a grant from Merck & Co., Inc. (to M.J.B.).

References

1. Burgoyne, R.D. (1984) *Biochim. Biophys. Acta 779*, 201-216.
2. Burgoyne, R.D., Morgan, A. & O'Sullivan, A.J. (1988) *FEBS Lett. 238*, 151-155.
3. Knight, D.E. & Baker, P.F. (1983) *FEBS Lett. 160*, 98-100.
4. Burgoyne, R.D. & Cheek, T.R. (1987) *Biosci. Rep. 7*, 281-288.
5. Knight, D.E. & Baker, P.F. (1982) *J. Membr. Biol. 68*, 107-140.
6. Dunn, L.A. & Holz, R.W. (1983) *J. Biol. Chem. 258*, 4989-4993.
7. O'Sullivan, A.J. & Burgoyne, R.D. (1989) *Biosci. Rep. 9*, 243-252.
8. Bunn, S.J. & Marley, P.D. (1989) *Neuropeptides 13*, 121-132.
9. Gill, D.L. (1989) *Nature 342*, 16-18.
10. Cheek, T.R., O'Sullivan, A.J., Moreton, R.B., Berridge, M.J. & Burgoyne, R.D. (1989) in *Biochemical Approaches to Cellular Calcium* [Vol. 19, this series] (Reid, E., Cook, G.M.W. & Luzio, J.P., eds.), Roy. Soc. of Chemistry, Cambridge, pp. 319-320.
11. Berridge, M.J. (1987) *Annu. Rev. Biochem. 56*, 159-193.
12. Cheek, T.R. & Burgoyne, R.D. (1985) *Biochim. Biophys. Acta 846*, 167-173.
13. Kao, L-S. & Schneider, A. (1985) *J. Biol. Chem. 260*, 2019-2022.
14. Eberhard, D.A. & Holz, R.W. (1987) *J. Neurochem. 49*, 1634-1643.
15. Cheek, T.R., O'Sullivan, A.J., Moreton, R.B., Berridge, M.J. & Burgoyne, R.D. (1989) *FEBS Lett. 247*, 429-434.
16. Penner, R. & Neher, E. (1988) *J. Exp. Biol. 139*, 329-345.
17. Forsberg, E.J., Rojas. E. & Pollard, H.B. (1986) *J. Biol. Chem. 261*, 4915-4920.
18. O'Sullivan, A.J., Cheek, T.R., Moreton, R.B., Berridge, M.J. & Burgoyne, R.D. (1989) *EMBO J. 8*, 401-411.

19. Cheek, T.R., Jackson, T.R., O'Sullivan, A.J., Moreton, R.B., Berridge, M.J. & Burgoyne, R.D. (1989) *J. Cell Biol. 109*, 1219-1227.

20. Jacob, R. (1990) *J. Physiol. 421*, 55-77.

21. Stauderman, K.A. & Pruss, R.M. (1989) *J. Biol. Chem. 264*, 18349-18355.

22. Sage, S.O., Merritt, J.E., Hallam, T.J. & Rink, T.J. (1989) *Biochem. J. 258*, 923-926.

23. Grynszpan-Winograd, O. (1971) *Phil. Trans. R. Soc. London B. Biol. Sci. 261*, 291-292.

24. Cheek, T.R., O'Sullivan, A.J., Moreton, R.B. Berridge, M.J. & Burgoyne, R.D. (1990) *FEBS Lett. 266*, 91-95.

25. Marley, P.D., Bunn, S.J., Wan, D.C.C., Allen, A.M. & Mendelsohn, F.A.O. (1989) *Neuroscience 28*, 777-787.

26. Burgoyne, R.D., Cheek, T.R., Morgan, A., O'Sullivan, A.J., Moreton, R.B., Berridge, M.J., Mata, A.M., Colyer, J., Lee, A.G. & East, J.M. (1989) *Nature 342*, 72-74.

27. Kim, K-T. & Westhead, E.W. (1989) *Proc. Nat. Acad. Sci. 86*, 9881-9885.

28. Lawson, D., Fewtrell, C. & Raff, M. (1978) *J. Cell Biol. 79*, 394-400.

#D-3

ANALYSIS OF CELLULAR CALCIUM WITH FLUO-3 AND FLOW CYTOMETRY, IN GUINEA-PIG EOSINOPHILS

Harm J. Knot[+], Kurt Müller and Urs T. Rüegg

Preclinical Research (Bldg. 386/422),
Sandoz Pharma Ltd., CH-4002 Basel, Switzerland

A method is described for measuring $[Ca^{2+}]_i$[*] in real time by combining the new Ca^{2+}-sensitive dye fluo 3 with FACS in flow-cytometers. Since this technique uses relatively few cells, one can now undertake extensive measurements on Ca^{2+} signalling in cells such as eosinophils, T-lymphocytes and other cells which, due to their limited availability, are too valuable for 'sacrifice' of one million per experimental point as needed in classical suspension measurements.

The measurement of changes in $[Ca^{2+}]_i$[*] is vital for revealing activation of cells by hormonal stimulation in almost all excitable and non-excitable cells. The development of the new fluorescent dye fluo 3 [1], which is excitable at the 488 nm wavelength emitted by argon-lasers in standard flow-cytometers, has extended their value for studying changes in $[Ca^{2+}]_i$ in many cell types [2]. *(Dye formula: see later.)*

The value of this technique is that fluorescence data are obtainable on quite a large number of single cells; thus each cell's behaviour can be assessed individually within the measured population, and statistics are acquirable as if a whole population were being studied. This assay uses much fewer cells than needed for measuring a suspension in a cuvette, e.g. with fura 2, in a spectrofluorimeter, and is therefore more economical with eosinophils or other biological material that is hard to procure or limited in amount. The technique increases by ~50-fold the number of experiments that can be done with a given number of cells. Common FACS analyzers are equipped with a 488 nm argon laser and at least the fluorescein emission filter sets, so allowing $[Ca^{2+}]_i$ measurement with almost any flow-cytometer.

[+]addressee for any correspondence

[*]*Abbreviations.* - $[Ca^{2+}]_i$, intracellular free Ca^{2+} (concentration); AM, acetoxymethyl ester; DMSO, dimethylsulphoxide; FACS, fluorescence-activated cell sorting; LT, a leukotriene; PAF, platelet-activating factor; PSS, physiological salt solution; i.p., intraperitoneal. [*Note by Ed. on* FACS, including instrumentation: consult Vol. 17 (Ormerod; his p. 163 cites Vol. 8 & 11 also.]

Most flow-cytometers allow at least 3 cell-associated fluorescence measurements besides cell-size and granularity assessments. The latter are directly proportional to the forward and right-angle scatter data.

This article gives a detailed description of such a $[Ca^{2+}]_i$ measurement assay. We used guinea-pig eosinophils as an example of a cell which has not been studied extensively.

METHODS

Animals.- The guinea-pigs (Dunkin-Hartley; 300-500 g) were actively sensitized with ovalbumin:- on the day following a cyclophosphamide dose (100 mg/kg, i.p.), 1 ml of an emulsion of $Al(OH)_3$ (2 mg), pertussin vaccine (0.25 ml) and ovalbumin (10 μg) was given i.p. A supplementary injection was given after 3 weeks and, starting after a further week, polymixin B was given weekly (2 mg in 1 ml, i.p.) for 5 weeks, to both sensitized and non-sensitized animals as distinguished by whether a wheal-and-flare reaction occurred after intradermal injection of ovalbumin (250 μg in 100 μl).

Cell preparation.- Cells were harvested by washing the peritoneal cavity of sensitized or non-sensitized animals twice with 2 x 20 ml of Ca^{2+}- and Mg^{2+}-free Hank's balanced salt solution containing bovine serum albumin (0.3%), EDTA (10 mM) and Hepes (10 mM). Lavage fluid was centrifuged (180 **g**, 10 min, 20°), the cell pellet resuspended in 2 ml of Ca^{2+}- and Mg^{2+}-free Minimal essential medium (MEM), and a total cell count was made using a Sysmex automatic cell counter. Cells (maximally 4×10^7/ml) were layered onto discontinuous gradients comprising 2 x 3 ml Percoll (densities 1.09 and, for the overlay, 1.08 g/ml). The tubes were centrifuged (700 **g**, 10 min, 20°). Eosinophils of >95% purity and viability were found at the 1.08/1.09 interface and at the bottom of the tube. These fractions were pooled, washed twice in the MEM, and finally suspended (10^7/ml) in MEM containing Ca^{2+} (1 mM) and Mg^{2+} (1 mM).

Dye loading.- A stock solution of fluo 3* (1 mM) in dry DMSO was mixed 1:1 with 0.25% Pluronic F-127 (w/w; non-ionic detergent) before adding to the cell suspension. Cells (10^7/ml) were exposed at 37° for 20 min to 4 μM fluo 3/AM in PSS containing (mM) 135 NaCl/5 KCl/1.5 $MgCl_2$/1 $CaCl_2$/10 Hepes/5.5 glucose at pH 7.4. Then the cells were diluted x 15 with cold PSS, centrifuged at 1000 rpm for 5 min, resuspended at 10^6/ml in PSS, and left at room temperature in the dark for 1 h prior to the experiments. Finally aliquots of the cell suspension, after equilibration for 5 min at 37°, were diluted with buffer into standard flow-cytometer tubes so that the cell-flow was at least 200 cells/sec. In an average experiment we

*from Molecular Probes; likewise Br-A23187 (other chemicals: Sigma).

used ~25,000 cells.   Solutions and reagents were injected
directly into the plastic tube using a normal syringe and
needle to mix the suspension in the tube.   The lag-time
between mixing and measurement was ~2 sec.   If cells have
a high anion efflux, loss of dye from the cells can be
a major problem, which Merritt [3] found could be eliminated
by use of probenicid.

Data acquisition and analysis

Cell fluorescence was analyzed on a FACScan Flow Cytometer
(Beckton Dickinson).   The cells were continuously taken out
of the suspension and injected into a laminar fluid stream.
Through a 'hydrodynamic focusing' process, the cells entered
one-by-one a quartz cuvette where they were excited by the
laser beam, which was hardly larger than the cell itself.
This minimizes disturbance from extracellular light to the
equivalent of 1 µM fluorescent dye in the medium.   The whole
transit from the tube to the cuvette took ~2 sec. Light-scatter
and fluorescence were collected and separated with the help
of dichroic mirrors and bandpass filters.   The forward and
right-angle scatters were used to 'gate' selectively on the
eosinophil population.   This excludes possible contaminating
cells and particles from the preparation (macrophages or
cellular aggregates).

Excitation was from the 488 nm argon laser.   Emission
via 'Channel 1', the fluorescein bandpass filter at 525 nm,
was measured on a linear scale.   For non-stimulated cells
the baseline was set arbitrarily at 'Channel 200'. The other
two longer wavelength channels were used only exceptionally,
e.g. when the pH-sensitive dye SNARF-1 in the ratio emission
mode was combined with fluo 3, enabling data to be collected
in these channels (unpublished work).

The events were acquired and analyzed in the continuous
mode for maximally 3 min (limited by the computer's RAM
memory) with the CHRONYS-software program (Becton Dickinson).
For a given time slice of at least 1 sec the mean fluorescence,
in arbitrary units with 1024 channels resolution, was calculated
by the software and subsequently plotted *vs*. time.   In every
1 sec time interval a histogram of fluorescence *vs*.   cell
number is obtained (Fig. 1).   Total cell number depends
on the flow speed and can be adjusted by changing the initial
dilution of the cell suspension in the tube.   Usually a
rate of 250 cells/sec was used, in order to achieve stable
measurements.

The software creates listmode datafiles on the computer's
hard-disk.   During analysis the mean fluorescence over any
time-slice, from a minimum of 1 sec, can be calculated and

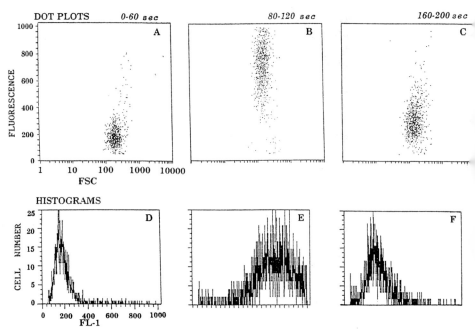

Fig. 1 Three time intervals of a typical calibration experiment. DOT PLOTS: 1000 cells in resting state (A), after adding
the Ca^{2+} ionophore Br-A23187 to 2 μM (B), and then after adding,
at 2 min, $MnCl_2$ to 10 mM (C). The x-axis represents forward
scatter of the cells ($\propto$ cell size). Fluorescence is in arbitrary units. HISTOGRAMS: distribution of fluorescence (= Ca^{2+})
vs. no. of cells, of the same fluorescence level, for the same
time-slices as in the dot plots. The mean fluorescence of
these histograms for every 12.5 sec vs. time is directly proportional to $[Ca^{2+}]_i$, and Fig. 2 shows the result.

plotted vs. time. How these calculations lead to the response
vs. time plot is illustrated in Fig. 2, showing a calibration
experiment.

   Calibration.- Since fluo 3, unlike indo 1, does not have
a Ca^{2+}-induced spectral shift in its emission wavelength,
one should calibrate at the end of each experiment. Using
eosinophils, the loss of fluorescence during 4 h after loading
was very small. Accordingly it sufficed, for calculating
absolute Ca^{2+} levels, to perform only 3-4 calibrations per
day, evenly spread between the experiments. $[Ca^{2+}]_i$ was
calculated using the equation [4][*]:

$$[Ca^{2+}]_i = Kd \cdot \frac{F - F_{min}}{F_{max} - F}$$

where Kd represents the dissociation constant for Ca^{2+}-bound
fluo 3, viz. 450 nM at 22° and ~860 nM at 37° [3] (obviously
very good temperature control during the experiments is needed).

[*]Cf. expression in #D-1 (Hallett); also Vol. 19 (Lückhoff particularly)

Fig. 2. Change in mean fluorescence when, as represented in
Fig. 1, Br-A23187 was introduced
at 60 sec and MnCl$_2$ at 120 sec.
Upper x-axis: no. of cells used
to calculate the mean within
the 12.5 sec plotted in the trace.
Lower x-axis gives time (sec).
Fluorescence is in arbitrary
units.

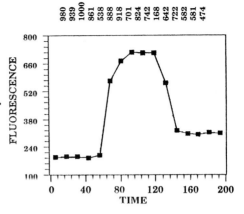

F_{max} represents the maximum fluorescence when all dye is
saturated with Ca^{2+}; this value was obtained by rendering
the cells permeable to Ca^{2+} using Br-A23187, a non-fluorescent derivative of the Ca^{2+} ionophore A23187. F is the
mean fluorescence of the samples. F_{min} represents the minimum
fluorescence which is found thus [4].- MnCl$_2$ (to 10 mM) is
added to the ionophore-treated cells.   The Mn^{2+} ion will
displace Ca^{2+} from the dye. However, the Mn^{2+}/fluo 3 complex
still has a fluorescence 8 times that of the free dye but
only 0.2 times that of the Ca^{2+}/fluo 3 complex [2].   Therefore
F_{min} can be calculated by the formula:

$$F_{min} = F_{max} - 1.25 \cdot (F_{max} - F_{MnCl_2})$$

RESULTS AND DISCUSSION

Fluo 3 dye measurements (see Fig. 3 for some dye properties).-
Cells are permeable to the AM ester of fluo 3 [5].   This
dye has major advantages over its predecessors.- (1) Its
excitation and emission spectra lie at wavelengths in the
visible range and therefore do not interfere with cellular
autofluorescence (arising mainly from pyridine nucleotides).
This enables the extension of Ca^{2+} measurements to flow-cytometry
and laser confocal microscopy, which usually incorporate the
488 nm argon laser and filter settings for fluorescein [6].
(2) The dye has improved Ca^{2+}-sensitivity, increasing its
fluorescence 40-fold upon binding Ca^{2+}; without Ca^{2+} the dye
is almost non-fluorescent.   (3) The Ca^{2+} affinity is much
lower compared to indo 1 and fura 2 [7].   This enables low
[Ca^{2+}]$_i$ values to be measured at lower intracellular dye
concentration, and also is advantageous for avoiding Ca^{2+}-
buffering effects observed with other dyes [8]. The measurement
remains accurate up to 5-10 µM [Ca^{2+}]$_i$.   Beekman [8] has
shown with fura 2 a very good example of this buffering
effect, interfering with the biological functioning of Ca^{2+}. (4) The
dye has superior binding kinetics [9]: very fast association

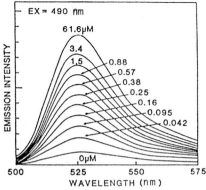

Mol. wt.:
free acid, 781
AM ester: 1141
For Kd's *see text.*

Fig. 3. Some features of fluo 3 (free acid), including the fluorescence emisssion spectrum (max: 526 nm) for Ca^{2+}-fluo 3 at different concentrations (490 nm excitation; max = 506 nm).

and dissociation rate-constants at physiological pH and temperature ensure proper reflection of actual changes in $[Ca^{2+}]_i$.

A calibration experiment was shown earlier in Fig. 1. The traces give an impression of how the software calculates the final response. The three time intervals represent non-stimulated, ionophore-treated and Mn^{2+}-addition situations. Fig. 2 showed, as the calibration outcome, mean fluorescence *vs.* time.

Effects of agonists, PAF and LTB_4, on guinea-pig eosinophils are illustrated in Fig. 4. For a 3-min exposure, Fig. 4A shows concentration-response curves. Both agonists raised $[Ca^{2+}]_i$, most likely from $InsP_3$-sensitive pools. The rise was rapid but transient, yet sustained with PAF (cf. LTB_4; Fig. 4B) if the external buffer contained Ca^{2+}, influx of which may explain the sustained rise. Calculation for PAF (Fig. 4A) showed that this rise, ~2 times the basal level, is more agonist-sensitive than the pre-peak transient rise in $[Ca^{2+}]_i$.

For all traces in Fig. 4 we needed ~1.5×10^6 cells. One guinea-pig can yield up to $1-2 \times 10^7$ cells. This would enable 20-40 dose-response curves to be acquired with the cells from one animal.

CONCLUDING COMMENTS

In this article we have shown that the combination of fluo 3 with flow-cytometry is a powerful tool for studying intracellular Ca^{2+} in eosinophils. The technique can easily be adapted to other cells. Additional experimental possibilities are the combination with other dyes, such as SNARF-1 for pH measurement, propidium iodide for viability, or a

Fig. 4. Concentration-response curves for $[Ca^{2+}]_i$ in eosinophils treated with LTB_4 or PAF.
A: 3-min treatment (pooled results of 3 expts.); the sustained Ca^{2+} level is that measured with 1.2 mM external Ca^{2+} present, and the peak level was that in Ca^{2+}-free buffer.
B: time-course traces with different agonist concentrations.

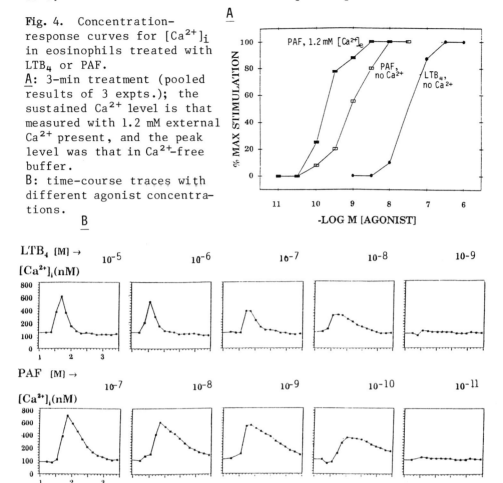

dye which reflects membrane potential. When experiments are done in a sorter (FACS) instead of an analyzer, as described in this article, one gets the opportunity to physically sort out the cells according to any combination of their responses as manifested by a fluorescent marker. Also, the visible-light excitation offers the possible combination with photochemically reactive probes such as the caged nucleotides or caged inositol phosphates [10-12].

Acknowledgements

We thank Chris Poll and Marko Rehn for providing the cells, Sibile Doumont for preparing the manuscript, and Becton Dickinson (Basel) for providing the CHRONYS software.

References

1. Minta, A., Kao, J.P. & Tsien, R.Y. (1989) *J. Biol. Chem.* *264*, 8171-8178.
2. Vandenberghe, P.A. & Ceuppens, J.L. (1990) *J. Immunol. Meth.* *127*, 197-205.
3. Merritt, J.E., McCarthy, S.A., Davies, M.P.A. & Moores, K.E. (1990) *Biochem. J. 269*, 513-519.
4. Grynkiewicz, G., Poenie, M. & Tsien, R.Y. (1985) *J. Biol. Chem. 260*, 3440-3450.
5. Tsien, R.Y. (1981) *Nature 290*, 127-128.
6. Niggli, E. & Lederer, W.J. (1990) *Cell Calcium 11*, 121-130.
7. Tsien, R.Y. (1989) *Meth. Cell Biol. 30*, 127-156.
8. Beekman, R.E., Van Hardeveld, C. & Simonides, W.S. (1990) *Biochem. J. 268*, 563-569.
9. Eberhard, M. & Erne, P. (1989) *Biochem. Biophys. Res. Comm. 163*, 309-314.
10. Gurney, A.M. & Lester, H.A. (1987) *Physiol. Rev. 67*, 583-617.
11. Walker, J.W., Somlyo, A.V., Goldman, Y.E., Somlyo, A.P. & Trentham, D.R. (1987) *Nature 327*, 249-252.
12. Kao, J-P., Harootunian, A.P. & Tsien, R.Y. (1989) *J. Biol. Chem. 264*, 8179-8184.

#D-4

ENGINEERING BIOLUMINESCENT PROTEINS: A NEW EXPERIMENTAL STRATEGY FOR MEASURING INTRACELLULAR SIGNALLING IN LIVE CELLS

Anthony K. Campbell and Graciela Sala-Newby

Department of Medical Biochemistry,
University of Wales College of Medicine,
Heath Park, Cardiff CF4 4XN, U.K.

A new experimental strategy has been established to measure intracellular signals and covalent modification of proteins in the cytosol and organelles of living cells. Firefly luciferase and aequorin have been engineered using the PCR*. The new proteins, synthesized in vitro, have been characterized in respect of activity, colour, pH profile and the effect of protein kinase A.

A universal feature of all eukaryotic and some prokaryotic cells is the ability of agents acting at the p.m.* to activate or impair structures and reactions within the cell [1-3]. The primary agents include physiological agonists such as hormones and neurotransmitters, pathogens (e.g. membrane pore-forming proteins), toxins, bacteria and viruses, and drugs. These agents first bind to a receptor on the cell surface, or insert into the p.m., and cause the release of a signal into the cytosol. This signal then initiates a sequence of events ending with a cell response. These responses include movement, secretion, transformation, division, defence, and death *via* lysis or apoptosis [1-3]. The primary intracellular signal may generate a secondary signal: e.g. $InsP_3$ releases Ca^{2+} [4], which has to reach its target site, where it induces covalent modification and structural reorganization of proteins and of other molecules such as DNA. It is these macromolecular changes that are directly responsible for provoking the ultimate end-response of the cell.

The experimental problem in elucidating the complete signalling sequence necessary for targeting a particular cell response is that each step in the sequence involves a threshold, or 'rubicon' [3], whose magnitude, timing and intracellular location varies by several seconds, minutes or even hours from cell to cell. The conventional 'grind and find' approach of homogenization destroys this heterogeneity, and also removes the ion gradients and localization of intracellular chemical

---

*Abbreviations.- $InsP_3$, inositol trisphosphate; PAGE, poly-acrylamide gel electrophoresis; PCR, polymerase chain reaction; p.m., plasma membrane. Aequorin and luciferase are 'probes' for measuring Ca^{2+} and ATP respectively.

processes that keep the cell alive. It is therefore essential
that methods be developed to measure, locate and manipulate
both the primary and the secondary signals, and the protein
modifications they induce, within the cytosol and organelles
of living cells.

This article aims to review a general strategy for
achieving this, applying protein engineering to generate bioluminescent proteins [5, 6].

Bioluminescence occurs in ~600 genera representing 16
phyla [2]. Because of the exquisite sensitivity of detecting
chemiluminescent analytes, down to 10^{-21} mol (1 tipomol) amounts,
several bioluminescent systems have been used to measure
reactions in cell extracts and live cells [1, 2, 7]. Furthermore, the isolation of the DNA coding for firefly and other
beetle luciferases [8], bacterial luciferase [9], aequorin
[10, 11] and *Vargula* luciferase has opened up a new era
for exploiting bioluminescent genes as reporters of promoter
and enhancer activity in live cells. In luminous beetles
a small number of charged amino acids vary between the
luciferases of different species, and thereby alter the colour
of the light emission from green to green-yellow, to yellow,
to orange or to red [12]. Similarly in imidazolopyridine
bioluminescence, which is found in animals from 6 distinct
phyla [13], the luciferase or photoprotein is responsible
for determining the spectrum of light emitted. In this
case the spectral range, without intermolecular energy transfer,
is likewise ~60 nm but falls in the blue region of the
spectrum, i.e. 430-460 nm. The new strategy we have established
is to incorporate cAMP, cGMP and $InsP_3$ binding sites, and
protein phosphorylation recognition sites [14] (Table 1), into
firefly luciferase and aequorin, such that binding of the
signal, or phosphorylation, induces a change in intensity
and/or colour of the light emitted [6, 15].

Protein engineering now achieved. - A procedure has been
established, using the PCR, to engineer the protein kinase A
recognition peptide, kemptide [6, 14], onto the N- and C-termini
of firefly luciferase and aequorin, and to mutate 2 bases
in the natural sequence of firefly luciferase, altering amino
acid 217 from V to R. This inserts the protein kinase
recognition site RRFS into the protein. We have also used
the PCR to remove the peroxisomal target signal from the
C-terminus of firefly luciferase [16], and to add various
organelle-targeting signal peptides to the N-terminus of firefly luciferase and aequorin. A crucial part of our strategy
is the ability to test *in vitro* the effect of the protein
engineering on the specific activity and colour of the light
emitted. The T7 RNA polymerase promoter was added to the

Table 1. Peptide recognition sequences for various kinases.
Phosphorylated residue is denoted by **bold** letter with . under.

Kinase	Recognition sequence
cAMP type I (kemptide)	LRRA**S**LG
calmodulin kinase II	HRQEE**T**VEC
myosin light chain	KKAKRRAAAEG**S**S
C	RKR**T**LRRL
pp60(src)	ARLIEDNE**Y**TARFG
insulin receptor	DI**Y**ETDYYRKGGK

5'-end of the PCR sense primer, so that the new DNA could
then be transcribed and the resulting mRNA translated *in
vitro* to form light-emitting protein, without the need to
sub-clone every variant [17-19].

METHODS

Isolation of bioluminescent genes.- Firefly luciferase
cDNA was isolated from a cDNA library constructed from firefly-
tail mRNA in pcDV1 primer + Honjo linker containing the SP6
RNA polymerase promoter [5, 17, 18]. The library was screened
using ^{32}P-oligonucleotides designed from the known cDNA sequence
[8], and light emission detected by X-ray film. The leak
of *E. coli* RNA polymerase activity with this vector results
in an overall expression of the luciferase of ~0.02% *vs.*
total soluble protein, detectable because of the high measurement sensitivity of bioluminescence.

Aequorin cDNA was isolated using the reverse transcriptase
PCR. Genomic RNA for firefly luciferase and aequorin were
isolated using the conventional PCR. The freeze-dried *Aequorea*
were a kind gift from Dr. C.C. Ashley (Univ. of Oxford).

Engineering bioluminescent genes.- Kemptide (LRRASLG) was
added to the 5'-end of the DNA encoding for firefly luciferase
or aequorin using a 2-step PCR procedure: the DNA coding
for kemptide was added in the first stage, and the T7 RNA
polymerase promoter in the second. A 2-stage PCR was also
used to alter 2 bases within the codon for amino acid 217
in firefly luciferase to form RRFS within the protein. The
normal PCR procedure involved 25 cycles in a Perkin-Elmer-Cetus
thermal cycler, each cycle consisting of 1 min at 94°, 1 min
at 56° and 1 min at 72°. An ATG site was incorporated
into the 5'-primer, and a stop codon + salI site and clamp
incorporated into the 3' anti sense primer. Successful PCR
execution was confirmed by manifestation, on agarose gel
electrophoresis, of a single band of 1650 bp for firefly
luciferase and 600 bp for aequorin + additions.

In vitro *synthesis of bioluminescent proteins.-* The PCR-generated DNA was extracted with phenol/CHCl$_3$/isoamyl alcohol (25:24:1 by vol.) and ethanol-precipitated. A sample redissolved in 10 mM Tris/1 mM EDTA pH 7.4 was then incubated for 1 h at 37° with T7 RNA polymerase, nucleotide triphosphates, ^{32}P-UTP in trace amount, and the mRNA cap m^{7}GpppG [description: 5, 18, 19]. The mRNA formed was phenol-extracted, ethanol-precipitated and redissolved as above. A 1-2 µl aliquot was translated in 5-10 µl of rabbit reticulocyte lysate (Amersham Internatl.) for 1 h at 37°; optimal translation needed KCl and MgCl$_2$, the amounts varying from one lysate batch to another. The luciferase activity attained by the normal protein was then determined as the chemiluminescent counts in the 10 sec after adding the luciferin at pH 7.8. The apo-photoprotein activity was determined by exposure to coelenterazine for up to 24 h and final measurement of chemiluminescent counts for 10 sec after adding Ca^{2+} [1, 2, 5, 20]. Specific activities of the proteins were measured by adding ^{35}S-methionine to the translation cocktail followed by PAGE and elution of the luciferase or aequorin. *Chemiluminescence was measured* in a home-built chemiluminometer [1,2]. The colour of the bioluminescent proteins was assessed by measuring the chemiluminescence in a chemiluminometer that allowed 2 wavelengths to be measured simultaneously and the ratio obtained [21].

RESULTS AND DISCUSSION

The PCR-generated DNA's were characterized using 3 criteria:-
(1) correct size of PCR product on agarose gel electrophoresis;
(2) ability to act as a template for synthesis of ^{32}P-mRNA of the right size on gel electrophoresis; (3) translation of mRNA to form light-emitting protein and ^{35}S-protein of the correct.M$_r$ on PAGE.

The transcription efficiency was 5-20 mol bioluminescent protein/mol mRNA. No PCR products were produced in the absence of primers or template DNA. No mRNA was formed when the T7 RNA polymerase promoter was omitted from the 5-primer.

The synthetic proteins from the rabbit reticulocyte lysate were characterized by 3 criteria:- (1) specific activity assessed as chemiluminescence counts/µg RNA or /µg protein; (2) pH profile of activity; (3) colour at various pH's by ratio measurements (545/603 nm) of chemiluminescence counts.

The PCR products from aequorin cDNA and gDNA transcribed to form mRNA which translated into fully active aequorin (Fig. 1). Apparently aequorin gDNA has no introns. However, although the PCR products from both cDNA and gDNA of firefly luciferase transcribed to form mRNA, only the mRNA generated from the cDNA formed active luciferase, since gDNA for luciferase has 6 introns [8].

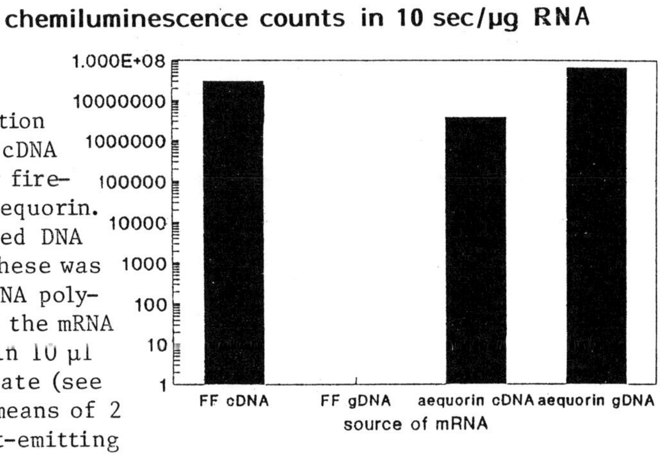

chemiluminescence counts in 10 sec/µg RNA

Fig. 1.
In vitro transcription
and translation of cDNA
and gDNA coding for fire-
fly luciferase and aequorin.
Each PCR-synthesized DNA
corresponding to these was
transcribed by T7 RNA polymerase, and 1 µl of the mRNA
formed translated in 10 µl
of reticulocyte lysate (see
text). Results are means of 2
estimations of light-emitting
capacity of the products.

Addition of the peptide kemptide and an endoplasmic
reticulum or mitochondrial signal peptide to the N-terminus
had little or no effect on the specific activity of firefly
luciferase and aequorin.   Removal of the last 12 amino
acids at the C-terminus of firefly luciferase, containing
the peroxisomal targeting signal, reduced activity by >99%
[18].   However, after removal of the last 3 amino acids, the
key residues for peroxisomal targeting, full activity survived.

Mutation of the valine at residue 217 of the luciferase
to arginine reduced its activity by >80% [19].   Furthermore
the mutant protein had a different pH profile and bioluminescent
spectrum from the native luciferase.   Phosphorylation of
the mutant appeared to drastically reduce the activity of
the mutant luciferase still further [19], but protein kinase A
(kindly gifted by Dr. K. Murray, SmithKline Beecham, U.K.) had
no effect on the activity of either the synthetic or the
native luciferase.

Addition of the luciferase to isolated liver mitochondria
resulted in a large increase in binding to the organelles
compared to the normal luciferase.   No evidence has yet
been obtained to show whether the luciferase was taken up
into the matrix of the mitochondria.

CONCLUSIONS

Evidently the PCR, coupled to an *in vitro* transcription/
translation system, can serve to ascertain whether activity
is affected by our novel method for engineering bioluminescent
proteins.   Addition of kinase recognition and signal peptides
to the N- or C-terminus appeared not to markedly impair
activity.   This strategy opens the door to a new technology
for measuring signals and protein phosphorylation in defined
cell compartments.   The proteins have been expressed in

E. coli, such that the live bacteria emit light.   These
new indicators of intracellular signalling and gene expression
must now be manifested in live eukaryotic cells, so that
there will be the novel ability to follow a complete sequence
of events from the p.m. to the end-response in a living cell.

Acknowledgements.- For financial support the Agricultural &
Food and Medical Research Councils and the Arthritis & Rheumatism
Council are thanked.

References

1.   Campbell, A.K. (1983) *Intracellular Calcium: its Universal Role
     as Regulator,* Wiley, Chichester, 556 pp.
2.   Campbell, A.K. (1988) *Chemiluminescence: Principles and Applica-
     tions in Biology and Medicine,* Horwood/VCH, Chichester & Weinheim,
3.   Campbell, A.K. (1990) in *Oxygen Radicals and Cellular*   [608 pp.
     Damage (Duncan, C.J., ed.), Camb. Univ. Press, Cambridge, pp. 184-217.
4.   Berridge, M.J. & Irvine, R. (1989) *Nature 341,* 197-205.
5.   Campbell, A.K., Sala-Newby, G., Aston, P., Jenkins, T. &
     Kalsheker, N. (1990) *J. Biolum. Chemilum. 4,* 131-139.
6.   Campbell, A.K. (1989) Br. Patent Applicn. 8916806.6.
7.   Campbell, A.K. (1989) *Essays in Biochemistry 24,* 41-81.
8.   De Wet, J.R., Wood, K.V., DeLuca, M., Helsinki, D.R. &
     Subramani, S. (1987) *Mol. Cell Biol. 7,* 725-737.
9.   Baldwin, T.O., Berends, T., Bunch, T.A., Holzman, T.F., Rausch, S.K.
     Shamansky, L., Treat, M.L. & Ziegler, M.M. (1984) *Biochemistry 23,*
     3663-3667.
10. Inoue, S., Noguchi, M., Sakaii, Y., Takagi, I., Miyata, T., Iwanaga, S.
     & Tsuji, F.I. (1985) *Proc. Nat. Acad. Sci. 82,* 3154-3158.
11. Cormier, M.J., Prasher, D.C., Longiaru, M. & McCann, R.O. (1989)
     Photochem. Photobiol. 49, 509-512.
12. Wood, K.V., Lam, Y.A. & McElroy, W.D. (1989) *Science 244,* 700-702.
13. Campbell, A.K. & Herring, P.J. (1990) *Marine Biology 104,* 219-225.
14. Cohen, P. (1988) *Proc. Roy. Soc. Lond. B, 234,* 115-144.
15. Jenkins, T., Sala-Newby, G. & Campbell, A.K. (1990) *Biochem.
     Soc. Trans. 18,* 563-565.
16. Gould, S.J., Keller, G.A. & Subramani, S. (1987) *J. Cell
     Biol. 105,* 2923-2931.
17. Sala-Newby, G., Kalsheker, N. & Campbell, A.K. (1990) *Biochem.
     Soc. Trans. 18,* 459-460.
18. Sala-Newby, G., Kalsheker, N. & Campbell, A.K. (1990)
     Biochem. Biophys. Res. Comm. 172, 477-482.
19. Sala-Newby, G. & Campbell, A.K. (1991) *Biochem. J.,* submitted.
20. Campbell, A.K., Patel, A.K., Razavi, Z.S. & McCapra, F. (1988)
     Biochem. J. 252, 143-149.
21. Campbell, A.K., Roberts, P.A. & Patel, A.K. (1985) in
     Alternative Immunoassays (Collins, W.P., ed.), Wiley,
     Chichester, pp. 153-183.
Added by Ed.- Campbell, A.K. (1989) in *Biochemical Approaches to
Cellular Calcium* [Vol. 19, this series; eds. & publisher as for prese
vol.], pp. 1-14 & 483-486.

#ncD

NOTES and COMMENTS relating to

INDIVIDUAL-CELL STUDIES, ESPECIALLY ON Ca²⁺

Comments on #D-1: M.B. Hallett - IMAGING STUDIES ON $[Ca^{2+}]_i$
       #D-2: T.R. Cheek - IMAGING STUDIES ON $[Ca^{2+}]_i$

   Question from S. Hourani.- Might the delayed responses you observe be due to stimuli released from cells which have already responded? **Hallett's reply.**- Quite possibly; imaging may reveal those types of interaction that could be hard to detect in cell populations. **A.K. Campbell, to Cheek.**- Do you agree that Ca^{2+} transit in neutrophils and adrenal chromaffin cells involves receptor-operated channels rather than $InsP_3$-releasable stores? **Reply: Yes!** **D. Scheller.**- Can you see how Ca^{2+} is removed from the cytoplasm once it has entered the cell? Is it internally stored or extruded to the outside? **Cheek's reply.**- To settle this in detail needs a particular experimental set-up. Some Ca^{2+} is extruded to the outside immediately, and some is stored in vesicles and removed with a certain latency. **Further reply to Scheller.**- We are already calculating the cytoplasmic diffusion coefficient for Ca^{2+}; diffusion is slower than in free solution. **Bruzzone** asked whether exocytosis needs not only particular localization of $[Ca^{2+}]_i$ rises but also a threshold level. **Reply.**- These two aspects are inter-linked. Possibly Ca^{2+} entry may generate a very high concentration only a few Å beneath the p.m. that we can't resolve with our technique, which has 1-2 μm resolution. **Remark by Berggren.**- The fact that you release ~30% of the total hormone content in your cells may imply that 30% of the granules are situated just beneath the p.m., or that there are different populations of granules released in different ways. **Reply to G.J. Barritt.**- The actions of nicotine on single cells show heterogeneity, but this seems to be in the phase that $[Ca^{2+}]_i$ has attained rather than its actual increase.

Comments on #D-3: H.J. Knot - USE OF FLOW CYTOMETRY
       #D-4: A.K. Campbell - BIOLUMINESCENT PROTEIN PROBES

   Knot, answering Berggren.- $[Ca^{2+}]_i$ can indeed be used as a sorting parameter. Knot asked Campbell whether introducing the actual high-affinity binding site for, say, cAMP or $InsP_3$ into the bioluminescent protein entails a risk of interfering with the physiological levels or function of the messenger itself. **Reply.**- We will have to check this when the probe is available. However, because of the high sensitivity of detecting light from the probe, the concentrations needed would be much lower than for the fluorophores currently used for Ca^{2+}, which clearly have this disadvantage. Reply to R. Heath.— It is indeed our expectation that the target site for protein insertion into the   photoprotein molecule will be universal for all proteins to be inserted.

===

SOME LITERATURE (not confined to individual-cell studies) on topics related to 'D' and, in respect of $[Ca^{2+}]_i$, to 'B' — noted by Senior Editor; attention also drawn to arts. (e.g. P.H. Cobbold: 'transients' in single cells) in Biochemical Approaches to Cellular Calcium *(Vol. 19, this series; eds. & publisher as for present book)*

Irvine, R.F. (1990) *FEBS Lett. 263*, 5-9.- A model for 'quantal' InsP-stimulated Ca^{2+} **mobilization** from intracellular stores and uptake into the cell, involving allosteric effects; besides having an $InsP_3$-binding site at its p.m. end (maybe abutting the $InsP_4$ receptor), the $InsP_3$ receptor may (spanning the p.m. and the e.r.) have a Ca^{2+}-binding site within the e.r. lumen. [Recent work, with W.T. Mason *et al.*: single–cell imaging.]
    Taylor, C.W. & Potter, B.V.L. (1990) *Biochem. J. 266*, 189-194.- Incompleteness of Ca^{2+} **mobilization** (both luminal Ca^{2+} stores and cytosol $InsP_3$ seem to be needed), relevant to Irvine's model.

Hepatocyte Ca^{2+} studies in *Biochem. Pharmacol.*: (1990) *40*, 2247-2257 (1), & (1991) *41*, 669-675 (2) & 1087-1090 (3).- (1) Nafenopin effects *in vitro* and *in vivo* (VP and amiloride for comparison); $[Ca^{2+}]_i$ increases by mobilization despite increased efflux, and pH_i transiently decreases, without InsP generation; cf. tumour promotion by chronic treatment with nafenopin.- Ochsner, M., *et al.* (2) Ca^{2+} pump in p.m.: myricetin and other flavonoids inhibit transport activity, probably through interaction with phenolic groups in the transporter.- Thiyagarajah, P., *et al.* (3) EEDQ (a dihydroxyquinoline) inhibits agonist-induced Ca^{2+} inflow across the p.m., probably by acting on the extracellular domains of the adrenaline and VP receptors.- Hughes, B.P. & Barritt, G.J.

Dugina, V.B., *et al.* (1987) *Proc. Nat. Acad. Sci. 84*, 4122-4125.- 'Special type of morphological reorganization induced by phorbol ester: reversible partition of cell into **motile and stable domains**' (fibroblast cell lines studied).
    Welling, L.W., *et al.* (1987) *Am. J. Physiol. 253*, F126-F140.- 'Morphometric analysis of distinct **microanatomy** near the base of **proximal tubule** cells'. In these there is axial heterogeneity of pH_i.- Pastoriza-Munoz, E., *et al.* (1987) *J. Clin. Invest. 80*, 207-215.   In **single cortical collecting tubules**, $[Ca^{2+}]_i$ was studied by fura 2 and a superfusion device (after microdissection).- Taniguchi, S., *et al.* (1985) *J. Biol. Chem. 260*, 3440-3450.

Tsunoda, Y., *et al.* (1991) *Exp. Cell Res. 193*, 356-363.- 'Cytosolic acidification leads to Ca^{2+} **mobilization** from intracellular stores in single and populational [gastric] parietal cells and platelets'.

===

#E

FIBRINOLYTIC, ONCOGENIC, JUNCTIONAL AND NEURAL PHENOMENA

#E-1

REGULATION OF THE FIBRINOLYTIC SYSTEM

J.W.C.M. Jansen[†] and J.H. Reinders

Duphar B.V., Department of Pharmacology,
P.O. Box 900, 1380 DA Weesp, The Netherlands

The fibrinolytic capacity of blood is determined by the balance between plasminogen activator(s), t-PA*, and their main inhibitor, PAI-1. That this balance is important is inferred from clinical studies which suggest that PAI-1 may play a crucial role in various pathological conditions associated with thrombosis. Both components are synthesized and released into blood by the vascular endothelium. Hepatocytes represent a second physiological plasma PAI-1 source, which can be modulated by insulin. As is reviewed, in vitro experiments show that the fibrinolytic balance is modulated in different ways by several essentially physiological and experimental agents, including thrombin, histamine, TNF, TGFβ, dexamethasone, phorbol ester and retinoic acid. Taking account of some supporting observations in vivo, it is suggested that the endothelial cells and hepatocytes play a key role in the regulatory system of fibrinolysis.

Haemostatic balance is based on finely tuned systems responsible for blood clot formation (the coagulation cascade plus platelet aggregation) and clot dissolution (the fibrinolytic system). The clinical consequences of an imbalanced haemostatic system are either a thrombosis or a bleeding tendency (Fig. 1). For example, an imbalanced fibrinolytic system is associated with an increased risk of bleeding or thrombotic events. This review focuses on the regulatory mechanisms, which may form the basis of a disturbed blood fibrinolytic activity. First, an outline is given of the vascular fibrinolytic system including its intrinsic regulation; then there is a discussion of regulatory aspects of fibrinolysis, *viz.* the cellular expression of factors determining the fibrinolytic balance. In addition, results from an *in vitro* model of fibrinolytic balance as well as those from an *in vivo* model are presented.[◊]

[†]addressee for any correspondence
[*]*Abbreviations.*- L-PA, tissue-type plasminogen activator (PAI-1, endothelial-type inhibitor of t-PA); α_2AP, α_2-antiplasmin; *(see text)* variant PA prefixes: u, urokinase; sc/tc, single-/two-chain. HMWK, high-mol. wt. kininogen; IL, interleukin; PG, prostaglandin; TGFβ, transforming growth factor β; TNF, tumour necrosis factor. cAMP, cyclic AMP. *See also Table 1.*
[◊]*Literature coverage has been slightly abridged.- Ed.*

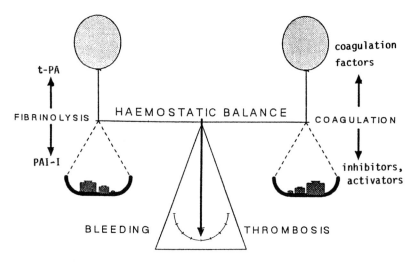

Fig. 1. Haemostatic balance. Numerous factors within the fibrinolytic and coagulation systems make up the delicate balance between bleeding and thrombosis.

Fibrinolytic system: intrinsic regulatory aspects

Fibrinolysis is an intricate system of several interacting proteolytic enzymes and their inhibitors (Fig. 2). Once a fibrin clot is formed it has to be removed sooner or later. This is performed by **plasmin** that is formed from its zymogen **plasminogen**. Plasmin digests **fibrin** into smaller polypeptides, the fibrin degradation products, which results in dissolution of the blood clot. The activation of plasminogen is by **plasminogen activators**, which are serine proteases with a high specificity for their natural substrate plasminogen. Physiological fibrinolysis is highly fibrin-specific and not associated with systemic activation of the fibrinolytic system. Specific inhibitors of plasmin and plasminogen activators are responsible for limiting proteolysis to the fibrin clot.

Two immunologically and biochemically distinct types of plasminogen activators are known at present: t-PA* (tissue type) and u-PA (urokinase type). t-PA is synthesized by endothelial cells in a single-chain form (sct-PA; M_r 72 kDa) and can be converted into a two-chain form (tct-PA) by limited proteolysis [1, 2]. The physiological relevance of this phenomenon remains unclear [3], since both activators show activities towards plasminogen that are comparable.

In purified systems without the presence of fibrin, t-PA is a poor activator of plasminogen: the K_m is high (≤ 8 μM) and k_{cat} low (0.01 sec^{-1}), while in the presence

*Abbreviations are listed at foot of previous p.

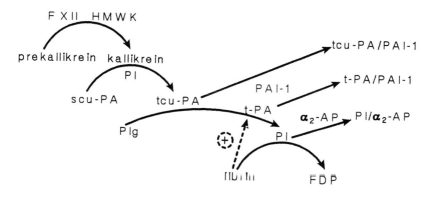

Fig. 2 The fibrinolytic system. Plasminogen (Plg) is
activated to plasmin (Pl) by either tcu-PA or t-PA. Both
plasminogen activators are inhibited by PAI-1. Pl is
inhibited by α_2AP; tcu-PA is formed from an inactive pro-
form scu-PA by Pl or kallikrein, which arises from pre-
kallikrein by the action of factor XII and HMWK. Fibrin
is a cofactor in the activation of Plg by t-PA. Finally,
fibrin is degraded by Pl into degradation products (FDP).
For abbreviations not defined here, see first p., footnote.

of fibrin the K_m decreases to <0.2 μM and k_{cat} becomes 0.1 sec^{-1}
[4, 5]. Because plasminogen circulates at a concentration
of 2 μM, plasminogen activation in the presence of fibrinogen
(i.e. on the clot surface) proceeds several orders of magnitude
more efficiently than in free circulation.

The specific fibrin binding of t-PA originates in the
so-called finger-and-kringle II structures of the t-PA molecule.
Based on experiments with t-PA mutants it appeared that
initial binding of t-PA to fibrin is mediated by the finger-
domain and that kringle II is required for lowering the
K_m towards plasminogen [6].

Single-chain u-PA (scu-PA, M_r 54 kDa) is the precursor
of the activator tcu-PA, which efficiently catalyzes the
conversion of plasminogen to plasmin but differing from t-PA
in affinity. Whereas scu-PA powerfully activates plasminogen
(K_m: 0.7 μM) and has a low k_{cat} (0.0011 sec^{-1}), tcu-PA has
a high K_m (60 μM) and a high k_{cat} (1.4 sec^{-1}) [7]. Whilst
scu-PA shows fibrin-specific thrombolytic activity in a plasma
clot-lysis system without systemic activation of the fibrinolytic system [8], it does not bind appreciably to fibrin
[9] – probably because u-PA lacks fibrin-binding domains,
having one kringle and no finger domain. Thus, the kinetics
of plasminogen activation are essential for confining fibrinolysis to the blood clot. Inhibitors of the proteases involved
are likewise essential for this.

The inhibitors concerned are α_2AP and (endothelial type) PAI-1. α_2AP (M_r 70 kDa) is a single-chain glycoprotein and is the major plasma inhibitor of plasmin. Upon formation of plasmin, α_2AP rapidly binds plasmin in plasma (rate constant $\sim 10^7$ $M^{-1}sec^{-1}$) and thereby protects against systemic proteolysis [10].   During blood clotting α_2AP is cross-linked to fibrin by factor XIIIa, a process that contributes to the resistance of clots to lysis [11, 12].

PAI-1 is a single-chain glycoprotein (M_r 52 kDa); it appears to be the primary physiological inhibitor of sct-PA, tct-PA and tcu-PA by reaction with their catalytic centres with second-order association constants $\sim 3 \times 10^7$ $M^{-1}sec^{-1}$ [13-15].   No complex formation has been observed between PAI-1 and scu-PA [15].   In plasma, PAI-1 is partly bound to **vitronectin** [16, 17], a complex which dissociates in the presence of t-PA [18].   The PAI-1/vitronectin complex is much more stable than PAI-1 as such, suggesting that the complex may stabilize PAI-1 in blood [17, 19].   *In vitro*, free PAI-1 rapidly changes its conformation into a latent form, which needs surface-active agents to restore its activity.   When this process is operational *in vivo*, negatively charged phospholipids exposed, for example, during platelet aggregation could be responsible for activating latent PAI-1 [20].

The normal concentrations of t-PA and PAI-1 in human plasma are in the respective ranges 2-13 [21, 22] and 6-60 ng/ml [22].   In various clinical conditions such as coronary artery disease and deep-vein thrombosis, increased plasma PAI-1 concentrations are found [21, 22]. This suggests that disordered regulation of the fibrinolytic system may be an independent risk factor for vascular diseases.   u-PA is also found in blood (9-27 ng/ml) but levels of u-PA are rather constant as compared to t-PA: e.g. venous stasis has no effect [18]. It is generally accepted that u-PA is involved in tissue degradation and tumour invasiveness, whereas t-PA is the plasminogen activator in thrombolysis [23].

CELLULAR SIGNALLING IN REGULATION OF FIBRINOLYSIS:
Studies *in vitro*

Blood being rich in plasminogen and its inhibitor, its fibrinolytic activity is determined by the balance between t-PA and PAI-1; the cell types relevant to the blood fibrinolytic balance are vascular endothelial cells, platelets and hepatocytes. Vessel endothelium of abdominal and breast skin has been shown histochemically to contain mainly t-PA, no u-PA being demonstrable [24] as likewise reported for primary cultures of endothelial cells from several sources [25, 26].

Depending on the tissue of origin, endothelial cells *in vitro* start to produce u-PA after one or more passages [26, 27]. Cultured renal endothelial cells have been shown to produce large amounts of u-PA, which suggests that the vascular bed of origin determines the expression of plasminogen activators [28].

So far the origin of plasma PAI-1 seems to be solely the vascular endothelial cell [29, 30]. Many cells in primary culture and cell lines including hepatocytes and hepatomas synthesize PAI-1 [31, 32]. PAI-1 is also found in the α-granules of platelets [33, 34] and is released during aggregation [35], probably to protect clots from premature lysis. As yet the contribution of non-endothelial cells to plasma PAI-1 is unclear.

Thus the endothelial cell produces both t-PA and PAI-1, factors which are associated with anti- and pro-thrombotic activities. Liver parenchymal cells are very efficient t-PA scavengers and clear t-PA rapidly from the circulation [36], and might also contribute to plasma PAI-1. Much has been learned from *in vitro* studies with these cell types and perfused organs, and many factors have been identified which regulate t-PA and PAI-1 synthesis by these cells.

PAI-1 inducing agents

Endothelial PAI-1 biosynthesis is, among other endothelial functions, an important target for cytokines such as TNF and IL-1, produced by activated monocytes/macrophages and other cells including smooth muscle cells [37]. IL-1 addition to the culture medium of endothelial cells results in high amounts of PAI-1 being secreted into the conditioned medium [e.g. 37, 38] and a decreased t-PA-antigen level [38]. Comparable results have been obtained with TNF [39] and endotoxin [37]. The effects of cytokines on PAI-1 biosynthesis in endothelial cells are at the level of PAI-1 gene expression [39, 40], but how these inflammatory mediators induce these prothrombotic effects is unclear. TNF does not affect PAI-1 production by human or rat hepatocytes [39].

Growth factors present in serum and possibly derived from platelets and white blood cells induce PAI-1 in both endothelial cells and hepatocytes (Table 1). For TGFβ, a potent inducing factor released from platelets during thrombus formation, it has been shown with cultured bovine endothelial cells that the PAI-1 increase reflects an increase in PAI-1 mRNA [41]. The PAI-1 promoter gene contains specific TGFβ-responsive elements [42]. TGFβ also activates hepatocytes (HepG2) to increase their PAI-1 synthesis [43]. EGF has no effect on PAI-1 expression in endothelial cells [44] or HTC-hepatoma (Table 1) cells. Likewise platelet-derived growth factor (PDGF) had no effect in either cell type [43].

Glucocorticoids, like the potent anti-inflammatory compound dexamethasone, suppress the fibrinolytic activity of both endothelial cells [45] and rat hepatocytes (HTC hepatoma;. [46] & Table 1). Dexamethasone increased PAI-1 levels as measured in cell lysates and the culture medium, decreasing net PA-activity [47, 48]. The dexamethasone depression of fibrinolytic activity is primarily caused by a 5- to 10-fold increase in PAI-1 activity and antigen level [48, 49]. A modest increase in t-PA antigen was also found [49]. Both effects are due to a direct effect on the expression of the t-PA and PAI-1 genes [50].

The PAI-1 secretion of endothelial cells is unaffected by insulin, but insulin induced a dose-dependent maximal 2-fold increase of PAI-1 secretion by the hepatoma cell lines HepG2 [51] and HTC (Table 1) and by primary cultures of human hepatocytes [52]. The increase in PAI-1 synthesis is preceded by a rise in PAI-1 mRNA levels [52].

On the PAI-1 promoter gene in the rat [53] and human [54], there are present the putative responsive elements for glucocorticoids, cAMP, TGFβ and phorbol ester, and that coding for the acute phase response.

t-PA inducing factors

Phorbol esters are activators of the protein kinase C pathway of signal transduction. Addition to endothelial cells of phorbol esters or activators of phospholipase C, such as histamine and thrombin, results in a stimulated t-PA release [55, 56], an effect which is preceded by an increase in t-PA mRNA levels [55]. With thrombin there is also a rise in PAI-1 [e.g. 55-57]. Retinoic acid raises t-PA [58], an effect probably mediated by activation of protein kinase C. Human endothelial cells in culture showed no effect of activators of the adenylate cyclase system or a slightly increased t-PA secretion [59, 60], but phorbol ester synergistically enhanced of t-PA synthesis [60], an effect preceded by a rise in t-PA mRNA levels [55].

The observation that intravenous epinephrine enhanced fibrinolytic activity suggested that the wide variety of stressful stimuli that have a similar effect might act *via* an adrenergic receptor [61]. t-PA in the perfusate of isolated rat hind-legs is elevated by epinephrine [62], and this can be blocked by the non-specific β-blocker propranolol

Table 1. Modulation of the fibrinolytic balance in HTC hepatoma cells: t-PA *vs*. PAI-1 activity as affected by various agents (including EGF = epidermal growth factor, and IBMX = *iso*butylmethylxanthine).

The cells were cultured till confluent in minimal Eagle's medium (MEM) with FCS and incubated with substances in MEM with 0.1% (w/v) bovine serum albumin for 24 h. Net PAI-1 activity was measured as t-PA-neutralizing activity in a chromogenic substrate assay using human t-PA, plasminogen, CNBr-treated fibrinogen and plasmin substrate 'S2251'. Each value is the mean of 3, and represents net PAI-1 activity (%) *vs*. untreated controls.- A negative value indicates that no net PAI-1 activity could be measured but there was a net t-PA activity (assuming that 1 u. t-PA can neutralize 1 u. PAI-1).

Modulator	% cont.[&]
Control	100
dibutyryl-cAMP (1 mM)	−71
IBMX (1 mM)	4
diB-cAMP + IBMX	−383
Butyrate (2 mM)	183
Butyrate + IBMX	50
Dexamethasone (0.1 μM)	1530
Dex + diB-cAMP	410
Dex + IBMX	361
EGF (10 ng/ml)	101
Foetal calf serum (1% v/v)	356
FCS + Dex	2799
FCS + diB-cAMP	104
Insulin (0.1 μM)	200
Insulin + diB-cAMP	19
Insulin + diB-cAMP + IBMX	−251

[&] PAI-1 activity

[63]. *In vitro* investigations with cultured endothelial cells do not show any increase in fibrinolytic activity in response to catecholamines [64]. The rat t-PA promoter gene possesses elements which mediate activation by phorbol esters and cAMP [65, 66].

HTC: model for *in vitro* fibrinolytic balance

A model for expression of PAI-1 + t-PA is HTC - a rat hepatoma cell line. Gelehrter and co-authors [46-48] have fully described this sytem. We evaluated the use of HTC as an *in vitro* model for the modulation of the fibrinolytic balance: the results are summarized in Table 1. Agents that raise cAMP, such as dibutyryl-cAMP and the phospho-diesterase inhibitor IBMX, induce PA activity, presumably by suppressing PAI-1 and inducing t-PA. The action of dibutyryl-cAMP was not due to the butyrate moiety of the molecule; in fact the differentiating agent butyrate induced PAI activity in these cells. The glucocorticoid dexamethasone and serum factors induced PAI, and cAMP-raising agents reduced their effects, likewise for insulin which was a moderate PAI inducer.

Table 2. Modulation of the fibrinolytic balance in rats (male Wistar), dosed orally for 5 days with dexamethasone (once daily) or retinoic acid (twice daily). Blood was taken (citrate) from the abdominal vena cava under nembutal. [^{125}I]fibrinogen was added, besides Ca^{2+} and thrombin to induce clot formation, with 1:6 final dilution. The rate of liberation of ^{125}I served as a measure of fibrinolytic activity, which was expressed as % of the placebo-treated animals (5 animals per group); values are mean ± S.E.M. (with no. of experiments).

| | Fibrinolytic activity, % | |
Dose, mg/kg	dexamethasone	retinoic acid
0.1	80 ±3 (3)	
0.3	73 ±4 (5)	127 ±12 (4)
1.0	62 ±4 (5)	167 ±11 (17)
3.0		200 ±23 (6)

MODULATION OF THE FIBRINOLYTIC SYSTEM: *IN VIVO* STUDIES

Some regulators have been tested for their effects on the fibrinolytic system *in vivo*. Intravenous administration of endotoxin [37, 67], IL-1 [37] or TNF [39] to rats or rabbits resulted after 2-4 h in decreased fibrinolytic activity in blood. TNF infusion to patients with refractory malignancies led to increased PAI-1 activity which peaked 4 h after finishing TNF infusion [e.g. 68]. Moreover, we found with dexamethasone treatment a fall in blood fibrinolytic activity in rats, and with retinoic acid [69] a rise; a stable PGI_2 analogue [70] also raised the activity.

Table 2 summarizes the effects we found in rats. The time frame of the activity change, and the lag phase, indicate an interaction of these agents at the level of gene expression of t-PA and PAI-1, rather than on acute release of fibrinolytic components.

Thrombin infusion in rabbits with high PAI-1 levels lowered the levels due to activation of protein C [71]. Although hyperinsulinaemia in humans correlates with high plasma PAI-1 levels, no significant increase occurs after insulin infusion [72].

CONCLUSIONS

Within the fibrinolytic system, t-PA and PAI-1 play a key role in the regulation of fibrinolytic activity. Both components are synthesized by the endothelial cell, but it cannot be excluded that other cell types such as hepatocytes affect the fibrinolytic activity of blood by synthesis of PAI-1. Insight into some of the factors and/or pathways important in controlling this activity has come from recent investigations. At the level of gene regulation the promoters for both t-PA and PAI-1 are characterized and will provide information about the molecular regulation of the fibrinolytic system. As a spin-off, these investigations will lead towards the development of anti-thrombotic agents whose mechanism of action is based on interaction with components of the fibrinolytic system.

Acknowledgements

We thank Dr T. Gelehrter for providing us with HTC cell. Peter Kaczmarek, Martina van der Neut, Janny Minkema, Gerardo Boon, Joost Brakkee and Hans van Giezen provided technical assistance and Janny Troost and Els Verschoof secretarial assistance.

References

1. Ichinose, A., Kisse, W. & Fugukawa, K. (1984) *FEBS Lett.* *175*, 412-418.
2. Pohl, G., Kallstrom, M., Bergsdorf, N., Wallen, P. & Jomvall, H. (1984) *Biochemistry 23*, 3701-3707.
3. Lijnen, H.R., Van Hoef, B. & Collen, D. (1990) *Thromb. Res. suppl. X*, 45-54.
4. Hoylaerts, M., Rijken, D.C., Lijnen, H.R. & Collen, D. (1982) *J. Biol. Chem. 257*, 2912-2919.
5. Ranby, M. (1982) *Biochim. Biophys. Acta 704*, 461-469.
6. Van Zonneveld, A.J., Veerman, H. & Pannekoek, H. (1986) *J. Biol. Chem. 261*, 14214-14218.
7. Stump, D.C., Lijnen, H.R. & Collen, D. (1986) *J. Biol. Chem. 261*, 1274-1278   (*see also* 1267-1273).
8. Zamarron, C., Lijnen, H.R., Van Hoef, B. & Collen, D. (1984) *Thromb. Haemostas. 52*, 19-23.
9. Gurewich, V. & Panell, R. (1987) *Blood 69*, 769-772.
10. Wiman, B. & Collen, D. (1978) *Eur. J. Biochem. 84*, 573-578.
11. Sakata, Y. & Aoki, N. (1980) *J. Clin. Invest. 65*, 290-297.
12. Jansen, J.W.C.M., Haverkate, F., Koopman, J., Nieuwenhuis, H.K., Kluft, C. & Boschman, Th.A.C. (1987) *Thromb. Haemostas. 57*, 171-175.
13. Hekman, C.M. & Loskutoff, D.J. (1988) *Arch. Biochem. Biophys. 262*, 199-210.
14. Thorsen, S., Philips, M., Selmer, J., Lecander, I. & Astedt, B. (1988) *Eur. J. Biochem. 175*, 33-39.
15. Andreasen, P.A., Nielsen, L.S., Kristensen, P., Grøndahl-Hansen, J., Shriver, L. & Danø, K. (1986) *J. Biol. Chem. 261*, 7644-7651.
16. Wiman, B., Almquist, A., Sigurdardottir, O. & Lindahl, W.T. (1988) *FEBS Lett. 242*, 125-128.
17. Declerk, P.J., De Mol, M., Alessi, M-C., Baudner, S., Pâques, E.P., Preissner, K.T., Müller-Berghaus, G. & Collen, D. (1988) *J. Biol.Chem. 263*, 15454-15461.

18. Nicoloso, G., Hauert, J., Kruithof, E.K.O., Van Melle, G.
    & Bachmann, F. (1988) *Thromb. Haemostas.* *59*, 299-303.
19. Lindahl, T.L., Sigurdardottir, O. & Wiman, B. (1989)
    Thromb. Haemostas. *62*, 748-751.
20. Lambers, J.W.J., Cammenga, M., König, B.W., Mertens, K.,
    Pannekoek, H. & Van Mourik, J.A. (1987) *J. Biol. Chem.*
    262, 17492-17496.
21. Wiman, B., Ljungberg, B., Chmielewska, J., Urdén, G.,
    Blombäck, M. & Johnsson, H. (1985) *J. Lab. Clin. Med.*
    105, 265-270.
22. Nguyen, G., Horellon, M-H., Kruithof, E.K.O., Conard, J.
    & Samama, M.M. (1988) *Blood 72*, 601-605.
23. Danø, K., Andreasen, P.A., Grøndahl-Hansen, J.,
    Kristensen, P., Nielsen, L.S. & Skriver, L. (1985) *Adv.
    Cancer Res. 44*, 139-266.
24. Kristensen, P., Larsson, L-I., Nielsen, L.S.,
    Grøndahl-Hansen, J., Andreasen, P.A. & Danø, K. (1984)
    FEBS Lett. 168, 33-37.
25. Levin, E.G. (1983) *Proc. Nat. Acad. Sci. 80*, 6804-6808.
26. Van Hinsbergh, V.W.M., Binnema, D., Scheffer, M.A.,
    Sprengers, E.D., Kooistra, T. & Rijken, D.C. (1987)
    Atherosclerosis 7, 389-400.
27. Booyse, F.M., Osikowics, G., Feder, S. & Scheinbuks, J.
    (1984) *J. Biol. Chem. 259*, 7198-7205.
28. Wojta, H., Hoover, R.L. & Daniel, T.O. (1989) *J. Biol.
    Chem. 264*, 2846-2852.
29. Emeis, J.J., Van Hinsbergh, V.W.M., Verheijen, J.H. &
    Wijngaards, J. (1983) *Biochem. Biophys. Res. Comm. 110*,
    392-398.
30. Loskutoff, D.J., Van Mourik, J.A., Erickson, L.A. &
    Lawrence, D. (1983) *Proc. Nat. Acad. Sci. 80*, 2956-2960.
31. Sprengers, E.D., Princen, H.M.G., Kooistra, T. &
    Van Hinsbergh, V.W.M. (1985) *J. Lab. Clin. Med. 105*,
    751-758.
32. Cwickel, B.J., Barouski-Millar, P.A., Coleman, P.L. &
    Gelehrter, T.D. (1984) *J. Biol. Chem. 259*, 6847-6851.
33. Booth, N.A., Anderson, J.A. & Bennett, B. (1985) *J. Clin.
    Path. 38*, 825-830.
34. Erickson, L.A., Ginsberg, M.H. & Loskutoff, D.J. (1984)
    J. Clin. Invest. 74, 1465-1472.
35. Kruithof, E.K.O., Tran-Cheng, C. & Bachmann, F. (1986)
    Thromb. Haemostas. 55, 201-205.
36. Kuiper, J., Otter, M. & Rijken, D.C. (1988) *J. Biol. Chem.
    263*, 18220-18224.
37. Emeis, J.J. & Kooistra, T. (1986) *J. Exp. Med. 163*,
    1260-1266 (*cf.* 1595-1600: Nachman & co-authors).
38. Bevilacqua, M.P., Schleef, R.R., Gimbrone, M.A. &
    Loskutoff, D.J. (1986) *J. Clin. Invest. 78*, 587-591.
39. Van Hinsbergh, V.W.M., Kooistra, T., Van den Berg, E.A.,
    Princen, H.M.G., Fiers, W. & Emeis, J.J. (1988) *Blood 72*,
    1467-1473.

40. Schleef, R.R., Bevilacqua, M.P., Sawdey, M., Gimbrone, M.A. & Loskutoff, D.J. (1988) *J. Biol. Chem. 263*, 5797-5803.

41. Sawdey, M., Podor, T.J. & Loskutoff, D.J. (1989) *J. Biol. Chem. 264*, 10396-10401.

42. Keeton, M., Van Zonneveld, A.J., Curriden, C. & Loskutoff, D.J. (1990) *Fibrinolysis 4, suppl. 3*, 175.

43. Fuijii, S., Lucore, C.L., Hopkins, W.E., Billadello, J.J. & Sobel, B.E. (1989) *Am. J. Cardiol. 63*, 1505-1511.

44. Mimuro, J. & Loskutoff, D.J. (1987) *Thromb. Haemostas. 58*, 1647.

45. Levin, E.G. & Loskutoff, D.J. (1982) *Ann. N.Y. Acad. Sci. 401*, 184-194.

46. Gelehrter, T.D., Barouski-Miller, P.A., Coleman, P.L. & Cwickel, B.J. (1983) *Mol. Cell. Biochem. 53/54*, 11-21.

47. Loskutoff, D.J., Roegner, K., Erickson, L.A., Schleef, R.R., Huttenlocker, A., Coleman, P.L. & Gelehrter, T.D. (1986) *Thromb. Haemostas. 55*, 8-11.

48. Coleman, P.L., Patel, P.D., Cwickel, B.J., Rafferty, U.M., Sznycer-Laszuk, R. & Gelehrter, T.D. (1986) *J. Biol. Chem. 261*, 4352-4357.

49. Gelehrter, T.D., Sznycer-Laszuk, R., Zeheb, R. & Cwickel, B.J. (1987) *Mol. Endocrin. 1*, 97-101.

50. Heaton, J.H. & Gelehrter, T.D. (1989) *Mol. Endocrin. 3*, 349-355.

51. Alessi, M.C., Juhan-Vague, I., Kooistra, T., Declerck, P.J. & Cwickel, B.J. (1988) *Thromb. Haemostas. 60*, 491-494.

52. Kooistra, T., Bosma, P.J., Töns, H.A.M., Van den Berg, A.P., Meyer, P. & Princen, H.M.G. (1989) *Thromb. Haemostas. 62*, 723-728.

53. Saksela, O., Moscatelli, D. & Rifhin, D.B. (1987) *J. Cell Biol. 105*, 957-963.

54. Riccio, A., Lund, L.R., Sartorio, R., Lania, A., Andreasen, P.A., Danø, K. & Blasi, F. (1988) *Nucl. Acids Res. 16*, 2805-2824.

55. Levin, E.G., Marotti, K.R. & Santell, L. (1989) *J. Biol. Chem. 264*, 16030-16036.

56. Hanss, M. & Cwickel, B.J. (1987) *J. Lab. Clin. Med. 109*, 97-104.

57. Dichek, D. & Quertermons, T. (1989) *Blood 74*, 222-228.

58. Inada, Y., Hagiwara, M., Kojima, S., Shimonaka, M. & Saito, Y. (1985) *Biochem. Biophys. Res. Comm. 13*, 182-187.

59. Thompson, E.A., Van Nuffelen, A., Nelles, L. & Collen, D. (1990) *Fibrinolysis 4, Suppl. 3*, 163.

60. Santell, L. & Levin, E.G. (1988) *J. Biol. Chem. 263*, 16802-16808.

61. Prowse, C.V. & Cash, J.D. (1984) *Sem. Thromb. Hemostas. 10*, 51-60.

62. Emeis, J.J. (1983) *Thromb. Res. 30*, 195-203.

63. Zhu, G.J., Abbadini, M., Donati, M.B. & Mussoni, L. (1989) *Am. J. Physiol. 256*, H404-H410.

64. Peracchia, F., Lipartiti, M., Fratelli, M., De Blasi, A., Donati, M.B. & Mussoni, L. (1989) *Haemostasis 19*, 235-240.

65. Van Zonneveld, A-J., Curriden, S.A. & Loskutoff, D.J. (1988) *Proc. Nat. Acad. Sci. 85*, 5525-5529.
66. Feng, P., Ohlsson, M. & Ny, T. (1990) *J. Biol. Chem. 265*, 2022-2027.
67. Colucci, M., Paramo, J.A. & Collen, D. (1985) *J. Clin. Invest. 75*, 818-824.
68. Silverman, P., Goldsmith, G.H., Spitzer, T.R., Rehmus, E.H. & Berger, N.A. (1990) *J. Clin. Oncol. 8*, 468-475.
69. Jansen, J.W.C.M. & Van Giezen, J.J.J. (1990) *Fibrinolysis 4, Suppl. 1*, 108.
70. Schneider, J. (1987) *Thromb. Res. 48*, 233-244.
71. Colucci, M., Triggiani, R., Calvallo, L.G. & Semeraro, N. (1989) *Blood 74*, 1976-1982.
72. Potter van Loon, B.J., De Bout, A.C.D., Radder, I.K., Frölich, M., Kluft, C. & Meinders, A.E. (1990) *Fibrinolysis 4, Suppl. 2*, 93-94.

ADDENDUM *(by request of Ed.)*: FIBRINOLYTIC METHODOLOGY

Experiments with cultured cells.- Endothelial cells (EC) are commonly isolated from human umbilical cord vessels. EC start to make PAI-1 immediately after isolation [73]. After some passages EC also start to produce u-PA [74]. Cell levels of t-PA and PAI-1 are difficult to measure because their secretion into the medium or blood is *via* binding protein-linked processes [e.g. 75]; hence effects are mostly measured in the medium, suitably by chromogenic substrate assay [76] (recent t-PA/u-PA/PAI-1 ELISA's apply only to the human). The cell lines requisite for hepatocyte culture [e.g. 31, 32, 49, 50, 51] can give aberrant results [51, 52, 77].

Animal experiments, e.g. with perfused rat hindlegs [62], entail assays such as the diluted blood clot lysis test [78] or the chromogenic substrate test. As yet t-PA, u-PA or PAI-1 levels cannot be measured quantitatively in animal species. There is a semi-quantitative method based on active components [79]: electrophoretic separation of the proteins is followed by incubation of the gel on a fibrin layer, which results in lysed zones (PA-activity) and lysis-resistant zones (PAI-1).

Additional references [T = Thromb(osis), H = Haemostas(is)]

73. Van den Berg, E.A., *et al.* (1988) *T H 60*, 63-67.
74. Van Hinsbergh, V.W.M. (1988) *H 18*, 307-327.
75. Hajhar, K.A. & Hamel, N.M. (1990) *J. Biol. Chem. 265*, 2908-2916.
76. Verheijen, J.H., *et al.* (1982) *T H 48*, 266-269.
77. Heaton, J.H., *et al.* (1989) *Mol. Endocrin 3*, 185-192.
78. Fearnley, G.R., *et al.* (1957) *Clin. Sci. 16*, 645-650.
79. Erickson, L.A., *et al.* (1984) *Anal. Biochem. 137*, 454-463.

#E-2

MEMBRANE-CYTOSKELETAL TARGETS FOR TYROSINE KINASES

Stuart Kellie[†], Andrea R. Horvath[⊗], Moira A. Elmore[†],
George Felice and Radhika Anand

Department of Biochemistry and Cell Biology,
The Hunterian Institute, Royal College of Surgeons of England,
Lincoln's Inn Fields, London   WC2A 3PN,   U.K.

TK's* may play a fundamental role in neoplastic transformation by several oncogenes and in ligand- or agonist-induced cell activation, but their intracellular targets are uncertain [1]. Recent data suggest that major targets for pp60[v-src] (RSV-generated TK) are membrane- or cytoskeletal-associated [2].   For two of these,   vinculin and the fibronectin receptor, the localization and phosphorylation have been investigated in CEF's transformed either by wild-type RSV, which induces a rounded, less adhesive phenotype with loss of surface fibronectin and gross cytoskeletal changes,   or by a variant which induces a flat, adhesive morphology with fibronectin survival and little cytoskeletal change.   Tyr phosphorylation occurred with vinculin in both transformed cell types, but with the fibronectin receptor only in the wild-type transformed cells.   Thus there was a correlation between integrin phosphorylation and loss of surface fibronectin. A Tyr phosphatase inhibitor, sodium orthovanadate, was needed for efficient recovery of phosphoTyr.   Since platelets are rich in the cellular pp60[c-src] [3],   a platelet fibronectin receptor, glycoprotein IIb/IIIa, was investigated for phosphorylation of membranes in vitro, disclosing an alkali-resistant phosphoprotein of ~100 kDa which was also immuno-precipitated by an anti-phosphoTyr Ab and co-migrated with gpIIIa.   IIIa phosphorylation was confirmed by comparing reduced and non-reduced SDS gels, and phospho-amino acid analysis showed that this protein was phosphorylated exclusively on Tyr. Our results suggest that Tyr-specific phosphorylation of matrix receptors by pp60 may play a regulatory role in both transformed and untransformed cell function.

MEMBRANE-CYTOSKELETAL SUBSTRATES FOR pp60[v-src] IN RSV-TRANSFORMED CELLS

Oncogenic transformation of cells by RSV* is mediated by the v-src gene-encoded TK, pp60[v-src] [4, 5] which is localized beneath the plasma membrane and is accumulated in the region

[†]now at Yamanouchi Research Institute, Littlemore Hospital, Oxford OX4 4XN. [⊗] now at Dept. of Clinical Chemistry, University School of Medicine, II 4012 Debrecen, P.O. Box 40, Hungary.

*Abbreviations (see also text & legends).- Ab, antibody; CEF, chick embryo fibroblasts; gp, glycoprotein; RSV, Rous sarcoma virus (see text for variants, e.g. rASV2234.3 and PrC);   PAGE, polyacrylamide gel electrophoresis; Tyr, tyrosine/tyrosyl; TK, Tyr kinase (pp60, a particular TK).

of transformed-cell structures responsible for adhesion [6, 7]. Several integral membrane proteins have been localized to these specialized regions of transmembrane linkages including extracellular matrix receptors (integrins) with multiple affinities for fibronectin, vitronectin, laminin and other matrix components [8-10], and the cytoskeletal proteins talin [11] and vinculin [12].* Interestingly, all of these components can act as substrates for pp60^{v-src}. Its localization in cell-cell and cell-matrix contacts suggested that phosphorylation of Tyr residues on target proteins such as vinculin and talin may result in the disassembly of focal adhesions [6]. However, later work has shown that Tyr phosphorylation of these cytoskeletal substrates appears to be irrelevant to the transformed phenotype [13-16].

Recently it has been suggested that Tyr-specific phosphorylation of integrin might mediate the reduction in adhesiveness and the loss of adhesion plaques of RSV-transformed cells. The Tyr phosphorylation site of the integrin β_1 subunit is in the cytoplasmic domain, in a region apparently involved in talin binding [17]. However, the functional relevance of Tyr-specific phosphorylation of integrin within whole cells remains to be determined. We have therefore explored the physiological significance of Tyr phosphorylation of integrins and its correlation with the altered phenotype in CEF's transformed by wild-type RSV or by a variant which induces a flatter, fusiform morphology.

The interaction of RSV-transformed cells with fibronectin

Cells were transformed by wild-type Prague C-strain RSV (PrC) and the non-conditional variant rASV2234.3 (Morf F) which confers a rather flat, fusiform morphology. Although rASV2234.3-transformed CEF possess high levels of pp60^{v-src} they retain stress fibres and a fibronectin meshwork on their surface (Fig. 1) [14]. We investigated whether these cells had the ability to interact with exogenous fibronectin. RSV-transformed CEF were plated onto a fibronectin substratum in serum-free medium and allowed to spread for 3 h, then fixed and stained for F-actin and vinculin. Whereas untransformed and rASV2234.3-transformed CEF spread and formed numerous microfilaments and bundles and vinculin-containing adhesion plaques, PrC-transformed CEF spread poorly and formed only limited numbers of microfilament bundles and adhesion plaques (Fig. 2). Therefore PrC-transformed cells are deficient in interacting with fibronectin whereas rASV2234.3-transformed cells can interact normally.

*In earlier vols. (Plenum; ed. E. Reid et al.) there is pertinent material, notably on fibronectin receptor (V. 13; Morré), on pp60-kinase substrates (V. 15; Owens et al.) and on the membrane-cytoskeletal axis (V. 17; Crawford, Holme & Kellie).- Ed.

Fig. 1.
Localization studies:
morphology (*right-hand panels*; phase
contrast) and
fibronectin (*left-hand panels*; anti-fibronectin immuno-fluorescence).

a, b: untransformed
CEF.
c, d: PrC-RSV-
transformed CEF.
e, f: rASV2234.3-
transformed CEF.

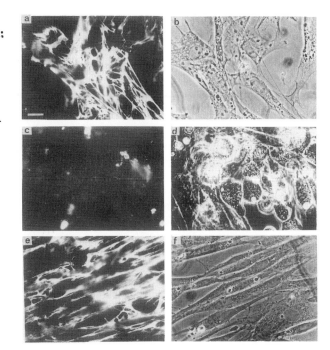

Localization of phosphotyrosyl proteins in RSV-transformed cells

Phospho-amino acid analysis of total phosphorylated cellular proteins revealed that the phosphoTyr content was elevated
in both PrC-transformed and rASV2234.3-transformed CEF to
similar levels, ~1%, showing that pp60^{v-src} is not only expressed
but also is active as a Tyr kinase in both cell types.
Immunofluorescence, using an anti-phosphoTyr Ab, revealed that
phosphoTyr-containing proteins in PrC-transformed CEF had a
discrete distribution corresponding to podosomes or point
contacts. In rASV2234.3-transformed CEF the proteins correlated
with surviving adhesion plaques or podosome-type adhesions,
or were aligned with portions of stress fibres (Fig. 3).
Therefore in both transformed cell types phosphoTyr proteins
were located in adhesion structures.

Phosphorylation of integrin in RSV-transformed CEF

Since many proteins in RSV-transformed cells appear to
be phosphorylated fortuitously, it is important to identify
those proteins whose phosphorylation is requisite for the
transformed process. Many phosphoTyr proteins are concentrated
in adhesion plaques; but there is no general correlation
between the presence of Tyr-phosphorylated proteins in adhesion
structures and the rounded morphology [2]. We therefore
investigated the phosphorylation of integrin in PrC-transformed

Fig. 2. Immuno-
fluorescence of
F-actin *(left-hand
panels)* and
vinculin *(right-
hand panels)* in
CEF plated onto a
fibronectin sub-
stratum.

A: Untransformed
CEF.
B: PrC-transfor-
med CEF.
C: rASV2234.3-
transformed CEF.

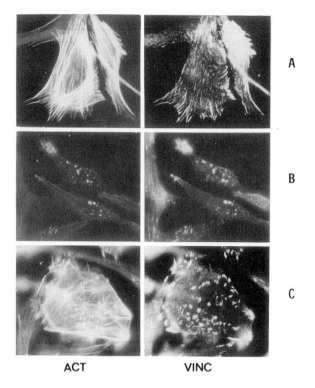

ACT   VINC

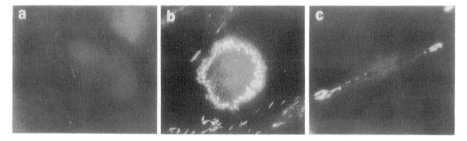

Fig. 3. Localization of phosphoTyr-containing proteins in RSV-
CEF, fixed and stained with monoclonal PY20: a, untransformed
CEF; b, PrC-transformed CEF; c, rASV2234.3-transformed CEF.

CEF which lose fibronectin or in rASV2234.3-transformed CEF
which retain a fibronectin matrix. Integrins were immunoprecipitated from cell lysates of untransformed and RSV-transformed
CEF labelled with [^{32}P]orthophosphate. The phosphorylation
of bands 2 and 3 (the β chains) of integrin in PrC-transformed
CEF was significantly increased compared to that of untransformed or rASV2234.3-transformed CEF (Fig. 4). The phosphorylation
of band 3 from rASV2234.3-transformed cells was only slightly

Fig. 4. Immuno-
precipitation of
integrin from ^{32}P-
labelled CEF (CELLS)
or *in vitro* phospho-
rylated CEF membranes
(MEM) using CSAT
monoclonal Ab.*
UN, untransformed
CEF; WT, PrC-RSV-
transformed CEF;
MF, rASV2234.3-
transformed CEF;
PRE, *same* but
using control non-
immune Ig.
Integrin bands:
1, α chain;
2, α chain;
3, β chain.
*See Acknowledgements

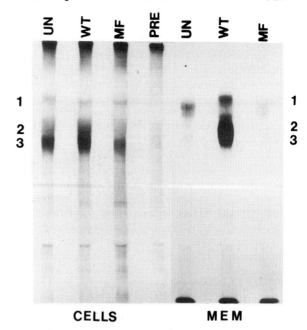

elevated compared with untransformed cells, and in some experiments there was no observable difference. *In vitro* phosphorylation of crude membrane preparations from normal and RSV-transformed CEF demonstrated results broadly similar to those
found *in vivo* (Fig. 4), with membranes from PrC-transformed
cells displaying elevated phosphorylation of integrin bands
2 and 3 compared with membranes from rASV2234.3-transformed
or untransformed cells.

In agreement with Tapley & co-authors [17] we found
that the level of phosphorylation was much higher in the
presence of sodium orthovanadate. Analysis for phosphoamino
acids revealed that integrin band 3 of PrC-transformed CEF
was phosphorylated almost exclusively on Tyr while hardly
any was detectable in untransformed or rASV2234.3-transformed
CEF. Thus the differential phosphorylation of integrin found
in our cell types was entirely due to Tyr phosphorylation.
While our experiments did not attempt to investigate the
correlation between the morphological changes, the ability
to interact with fibronectin and Tyr-phosphorylation of integrin
suggests that this post-translational modification of integrin
molecules is a physiologically important event in the induction
of the transformed phenotype.

A SUBSTRATE FOR pp60^{c-src} IN HUMAN PLATELETS

Matrix receptors on the platelet membrane play a major
role in activation [18]. Glycoprotein gpIIb-IIIa is the
major platelet receptor for fibronectin, although it will

also bind to von Willebrand factor, fibronectin and thrombo-
spondin [19], and loss of gpIIb-IIIa from platelets leads
to bleeding disorders such as Glanzmann's thrombasthenia [20].
GpIIb-IIIa is a complex of two transmembrane glycoprotein
subunits and belongs to the integrin superfamily of adhesion
receptors, other members of which have fundamental roles
in cell-matrix and cell-cell interactions [9]. In the resting
platelet, although the intact heterodimer is present on the
cell surface, gpIIb-IIIa does not bind fibronectin; however,
following activation the binding site becomes available.

The major TK in platelet membranes is $pp60^{c-src}$ which
represents 0.2-0.4% of the total protein [3]; but no substrates
for this enzyme within the platelet have yet been identified.
We have already shown that the β-chain of the fibronectin
receptor can be phosphorylated by the oncogenic homologue
of $pp60^{c-src}$, $pp60^{v-src}$. Since gpIIIa has extensive homology
with other integrin β-chains it is therefore a potential
target for the platelet enzyme, as we have investigated.

Purified platelet surface membranes were phosphorylated
in vitro and subjected to immunoprecipitation with monoclonal
anti-gpIIb-IIIa $MM_2/356$. A phosphopeptide of ~100 kDa having
the same electrophoretic mobility as gpIIIa was precipitated
(Fig. 5). Electrophoresis under reducing and non-reducing
conditions also revealed a shift in the M_r of the phosphorylated
band characteristic of that of gpIIIa. Coomassie Blue staining
of the gel showed that both gpIIb and gpIIIa were precipitated
by the Ab; therefore the IIIa subunit of this complex was
specifically phosphorylated in this membrane (Fig. 5). The
gpIIIa was phosphorylated exclusively on Tyr (data not shown).
Platelets contain large amounts of the cellular homologue
$pp60^{c-src}$, and the IIIa subunit contains a Tyr phosphorylation
site similar to that of the fibroblast β_1 integrin chain
[21]. Therefore it is likely that $pp60^{c-src}$ is the enzyme
responsible for phosphorylation of gpIIIa in these membrane
preparations. Although these experiments were performed on
isolated membranes, conceivably a similar situation may exist
within whole platelets, and our data suggest that the phosphorylation of gpIIIa by $pp60^{c-src}$ in platelet membranes may be
an important regulatory process in platelet function.

CONCLUSION

We have shown that integrins are substrates for both
the oncogenic TK $pp60^{v-src}$ and the non-oncogenic proto-oncogene
$pp60^{c-src}$. Tyr-specific phosphorylation of the integrin β-chain
may result in an alteration in the affinity of the receptor
for extracellular matrix components, leading to changes in
cell-matrix interactions which occur after neoplastic transformation or platelet activation.

Fig. 5. Immunoprecipi-
tation of ^{32}P-labelled
gpIIb–IIIa from platelet
surface membranes.
These were phosphoryla-
ted *in vitro* and gpIIb–IIIa
immunoprecipitated using
monoclonal Ab MM$_2$/356 and
subjected to SDS-PAGE
under non-reducing
(NONRED) or reducing
(RED) conditions (5%
2-mercaptoethanol).
PRE, immunoprecipitation
with non-immune Ig, and
IIb/IIIa with anti-
gpIIb–IIIa.

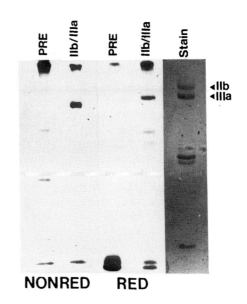

Acknowledgements

This work was supported by the Medical Research Council.
S.K. thanks the Smith & Nephew Foundation for core support.
We thank Drs. C. Buck and J.M. Wilkinson for gifts of monoclonal
Ab's CSAT and MM$_2$/356 respectively.

References

1. Wyke, J.A. & Stoker, A.W. (1987) *Biochim. Biophys. Acta 907*,
    47–70.
2. Kellie, S. (1989) *BioEssays 8*, 25–30.
3. Golden, A., Nemeth, S.P. & Brugge, J.S. (1986) *Proc. Nat.
    Acad. Sci. 83*, 852–856.
4. Bishop, J.M. (1983) *Annu. Rev. Biochem. 52*, 301–354.
5. Collet, M.S., Purchisio, A.F. & Erikson, R.L. (1980)
    Nature 285, 167–169.
6. Rohrschneider, L.R. (1980) *Proc. Nat. Acad. Sci. 77*, 3514–3518.
7. Nigg, E.A., Sefton, B.M., Hunter, T., Walter, G. &
    Singer, S.J. (1982) *Proc. Nat. Acad. Sci. 79*, 5322–5326.
8. Neff, N.T., Lowrey, C., Decker, C., Tover, A., Damsky, C.,
    Buck, C. & Horwitz, A.F. (1982) *J. Cell Biol. 95*, 654–666.
9. Hynes, R.O. (1987) *Cell 48*, 549–554.

10. Buck, C.A. & Horwitz, A.F. (1987) *Annu. Rev. Cell Biol. 3*, 179-205.
11. Horwitz, A., Duggan, K., Buck, C., Beckerle, M.C. & Burridge, K. (1986) *Nature 320*, 531-533.
12. Burridge, K. & Mangeat, P. (1984) *Nature 308*, 744-746.
13. Rohrschneider, L.R. & Rosok, M.J. (1983) *Mol. Cell. Biol. 3*, 731-746.
14. Kellie, S., Patel, B., Mitchell, A., Critchley, D.R. & Wyke, J.A. (1986) *Exp. Cell Res. 165*, 216-228.
15. Kellie, S., Patel, B., Mitchell, A., Critchley, D.R., Wigglesworth, N.M. & Syke, J.A. (1986) *J. Cell Sci. 82*, 129-141.
16. DeClue, J.E. & Martin, G.S. (1987) *Mol. Cell.Biol. 7*, 371-378.
17. Tapley, P., Horwitz, A., Buck, C., Duggan, K. & Rohrschneider, L. (1989) *Oncogene 4*, 325-333.
18. Fox, J.E.B. (1985) in *Platelet Membrane Glycoproteins* (George, J.N., Nurden, A.T. & Phillips, D.R., eds.), Plenum, New York, pp. 273-298.
19. Phillips, D.R., Charo, I.F., Parise, L.V. & Fitzgerald, L.A. (1988) *Blood 71*, 831-843.
20. Clemetson, K.J. & Luscher, E.F. (1988) *Biochim. Biophys. Acta 947*, 53-73.
21. Ruoslahti, E. (1988) *Annu. Rev. Biochem. 57*, 375-413.

#E-3

GAP JUNCTIONS AND CELL-TO-CELL SIGNALLING

[1]Roberto Bruzzone[†], [2]Marc Chanson, [1]Katherine I. Swenson,
[2]Domenico Bosco, [2]Paolo Meda and [1]David L. Paul

[1]Department of Anatomy and Cellular Biology,
Harvard Medical School,
25 Shattuck Street, Boston, MA 02115, U.S.A.

[2]Department of Morphology, University of Geneva,
1211 Geneva 4, Switzerland

Gap junctions are specialized membrane channels that enable the intercytoplasmic exchange of small molecules and ions between contacting cells. They share a basic structural similarity, but it is becoming evident that there exists a family of related proteins, connexins, *that form the cell-to-cell channel. The functional properties of individual Cx's* * can be studied by using the* Xenopus *oocyte expression system, allowing investigation of the ability of Cx's to form homomolecular and heteromolecular channels. Moreover, by constructing and expressing chimaeric Cx's, this system may allow delineation of the role of distinct domains of Cx's in the formation and regulation of the cell-to-cell pore.*

Although Cx's are ubiquitous and show molecular diversity, their functions are elusive except for excitable tissues where gap junctions conduct electrical impulses. As recently proposed, intercellular coupling may be involved in the regulation of the specific secretory activity of various cell types. The further characterization of the apparent correlation between the control of junctional permeability and that of secretion requires a combined biochemical and physiological approach. Pancreatic acinar cells represent an excellent model, since cell-to-cell communication between enzymatically isolated pairs of cells can be assessed by double whole-cell patch clamp, while secretion can be monitored by a reverse haemolytic plaque assay. Thereby one could test more directly whether dynamic adaptations of junctional communication provide a pathway for the fine regulation of the rate at which enzymes are released by pancreatic acinar cells.

In many tissues the response to an external stimulus involves the coordinated activity of groups of cells. So as to achieve a rapid transfer of signals, most cell types

[†]addressee for any correspondence
*Abbreviations.- [Ca^{2+}]$_c$, cytosolic free Ca^{2+} concentration; cAMP, cyclic AMP; Cx, connexin; InsP$_3$, inositol trisphosphate; KRB, Krebs-Ringer-bicarbonate medium; MB, Barth's saline without added Ca^{2+}; PKC, protein kinase C; p.m., plasma membrane.

are endowed with membrane channels that permit the direct exchange of low mol. wt. molecules and ions between adjacent cells without secretion into the extracellular space. These structures are clustered in specialized membrane structures, the gap junctions [1]. The junctional channel, the *connexon*, is a collection of 6 protein subunits spanning the p.m.*, which are paired with another hexamer of the adjacent cell to delineate a dodecameric channel across the intercellular gap [2]. Freeze-fracture analysis reveals an array of regularly packed membrane particles, each representing a connexon [3]. The molecular cloning of the major protein component of liver gap junctions [4-6] has opened the way to the characterization of a family of related proteins, the *connexins* (Cx's), that are expressed in different tissues (as reviewed: [7, 8]). The two best studied members of this protein family are Cx32, abundant in liver and exocrine pancreas [4, 5, 9] and Cx43, present in heart, smooth muscle and endocrine pancreas [9, 10].

A topological model of Cx's has been proposed on the basis of their deduced amino acid sequence, and its validity tested by a combination of biochemical and immunological experiments [11-13]. The predicted structure assumes that there are 4 hydrophobic domains which traverse the p.m. and that both the amino- and carboxy-termini are on the cytoplasmic side (Fig. 1). The cytoplasmic portions are very divergent, whereas the transmembrane and extracellular regions reveal a high degree of homology among members of the Cx family [7, 8]. The reason for the multiplicity of Cx's is not understood. Although many biological roles have been attributed to this direct form of cell-to-cell communication, only a few have received some experimental support, such as synchronization of myocardial cells, or regulation of cell growth, of glandular secretion and of cellular differentiation [1]. The demonstration that cAMP, $InsP_3$ and Ca^{2+} can diffuse through gap junctions and control biological responses in the coupled cells [14-18] suggests that second-messenger systems may use this direct pathway of cell-to-cell communication. In this perspective, the study of the properties and assembly of gap junction channels, and of the regulation of electrical conductance between cells, appears a requisite for understanding how a localized stimulus is translated into a multicellular response.

In this article we first summarize the characteristics of the *Xenopus* oocyte expression system, which has proved advantageous in defining the functional properties of individual Cx's. We also discuss the strategies that can be adopted to delineate the role of different protein domains

---

Abbreviations are at foot of previous p.

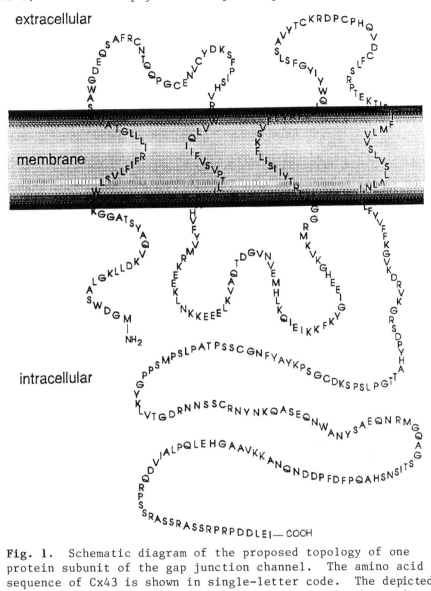

Fig. 1. Schematic diagram of the proposed topology of one
protein subunit of the gap junction channel. The amino acid
sequence of Cx43 is shown in single-letter code. The depicted
model predicts that Cx's have 4 transmembrane domains and
have both the N- and the C-termini on the cytoplasmic side of
the junctional membrane. The sequence of the 2 extracellular
and 4 transmembrane regions is highly conserved among
members of the Cx family, whereas both the central cytoplasmic
loop and the C-terminal tail are very divergent.

(Text, continued)

in the formation of the channel. Finally we review the
evidence suggesting a role of functional coupling in the
control of pancreatic enzyme secretion, and outline the
features and advantages of a two-cell system in order to
address these functional questions.

XENOPUS OOCYTES

The use of *Xenopus* oocytes has proved to be an invaluable tool for studying protein synthesis and function (review: [19]). When injected with foreign mRNA's, *Xenopus* oocytes faithfully carry out the synthesis and post-translational modifications of many proteins. Oocytes can target proteins to the correct membrane compartment and also export secretory proteins [19]. Because of its versatility, this system has been used to study the biosynthesis of complex receptor molecules, such as the nicotinic receptor, which involves interactions among subunits in order to achieve functional expression [20]. These features have been exploited by Dahl & co-workers [21] to induce the formation of gap junction channels between manually paired oocytes injected with uterine smooth muscle mRNA. Since their pioneer work, this approach has been successfully applied to investigate the functional characteristics of various members of the Cx family [22-25]. The main features of this experimental strategy are outlined below.

Preparation, microinjection and pairing of *Xenopus* oocytes

Mature *Xenopus* females are anaesthetized on ice, and pieces of the ovaries are removed following laparotomy. Because a typical experiment rarely requires >100 oocytes, the same animal can be used several times. Ovarian lobes are collected into Ca^{2+}-free saline (MB), cut into long strips and then digested (~20 min, room temp.) with collagenase (20 mg/ml in MB). After dilution with excess MB, oocytes are manually dissected away from the lobe and the surrounding follicular cell layers in MB supplemented with protease inhibitors: chymostatin, pepstatin-A, leupeptin and soybean trypsin inhibitor (each 10 µg/ml), and aprotinin (Trasylol; 10 Kallikrein u./ml) [23]. For optimum results this incubation with inhibitors should not exceed 2 h. Because mechanical dissociation inevitably causes some damage to cells, defolliculated oocytes are routinely incubated at 19° overnight in MB supplemented with $Ca(NO_3)_2$ (to 0.32 mM) and $CaCl_2$ (to 0.41 mM), whereafter the damaged ones can easily be identified.

For microinjection, defolliculated stage VI oocytes, distinguished by size, are transferred to a Petri dish and aligned along a ridge created by sticking a plastic bar onto the bottom of the dish. Oocytes are microinjected near the centres of the vegetal poles with ~40 nl of autoclaved water or mRNA, according to the standard protocols [19, 23]. Synthetic mRNA's are produced *in vitro* using the transcription vector SP64T [26] and dissolved (5-20 µg/ml) in sterile water. The efficiency of translation is enhanced

Fig. 2. A pair of oocytes, pre-
injected (water or mRNA) and freed
from vitelline membranes by dissection and then shrinkage by hyper-
tonicity; the oocytes, manipulated
into pairs with the vegetal poles
opposed, are incubated overnight
at 19°, then injected Cx's are
studied (see text). × 18.

by cloning the Cx cDNA's between the 5' and the 3' non-coding
sequences of the oocyte globin gene [23].   Finally, to
allow intimate contact between the oolemma of adjacent cells,
the vitelline membrane needs to be removed, following shrinkage
of oocytes by incubation in hyperosmotic medium [20].

Oocytes are transferred to culture dishes and then manipulated together to form pairs (Fig. 2).   So as to keep cells
in contact, pairs are positioned in the concavity of Teflon
tubing cut longitudinally.   Oocytes are always paired with
the vegetal poles opposed, to facilitate comparisons between
different experimental conditions, and are incubated overnight
at 19° in MB with added Ca^{2+}, before measurements of electrical
conductance are made.

Expression of connexins and formation of cell-to-cell channels

The functional properties of cell-to-cell channels are
assessed using a double voltage clamp procedure which enables
junctional conductance to be directly quantitated [27].   Each
cell of a pair is impaled with one voltage and one current
electrode (1-2 Mohms).   The two cells are voltage-clamped,
usually at -80 mV, and alternating symmetrical depolarizing
pulses of various amplitudes are imposed. The current supplied
by the voltage clamp to the cell not stepped is equal in
amplitude to the junctional current; junctional conductance
is the ratio of junctional current to the trans-junctional
voltage step. It should be recognized that *Xenopus* oocytes
form gap junctions with adjacent follicular cells *in vivo*
[28] and, therefore, may generate a background level of
junctional conductance when paired together.   In virtually
all water-injected pairs tested, however, endogenous electrical
coupling was undetectable (see below).   To cope with this
potential source of error, experimental and control conductances
were always measured in the same batch of oocytes.   It
appears that the incidence of spontaneous coupling can be
minimized by tightly maintaining the incubation temperature
at 19° throughout the electrophysiological experiments (pers.
comm. from L. Ebihara).

Oocytes injected with synthetic mRNA's encoding Cx32 and Cx43 express both proteins, as detected by analysis of total proteins and immunoprecipitates [23]. Pending a detailed study, preliminary results indicate that pairs injected with Cx32 mRNA form structures which exhibit some of the ultrastructural features of gap junctions [29]. Pairs of oocytes expressing different Cx's through injection of either Cx32 or Cx43 mRNA's develop large junctional conductances, the μS values (with S.E.M. for the stated no. of pairs) being as follows.-

H_2O/H_2O (n = 28):   not detectable;
Cx43/Cx43   (n = 19):   16.1 ±2.9;
Cx32/Cx32   (n = 10):   10.2 ±4.3;
Cx43/H_2O (n = 3):   8.2 ±2.8 (see comment below);
Cx32/H_2O (n = 6):   not detectable.

Evidently with either Cx there develop cell-to-cell channels. These are assumed to represent homomolecular channels. Gap junction channels between oocyte pairs composed of either Cx32 or Cx43 appear to be largely voltage-insensitive (Fig. 3) [23, but cf. 30].

By pairing heterologously-injected oocytes, it has been demonstrated that Cx32 and Cx43 can form heteromolecular channels. Proof of the expression of such channels has come from the discovery of a unique property of Cx43 [23]. While no conductance is detectable between pairs of oocytes injected with Cx32 mRNA and water, pairs composed of Cx43 mRNA- and water-injected oocytes develop junctional coupling (see values above). Because the levels of junctional coupling are similar to those observed with homologous Cx43/Cx43 channels, it can be inferred that Cx43 stimulates the assembly of endogenous oocyte proteins into gap-junction channels. The electrical characteristics of conductances developed by hetero-logous channels are easily distinguishable from those of homologous channels, since they display a markedly asymmetrical voltage-inactivation (Fig. 3B). Two members of the Cx family have been identified in *Xenopus* oocytes [25, 31]. The precise identity of the protein which forms gap-junction channels with Cx43 is not known. Channels consisting of *Xenopus* Cx38, however, display voltage inactivation similar in character to that observed for endogenous channels in Cx43/water pairs [23, 25].

Perspectives

Despite the unresolved issues, studies carried out with pairs of *Xenopus* oocytes have already defined unique character-istics of different Cx's [8, 23, 24]. For example, it has been demonstrated that the ability to form cell-to-cell

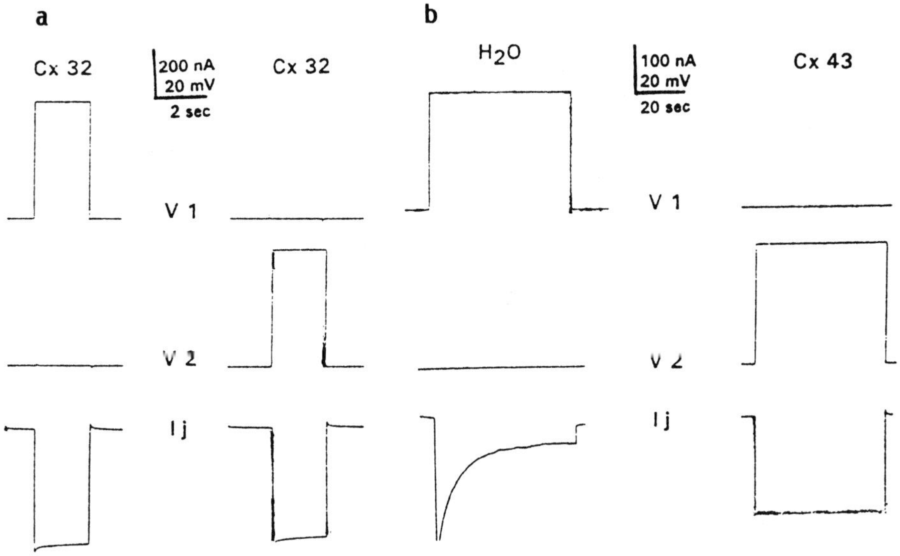

a   **b**

Cx 32   200 nA   Cx 32 H_2O 100 nA   Cx 43
             20 mV   20 mV
             2 sec   20 sec

V 1   _____   V 1   _____

V 2   V 2

I j   I j

Fig. 3. Electrical characteristics of the junctional conductance developed by pairs of oocytes injected with synthetic mRNA, coding for **(A)** Cx32 or **(B)** Cx43, and with water. Both cells are initially clamped at -80 mV to ensure zero trans-junctional potential. Depolarizing voltage steps (60 mV) are sequentially applied, and the resulting junctional current is recorded and calculated (see text). Under these conditions, junctional conductance in **(A)** displays no significant voltage-dependent inactivation, and that in **(B)** is asymmetrically voltage-dependent such that depolarization of cell #1, which was injected with water, evokes a current which decreases with time. Each record is representative of several obtained.

channels is not limited by differences in the Cx's expressed by two individual cells. The next goal is to delineate the molecular mechanisms underlying the assembly and regulation of gap-junction channels between neighbouring cells.

The identification of protein domains which determine these characteristics may be achieved by expressing chimaeric proteins in *Xenopus* oocytes. Fig. 4 outlines the strategy adopted to obtain chimaeric Cx's. In this example, the cDNA's for Cx32 and Cx43 are subcloned in tandem into the same vector, and then single-stranded DNA is rescued with the helper phage R408. An artificial splice site is created by annealing an oligonucleotide (32-mer) complementary to both a region of Cx32, in this example the last 16 nucleotides of the fourth transmembrane domain, and a region within Cx43, here the first 16 nucleotides of the C-terminal cytoplasmic tail. Double-stranded DNA is eventually produced by polymerase/ligase treatment, followed by nuclease digestion to remove

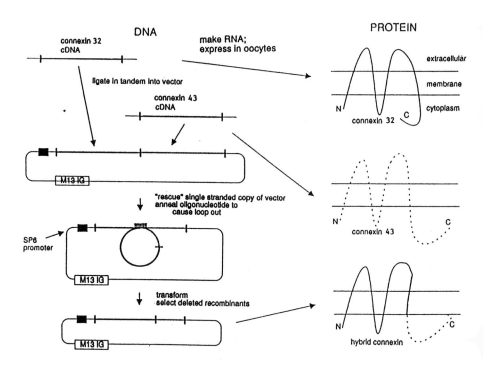

Fig. 4. Schematic representation of the strategy adopted to construct chimaeric proteins composed of combinations of Cx32 and Cx43. Chimaeric proteins are prepared by exchanging corresponding domains between the two Cx's. In this example the cytoplasmic C-terminus of Cx32 has been deleted and replaced by the corresponding portion of Cx43. Similarly, progressive deletions/exchanges towards the N-terminus can be created in order to identify the protein domains which confer different functional characteristics on members of the Cx family.

the portion of single-stranded DNA gapped by the synthetic oligonucleotide. Thus, the new construct is a hybrid composed of Cx32 until the end of the fourth transmembrane domain, with the C-terminus of Cx43 (Fig. 4).

Expression of chimaeric Cx's will enable one to determine which domains of Cx43 are necessary to confer on Cx32 the ability to form channels with the endogenous protein. The use of hybrids, in combination with site-directed mutagenesis, will allow detailed study of the structure-function relationships of Cx's and may contribute to a general understanding of the mechanisms of regulation of membrane channels.

EXOCRINE ACINAR CELLS

The functional unit of the exocrine pancreas, the *acinus*, is a cluster of apparently alike cells oriented around a common central lumen into which enzymes are discharged. These cells act mainly to synthesize, store and release digestive enzymes in response to appropriate stimuli such as hormones and neurotransmitters [32]. Since the final secretory product is the result of the coordinated activity of numerous cells, it is not surprising that they may exchange information in order to sense the functional state of their neighbours. Thus, acinar cells are connected to each other by large gap junctions [33, 34], each comprising 60 to 138,000 particles (median value = 2,400). Pending detailed study of molecular composition, it appears from Northern and Western blot analyses and from immunohistochemistry that the major protein component of gap-junction channels is Cx32 [9, 35-38]. On average there are 20-25 junctions per cell, which allow for the electrical and dye coupling demonstrable in both lobules and individual acini [34, 36, 39-41].

This extensive intra-acinar coupling under control conditions raises the question of its functional role in highly differentiated secretory epithelia. The recognition that pancreatic secretagogues can modulate junctional permeability has led to the suggestion of some role of cell-to-cell communication in secretion [41-45]. To test this hypothesis we have used alcohols such as heptanol, octanol and nonanol which have been shown to block junctional coupling by an as yet undetermined mechanism [34, 45, 46]. These alkanols were found to markedly reduce both dye and electrical coupling between pancreatic acini. The blockage of cell-to-cell communication was associated with a 2- to 4-fold increase in amylase release from both isolated acini and *in situ* perfused pancreas [34, 36, 43]. (Acinar secretion studies featured earlier in this series [47] - *Ed.*)

The molecular mechanisms linking uncoupling to increased enzyme release are unclear. However, the close correlation of the two phenomena, for both onset and reversibility, together with the lack of activation of any known second-messenger system, suggests a causal relationship, as discussed elsewhere [43-45]. In an attempt to further characterize the relationships between the control of cell coupling and that of secretion, we have turned to a simpler system, now outlined, consisting of pairs of acinar cells.

Preparation of pairs of acinar cells entails firstly the isolation of acini by collagenase digestion (full details in [48, 49]). To prepare cell pairs, acini are suspended at room temperature in KRB without added Ca^{2+} and Mg^{2+},

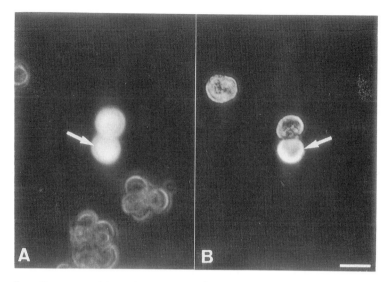

Fig. 5. Dye coupling in pairs of pancreatic acinar cells.
A: With secretagogues absent, injection of Lucifer Yellow into
an exocrine pancreatic cell *(arrow)* is usually followed by
the rapid transfer of the tracer into the adjacent cell,
indicating junctional coupling. **B:** In contrast, in the
presence of secretory stimuli (e.g. muscarinic agents, some
alkanols) Lucifer Yellow often remains within the injected
cell *(arrow)*, indicating cell uncoupling. *Bar represents 20 μm.*

supplemented with EGTA (to 3 mM).   Acini are dissociated
by aspirating them repeatedly, through first a 18-gauge and
then a 22-gauge needle. After adding excess KRB supplemented
with bovine serum albumin (to 0.1%) and buffered to pH
7.4 with 12.5 mM HEPES (giving 'control KRB'), the undissociated
acini are allowed to sediment under gravity during 1-2 min
and then discarded.   The supernatant is centrifuged at 100 g
for 3 min.   The cells present in the pellet are resuspended
in control KRB and resedimented. This procedure is repeated
twice before the final supernatant – containing mainly single
cells, pairs (Fig. 5) and small clusters – is plated onto
dishes which do not permit cell attachment and which contain
RPMI 1640 supplemented with foetal calf serum (10% v/v).
Dishes are kept at 37° in a humidified air-CO_2 incubator
until just before use.   To study dye and electrical coupling,
aliquots are rinsed in control KRB without added albumin,
and attached (15-30 min) to dishes pre-coated with poly-L-
lysine (M_r 150,000-300,000;   0.5 mg/ml in phosphate-buffered
saline).   Most experiments were performed within 4-5 h from
isolation [50].

Dye coupling

Dishes with attached cells are transferred to the heated (37°) stage of an inverted microscope. One cell of a pair is impaled with a high-resistance (150-200 Mohms) glass micro-electrode filled with a 4% solution of Lucifer Yellow CH in 150 mM LiCl, buffered to pH 7.2 with 10 mM HEPES. Following iontophoretical injection of the dye, the occurrence of coupling between the two cells of a pair is assessed by photographing the cells under fluorescent illumination [34]. Under control conditions virtually all pairs tested show dye coupling (Fig. 5A) [50]. Addition of heptanol to the medium always uncouples acinar cells (Fig. 5 B), indicating that pairs react to alcohols similarly to intact acini [34, 50].

Electrical coupling

To determine whether the heptanol-induced dye uncoupling reflects a complete closure of gap-junction channels, the electrical coupling of pairs of acinar cells can be studied under dual whole-cell patch clamp [51]. Experiments are usually peformed at room temperature, but occasionally at 37° without noticeable differences in the results obtained. The two cells of a pair are voltage-clamped, generally at a holding potential of -40 mV with a single patch electrode (6-10 Mohms). Changes of 20 pS can be detected using a chopping procedure [50]. Under these conditions, the actual voltage of a cell is measured at every cycle, allowing for a control of the membrane potential value without requiring a correction of series resistances. To estimate junctional conductance, the current evoked into one cell, e.g. cell #1, is divided by the voltage jump applied to cell #2. Voltage steps are alternately applied to each cell of a pair to allow constant control of the voltage in both cells and to exclude a possible rectification. Non-junctional membrane currents, negligible under these experimental conditions in our cell system [50, 52], are evaluated by subtracting the junctional current from the current recorded in the pulsed cell, which represents the sum of both junctional and non-junctional currents.

Pairs of acinar cells are found to be electrically coupled, with junctional conductance values ranging from 3 to 90 nS (median value = 15 nS). Because channels composed of Cx32, which represents the main junctional protein in this tissue, have a unitary conductance of 120-150 pS [30], this figure suggests that there are 25-750 fully open junctional channels at the interface between two acinar cells. These calculations are consistent with the ultrastructural data, assuming that no more than 10% of channels are open at any given time [34, 36, 45]. In contrast to reports from

Fig. 6. Effect of heptanol on the junctional conductance of a pair of acinar cells, evaluated every 6 sec by dividing the current measured in one cell by the voltage jump applied to the other cell of the pair. Heptanol (3.5 mM) induces a complete and fully reversible uncoupling of the two cells. This record is typical of several obtained. Duration of the alcohol infusion indicated by the bar.

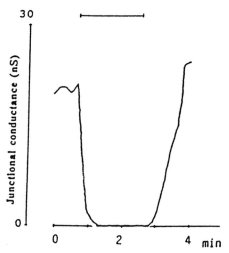

other groups [42, 51, 53], the initial conductance levels are generally maintained for a long recording period (up to 30 min) and are not affected by the variable time that elapses between the isolation of cell pairs and the electrical measurements [50, 54].

Millimolar concentrations of heptanol (Fig. 6), octanol and nonanol block the electrical coupling of acinar cells in the absence of detectable changes in the conductance of the non-junctional membrane [50]. Blockage of cell-to-cell coupling occurs within seconds, is always complete as long as the alkanol is present, and is fully reversible upon superfusion with control KRB and subsequent wash-out of the alkanol (Fig. 6). Although the mode of action of alcohols is still mysterious, it does not involve detectable changes in $[Ca^{2+}]_c$, pH, cAMP or PKC activity [34, 43, 44, 54, 55].

Clearly, it is uncertain whether the effect of alcohols has any relevance to the physiological gating of gap-junction channels. With this in mind, we have used the two-cell system to re-evaluate the effects on electrical coupling of the neurotransmitter acetylcholine (ACh), which had previously been shown to uncouple acinar cells [39, 41, 43]. Dual whole-cell patch clamp recordings demonstrate that μM concentrations of ACh decrease junctional conductance (Fig. 7). The extent of this effect is variable, but most pairs show a 50% inhibition of electrical coupling, in agreement with the quantitative estimates obtained in dye-coupling experiments [43]. Muscarinic receptors of pancreatic acinar cells, which are of the 'm3' type, are coupled to the activation of the inositol lipid signalling system [56]. Hence the same high concentrations of ACh that inhibit junctional coupling raise $[Ca^{2+}]_c$ and activate PKC [57], both effects being known to block cell-to-cell communication in several systems [1].

Fig. 7. Effect of acetylcholine on the junctional
conductance of a pair of
acinar cells. As evaluated
by on-line dual whole-cell
patch clamp recording, 5 µM
ACh decreases by 60% this
pair's initial conductance
(Gj), *viz.* ~3.5 nS. This
inhibition takes place
within 1-2 min and is fully
reversible, with a similar
time course, after ACh is
removed from the perfusion
medium.

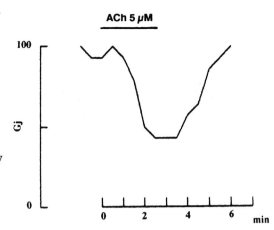

The uncoupling effect of ACh, however, is maintained when
the electrode-filling solution is supplemented with 11 mM
EGTA [45]. Under these extreme conditions, it is very
unlikely that even local changes in $[Ca^{2+}]_c$ might have occurred,
since the well-known effects of ACh on Ca^{2+}-dependent Cl^-
and cation channels in the non-junctional membrane are completely
abolished [45]. Because phorbol esters and other powerful
pharmacological activators of PKC in general do not perturb
coupling between pancreatic acinar cells [54, 55; but cf.
58], it is concluded that it is still unknown what molecular
mechanism underlies the effect of muscarinic activation on
junctional-channel conductance and permeability. As needs
study, maybe a GTP-binding protein is involved in the gating
of this channel [59], as suggested for K^+-channel activation
in cardiac myocytes [60].

Amylase secretion

    Studies using intact pancreatic acini have suggested
a role of cell coupling in the control of regulated secretion
of pancreatic enzymes [44, 45]. A more direct demonstration
of the causal relationship between the two events may be obtained
by combining the dual whole-cell patch clamp of acinar cell
pairs with a reverse haemolytic plaque assay, a technique
which permits the direct visualization of secretion from
individual acini and acinar cells [61]. In this assay,
the cell preparation is mixed with protein A-coated red blood
cells and incubated in the presence of a specific anti-amylase
serum. Under these conditions, the secreted amylase binds
to the anti-amylase serum and the resulting complex becomes
attached to protein A on red cells. When complement is
added to the medium, red cells bearing the amylase/anti-amylase
complex are lysed, and amylase release is revealed by the
formation of haemolytic plaques around the secreting cells.

Pairs of acinar cells prepared as described above retain
their ability, similarly to individual acini, to secrete
enzymes in response to secretagogues (Fig. 8). Thereby it
was found that the uncoupler heptanol stimulates amylase
secretion above basal from acini [61] and pairs of cells,
but not single acinar cells. Furthermore, preliminary experiments indicate that electrical coupling and amylase secretion
can be monitored simultaneously in the same pair of cells,
by combining the dual patch-clamp technique with a reverse
haemolytic plaque assay.

Perspectives

Pancreatic acinar cells have proved to be a useful
model to study the functional role of cell coupling in
secretory epithelia. Despite the demonstration that inhibition
of coupling is parallelled by an increase in the rate of enzyme
release, several critical questions remain unanswered. Thus,
it is not known which are the physiological uncoupling agents
in vivo, since not all pancreatic secretagogues decrease
junctional conductance [43, 52]. In addition, neither the
intracellular messengers involved in the gating of channels,
nor the mechanism linking the blockage of gap junctions to
changes in secretion, have been elucidated. In this perspective, the system of pairs of acinar cells described here may
offer significant advantages in addressing these issues in
the near future. These studies are likely to provide valuable
information pertaining to the more general field of signal
transduction.

Acknowledgements

This work was supported by grants from the Swiss
National Science Foundation (# 83.627.0.88 to R.B. and
#31-26625.89 to P.M.) and from the National Institutes of
Health, U.S.A. (#GM37751 to D.L.P.).

....

Legend to **Fig. 8,** *opposite*

Pancreatic acinar cell pairs used for coupling and secretion
studies. **A)** This phase-contrast view demonstrates that, after
isolation, the cells retain their characteristic polarity, as
indicated by the accumulation of dense secretory granules at
the apical pole and by the intermediate position of the Golgi
apparatus (the *arrowed* clear areas). Most pairs show a rather
flat membrane interface all along the area of cell contact.
B) The amylase secretion of exocrine pancreatic cell pairs
can be studied in the presence of antibodies against amylase,
by monitoring the development of complement-induced haemolytic
plaques (**hp**) in a monolayer of protein A-coated red blood
cells (**rbc**). After a 30-min stimulation by 10 µM carbachol,
[continued opposite

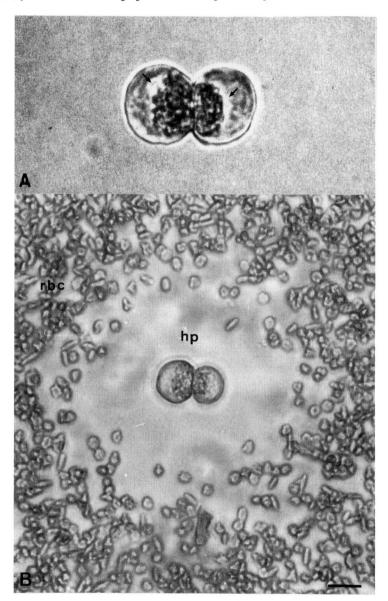

Fig. 8 *legend, continued from opposite*

individual acinar cell pairs secreting amylase can easily be
identified, under microscopic control, by the presence of a
surrounding haemolytic plaque. Scoring of these plaques and
measurement of their areas provide reliable quantitative
estimates of amylase secretion for individual acinar cell
pairs.   *The bar represents 7.5 μm in* **A** *and 15 μm in* **B**.

References

1. Hertzberg, E.L. & Johnson, R.G., eds. (1988) *Gap Junctions*, Alan R. Liss, New York, 548 pp.
2. Caspar, D.L.D., Goodenough, D.A., Makowski, L. & Phillips, W.C. (1977) *J. Cell Biol. 74*, 605-628.
3. Goodenough, D.A. & Revel, J.P. (1970) *J. Cell Biol. 45*, 272-290.
4. Paul, D.L. (1986) *J. Cell Biol. 103*, 123-134.
5. Kumar, N.M. & Gilula, N.B. (1986) *J. Cell Biol. 103*, 767-776.
6. Heynkes, R., Kozjek, G., Traub, O. & Willecke, K. (1986) *FEBS Lett. 205*, 56-60.
7. Stevenson, B.R. & Paul, D.L. (1989) *Current Opinion in Cell Biology 1*, 884-891.
8. Beyer, E.C., Paul, D.L. & Goodenough, D.A. (1990) *J. Membr. Biol. 116*, 187-194.
9. Meda, P., Chanson, M., Pepper, M., Giordano, E., Bosco, D., Traub, O., Willecke, K., El Aoumari, A., Gros, D., Beyer, E.C., Orci, L. & Spray, D.C. (1991) *Exp. Cell Res. 192*, 469-480.
10. Beyer, E.C., Paul, D.L. & Goodenough, D.A. (1987) *J. Cell Biol. 105*, 2621-2629.
11. Hertzberg, E.L., Disher, R.M., Tiller, A.A., Zhou, Y. & Cook, R.G. (1988) *J. Biol. Chem. 263*, 19105-19111.
12. Milks, L.C., Kumar, N.M., Houghten, R., Unwin, N. & Gilula, N.B. (1988) *EMBO J. 7*, 2967-2975.
13. Goodenough, D.A., Paul, D.L. & Jesaitis, L. (1988) *J. Cell Biol. 107*, 1817-1824.
14. Lawrence, T.S., Beers, W.H. & Gilula, N.B. (1978) *Nature 272*, 501-506.
15. Murray, S.A. & Fletcher, W.H. (1984) *J. Cell Biol. 98*, 1710-1719.
16. Dunlap, K., Takeda, K. & Brehm, P. (1987) *Nature 325*, 60-62.
17. Sáez, J.C., Connor, J., Spray, D.C. & Bennett, M.V.L. (1989) *Proc. Nat. Acad. Sci. 86*, 2708-2712.
18. Sanderson, M.J., Charles, A.C., & Dirksen, E.R. (1990) *Cell Regul. 1*, 585-596.
19. Colman, A. (1984) in *Transcription and Translation - A Practical Approach* (Hames, D. & Higgins, S., eds.), IRL Press, Oxford, pp. 271-301.
20. Methfessel, C., Witzemann, V., Takahashi, T., Mishina, M., Numa, S. & Sakmann, B. (1986) *Pflügers Arch. 407*, 577-588.
21. Werner, R., Miller, T., Azarnia, R. & Dahl, G. (1985) *J. Membr. Biol. 87*, 253-268.
22. Dahl, G., Miller, T., Paul, D., Voellmy, R. & Werner, R. (1987) *Science 236*, 1290-1293.
23. Swenson, K.I., Jordan, J.R., Beyer, E.C. & Paul, D.L. (1989) *Cell 57*, 145-155.
24. Werner, R., Levine, E., Rabadan-Diehl, C. & Dahl, G. (1989) *Proc. Nat. Acad. Sci. 86*, 5380-5384.

25. Ebihara, L., Beyer, E.C., Swenson, K.I., Paul, D.L. & Goodenough, D.A. (1989) *Science 243*, 1194-1195.
26. Krieg, P.A. & Melton, D.A. (1984) *Nucl. Acids Res. 12*, 7057-7070.
27. Spray, D.C., Harris, A.L. & Bennett, M.V.L. (1981) *J. Gen. Physiol. 77*, 77-93.
28. Gilula, N.B., Epstein, M.L. & Beers, W.H. (1978) *J. Cell Biol. 78*, 58-75.
29. Dahl, G., Werner, R. & Levine, E. (1988) in *as for* 1., pp. 183-197.
30. Eghbali, B., Kessler, J.A. & Spray, D.C. (1990) *Proc. Nat. Acad. Sci. 07*, 1328-1331.
31. Gimlich, R.L., Kumar, N.M. & Gilula, N.B. (1990) *J. Cell Biol. 110*, 597-605.
32. Go, V.L.W., Gardner, J.D., Brooks, F.P., Lebenthal, E., Di Magno, E.P. & Scheele, G.A., eds. (1986) *The Exocrine Pancreas: Biology, Pathobiology, and Diseases*, Raven Press, New York.
33. Friend, D.S. & Gilula, N.B. (1972) *J. Cell Biol. 53*, 758-776.
34. Meda, P., Bruzzone, R., Knodel, S. & Orci, L. (1986) *J. Cell Biol. 103*, 475-483.
35. Paul, D.L. (1985) in *Gap Junctions* (Bennett, M.V.L. & Spray, D.C., eds.), Cold Spring Harbor Laboratory, New York, pp. 107-122.
36. Bruzzone, R., Trimble, E.R., Gjinovci, A., Traub, O., Willecke, K. & Meda, P. (1987) *Pancreas 2*, 262-271.
37. Dermietzel, R., Leibstein, A., Frixen, U., Janssen-Timmen, U., Traub, O. & Willecke, K. (1984) *EMBO J. 3*, 2261-2270.
38. Hertzberg, E.L. & Skibbens, R.V. (1984) *Cell 39*, 61-69.
39. Iwatsuki, N. & Petersen, O.H. (1978) *J. Cell Biol. 79*, 533-545.
40. Iwatsuki, N. & Petersen, O.H. (1979) *Pflügers Arch. 380*, 277-281.
41. Findlay, I. & Petersen, O.H. (1982) *Cell Tissue Res. 225*, 633-638.
42. Neyton, J. & Trautmann, A. (1986) *J. Exp. Biol. 124*, 93-114.
43. Meda, P., Bruzzone, R., Chanson, M., Bosco, D. & Orci, L. (1987) *Proc. Nat. Acad. Sci. 84*, 4901-4904.
44. Bruzzone, R. & Meda, P. (1988) *Eur. J. Clin. Invest 18*, 444-453.
45. Meda, P., Bosco, D., Giordano, E. & Chanson, M. (1991) in *Biophysics of Gap Junction Channels* (Peracchia, C., ed.), CRC Press, Boca Raton, FL, in press.
46. Johnston, M.F., Simon, S.A. & Ramón, F. (1980) *Nature 286*, 498-500.
47. Dormer, R.L. & Al-Mutairy, A.R. (1987) in *Cells, Membranes and Disease, including Renal* [Vol. 17, this series] (Reid, E., Cook, G.M.W. & Luzio, J.P., eds.), Plenum, New York, pp. 273-283.

48. Bruzzone, R., Halban, P.A., Gjinovci, A. & Trimble, E.R. (1985) *Biochem. J.* *226*, 621-624.
49. Bruzzone, R. (1989) *Eur. J. Biochem.* *179*, 323-331.
50. Chanson, M., Bruzzone, R., Bosco, D. & Meda, P. (1989) *J. Cell. Physiol.* *139*, 147-156.
51. Neyton, J. & Trautmann, A. (1985) *Nature 317*, 331-335.
52. Petersen, O.H. & Findlay, I. (1987) *Physiol. Rev. 67*, 1054-1116.
53. Somogy, R. & Kolb, H-A. (1988) *Pflügers Arch. 412*, 54-65.
54. Chanson, M., Bruzzone, R., Spray, D.C., Regazzi, R. & Meda, P. (1988) *Am. J. Physiol.* *255*, C699-C704.
55. Chanson, M., Meda, P. & Bruzzone, R. (1989) *Exp. Cell Res.* *182*, 349-357.
56. Berridge, M.J. & Irvine, R.F. (1984) *Nature 312*, 315-321.
57. Bruzzone, R. (1990) *Gastroenterology 99*, 1157-1176.
58. Randriamampita, C., Giaume, C., Neyton, J. & Trautmann, A. (1988) *Pflügers Arch. 412*, 462-468.
59. Somogy, R. & Kolb, H-A. (1989) *FEBS Lett. 258*, 216-218.
60. Codina, J., Yatani, A., Grenet, D., Brown, A.M. & Birnbaumer, L. (1987) *Science 236*, 442-445.
61. Bosco, D., Chanson, M., Bruzzone, R. & Meda, P. (1988) *Am. J. Physiol. 254*, G664-G670.

#E-4

GROWTH CONE COLLAPSE: A SIMPLE ASSAY FOR MONITORING CELL-CELL REPULSION

Geoffrey M.W. Cook, Jamie A. Davies and Roger J. Keynes

Department of Anatomy, University of Cambridge,
Downing Street, Cambridge CB2 3DY,   U.K.

The development of a growth cone collapse assay [1] has provided an important experimental strategy for identifying molecules that inhibit the motility of developing neurites. This bioassay, as well as identifying specific collapse-initiating molecules, may be used for monitoring purification procedures.   In this article, the utility of the method is illustrated with respect to molecules involved in peripheral nerve segmentation.

How axons manage to find their correct targets during neuronal development is a major question, as yet unanswered. It has been recognized for some years that the growth cone[*], a highly motile and dynamic structure at the tips of growing neurites, is involved in making appropriate decisions that direct outgrowth.   Clearly, attractive or permissive cues may play a role in influencing growth cone guidance, though evidence has recently been accumulating which indicates that inhibitory or repulsive factors are also likely to play a critical role in axon guidance.

One system which has contributed to the emergence of the latter concept stems from studies on peripheral nerve segmentation in the developing chick embryo [2, 3].   During the course of normal development of vertebrate embryos, motor and sensory neurons emerge in register with the somites which are arranged in a repeating pattern along the longitudinal axis of the embryo.   Recently Keynes & Stern [2] have shown that, in the chick embryo, somites do not present a physical barrier to axon growth, but rather that motor axons grow out in the anterior half of the somite, even after it has been surgically displaced.   The question therefore arises: are there permissive cues in the anterior half and non-permissive cues in the posterior half of the somite?   This article addresses the latter possibility, and in a volume dedicated to examining experimental strategies in cell signalling presents a simple and reliable method which has proved valuable in assaying for inhibitory interactions between cells in the developing nervous system.

---

[*]*This and other structures are explained in an ADDENDUM (p.366)*

GROWTH CONE COLLAPSE

When *in vitro* growth cones derived from peripheral neurons make contact with axons derived from central neurons they collapse, that is the growth cone undergoes a dramatic change in shape from a spread morphology to a retracted state (Fig. 1). The same happens when growth cones from central neurons make contact with peripheral axons. This phenomenon was first described clearly by Kapfhammer & Raper [4]. These authors have a working hypothesis that certain neurites have labels associated with their membranes that initiate the collapse of those growth cones able to detect them, and point out that this is analogous to an explanation for the contact paralysis between motile non-neuronal cells in culture.

On the basis of this hypothesis, Raper & Kapfhammer [1] have recently devised a simple bioassay in order to allow the identification of those molecules that are responsible for initiating growth cone collapse. Growth cones are grown on a generally permissive substratum. Normally this comprises laminin-coated glass coverslips, as used here. (An alternative [1] is use of immunopurified G4, a cell surface glycoprotein, which is added to coverslips pre-dipped in a solution of Schleicher & Schnell BA 83 nitrocellulose in methanol. After drying, the cells may be added in appropriate tissue culture media; to prevent 'non-specific' binding of proteins from culture media to the nitrocellulose the coverslips are soaked in a 3 mg/ml solution of haemoglobin and extensively rinsed in medium before use. We have not tried this procedure.)

Test membranes in suspension, or test molecules in solution or as liposomes, are then introduced into the culture, and after a set period (usually 1 h) the cultures are fixed (4% formaldehyde in phosphate-buffered saline (PBS) containing 10% sucrose) and the morphology of the tips of the extending neurites is examined under phase-contrast microscopy. The spread growth cones (Fig. 1b) with associated lamellipodia and filopodia are readily distinguishable from the collapsed growth cones where the tips of the neurons exposed to the appropriate agent are bullet-shaped and lack lamellipodia or filopodia (Fig. 1c). Raper & Kapfhammer [1] point out

Fig. 1 *(opposite)*. Growth cone collapse. **a** *(amplified alongside)*: ganglia grown in culture and treated with liposomes for testing sclerotome proteins. **b, c** (phase-contrast micrographs, × 630; *from [5], by permission*): ganglion growth cones on a laminin substrate, treated with liposomes, control (**b**) or test (**c**; *see later in text*). The cones (growing neuron tips) show characteristic 'spread' morphology in **b**, and are collapsed in **c**.

Amplification of
Fig. 1a legend at foot
of opposite page.- Note
the marked % increase in
growth cones with a collapsed morphology after treatment with the liposomes
containing sclerotome
proteins (extract from
stage 17-19 chick embryo
trunks). The error bars
represent 95% confidence
limits based upon
sampling error.

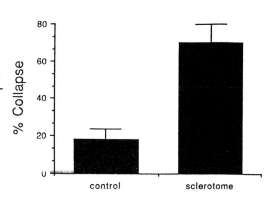

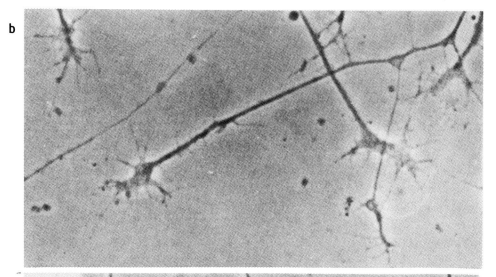

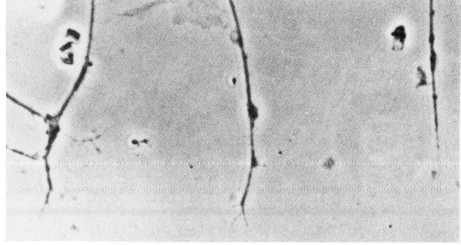

that only single uncrowded neurite tips should be scored and recommend that counts be limited to the longest neurites. This latter recommendation was based on their finding that weak collapsing activity has a greater effect on those neurites that are most highly extended; thus by following this procedure both reproducibility and sensitivity of the assay procedure are increased.

APPLICATION OF THE GROWTH CONE ASSAY TO THE SOMITE SYSTEM

The application of the collapse assay to the problem of peripheral nerve segmentation has served to strengthen the view that the collapse phenomenon is of relevance to explaining a normal biological process. In a search for molecules that guide growing axons during vertebrate development, Davies and co-workers [5] have been able to apply this simple assay to advantage. Whilst outgrowing spinal axons in higher vertebrates traverse exclusively the anterior-half somite, there are relatively few molecular markers which distinguish the cells of the two halves of the somite. Histochemical studies using a variety of plant lectins [6] have revealed a difference between anterior-half and posterior-half sclerotome: namely, peanut agglutinin binds only to the posterior-half sclerotome. Exploiting this finding, Davies and co-workers [5] have isolated from chick somites a glyco-protein that binds peanut agglutinin, and have demonstrated that it has potent growth cone collapse-inducing activity. Furthermore, they have made a rabbit polyclonal antiserum to this fraction and shown immunohistochemically that the antibody binding is only to cells of the posterior-half sclerotome. In addition, when the immunoglobulin fraction is immobilized on Sepharose 4B, it can be used to remove collapse-inducing activity, whilst pre-immune globulin when immobilized has no such capacity. Combining these findings, the authors [5] suggest that spinal nerve segmentation is produced by inhibitory interactions between the peanut agglutinin-binding glycoproteins on the surfaces of the posterior-half sclerotome and the advancing growth cones.

EXPERIMENTAL PROCEDURES (and particular conditions for Fig. 1)

Fig. 2 outlines our collapse-assay experimental procedure using dorsal root ganglion (DRG) growth cones with adoption, for coating the coverslips (see above), of the laminin procedure.

Cultures.- Acid-washed, sterile, glass coverslips (13 mm diam.) are coated in pairs by sandwiching 40 µl of 30 µg/ml laminin (Sigma) in Hank's balanced salt solution (BSS) between the two coverslips and incubating them at 37° for 1 h in 6% CO_2. After being washed in Hank's BSS, each coverslip

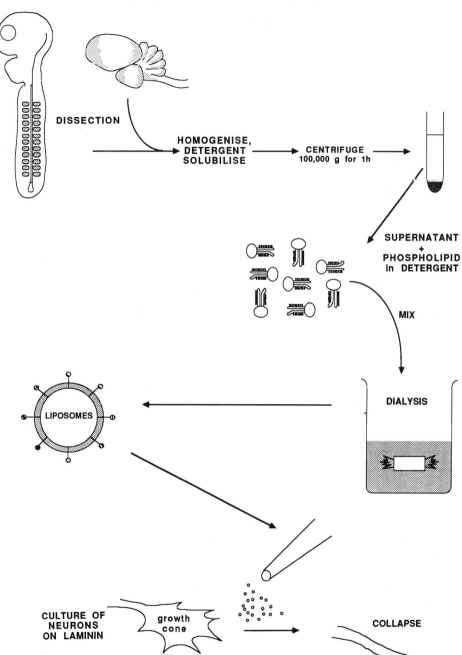

g. 2. Growth cone collapse assay based on method of Raper & pfhammer [1] *(from [7], by permission)*. Detergent extracts of the test terial (brain or somites) are mixed with phospholipids, and dialyzed to move detergent. The liposomes, incorporating the putative inhibitory ent, are added to neuron explants on a laminin substratum. Cultures are xed 1 h later and scored for growth cone collapse (phase contrast).

is placed in 1 ml of a culture medium consisting of 5 µg/L
7S Nerve Growth Factor (Sigma) in F12 medium (90%, v/v) and
foetal-calf and chick serum (Flow Labs.; each 5% v/v),
supplemented with 100 u./ml penicillin, 0.8 mg/ml streptomycin
and 0.28 µg/ml amphotericin (Sigma). A 24-well plate (Flow
Labs.) is preferable. Isolated DRG (from stage 33-35 chick
embryos) are divided into 2 approximately equal portions and
then transferred to the laminin-coated substratum and grown
for up to 20 h.

After this period the cultures are inspected with an
inverted phase contrast microscope, and those showing
substantial growth with few migrating non-neuronal cells are
selected for the collapse assay.

Liposome treatment, and assessment. - The test samples,
routinely added to the cultures as liposomes, are detergent-
extracted with 2% (w/v) CHAPS (Sigma) in PBS and centrifuged
for 1 h at 100,000 g_{av}. To 100 µl, 100 µg phosphatidylcholine
(Sigma) and 20 µg phosphatidylserine (Sigma) in 20 µl 4%
CHAPS in PBS are added. The liposomes are formed by extensive
dialysis against PBS at 4°.

Each culture receives 100 µl of liposome suspension.
After 1 h at 37° the cultures are fixed by adding 2 ml
formaldehyde (4% w/v) in PBS containing 15% (w/v) sucrose.
The top 2 ml of liquid, containing most of the original
medium, is removed from each well, and the cultures left
at least 6 h at room temperature before being scored by
phase-contrast microscopy. The cultures are scored as the
% ratio of collapsed growth cones to the total number of
growth cones (spread and collapsed) examined. Ideally the
samples should be blind-coded before viewing to eliminate
any operator bias.

With DRG, control values of 10% collapsed growth cones
in untreated cultures are seen. Controls employing plain
liposomes (i.e. liposomes prepared in the absence of test
material) may attain 20% collapsed growth cones but this
is entirely acceptable where test material manifests values of ~75%
collapsed growth cones.

COMMENTS ON THE METHOD

The incorporation of test materials into the liposomes
can be monitored using radiolabelled samples. In the case
of detergent-solubilized affinity-purified embryonic material
the efficiency of incorporation into liposomes of somite
polypeptides, metabolically labelled with ^{35}S-methionine, is
>80%. In the method as described, 200 µg of protein per

culture was routinely used in the test samples, but in practice the amount required will depend on the system under examination.   In those cases where the assay is being used to monitor a purification procedure, a series of different protein concentrations will need monitoring so that accurate dose-response measurements can be made.

The cellular mechanism that leads to growth cone collapse is still not known, though it seems highly likely that a collapse-initiating signal induces a reorganization of the cytoskeleton of the growth cone.   It is unlikely that growth cone collapse as measured by Raper & Kapfhammer [1] and Davies & co-authors [5] is due to some test-material constituent that modifies the substratum.   Collapse induced by extracts of embryonic brain occurs both on laminin and on the cell-surface glycoprotein G4 [1], and collapse induced by extracts of chick somites occurs on laminin, polylysine and Type 1 collagen (R.J. Howells, G.M.W. Cook & R.J. Keynes, unpublished observations).   For the same reasons it is also unlikely to be due to the release of some cellular component that has a general toxic effect, since the phenomenon is entirely reversible.   With the somite system the collapse occurs mainly in the first 20 min of adding the liposomes and has reached a plateau within 1 h; on removing the liposomes by washing the culture, a return to a spread morphology is achieved within 4 h.

Acknowledgements

This work was supported by a project grant (G86044393) from the Medical Research Council.   J.A.D. was in receipt of an Elmore Research Studentship, Gonville & Caius College, Cambridge, and G.M.W.C. is a Member of the External Research Staff of the Medical Research Council.

References

1. Raper, J.A. & Kapfhammer, J.P. (1990) *Neuron 4*, 21-29.
2. Keynes, R.J. & Stern, C.D. (1984) *Nature 310*, 786-789.
3. Keynes, R.J. & Stern, C.D. (1985) *Trends Neurosci. 8*, 220-223.
4. Kapfhammer, J.P. & Raper, J.A. (1987) *J. Neurosci. 7*, 201-212.
5. Davies, J.A., Cook, G.M.W., Stern, C.D. & Keynes, R.J. (1990) *Neuron 4*, 11-20.
6. Stern, C.D., Sisodiya, S.M. & Keynes, R.J. (1986) *J. Embryol. Exp. Morph. 91*, 209-226.
7. Keynes, R.J., Johnson, A.R. & Cook, G.M.W. (1991) in *Seminars in Neurosciences*, Cell Adhesion Molecules in the Nervous System (Rathjen, F., ed.), Saunders, London, in press.

ADDENDUM (by request of Senior Editor):

EXPLANATORY NOTE ON GROWTH CONES, AND OTHER STRUCTURES

The term refers to the terminal enlargement at the ends of axons, which contains mitochondria, vesicles, microfilaments and microtubules. Extending beyond the growth cone are fine outgrowths termed *filopodia* (micro-spikes), which contain microfilaments that react with antibodies raised against actin and myosin. The term *sclerotome* refers to the ventro-medial part of the somite which forms the vertebral column.

#ncE

NOTES and COMMENTS relating to

FIBRINOLYTIC, ONCOGENIC, JUNCTIONAL AND NEURAL
PHENOMENA *and to signalling phenomena in some other areas*

This subsection opens with supporting ('**nc**') articles.

From p. 381 there is Forum discussion material.

From p. 384 there is supplementary material provided by
the Editor, starting with **neural** items and including:
    inter-cell phenomena: p. 385
    proliferation/differentiation & **growth factors**: p. 387
      (p. 386: Forum material on proliferation)
    endothelins, eicosanoids[*]**, tyrosine kinases**, etc.: p. 388.
Some material in these areas is to be found in other #**nc**'s.

Consult the start of the main Contents list concerning
the book structure and inescapable compromises in the section
assignment for certain contributions whose wide-ranging
subject-matter does not fall neatly into one section.

---

[*]*For guidance on the 'jungle' of* **eicosanoids** *(including
abbreviations), consult the Fig. and legend on p. 130,
as now amplified (a,b, etc., refer to depicted routes):-*

```
5-HETE                        PL-A₂ action
   ↑                            furnishes
5-HPETE ←————————————— arachidonic acid, aa —(c)→ 15-HPETE —(f)→ LX's
   │      (a)  lipoxygenase      (b) │cyclo-oxygenase
   ↓
 LT's
[(d) &          endoperoxides, e.g. PGG₂
 (e) by-   PG's ←——————————||L——————————————→ TX's
 pass                      PGI₂
 5-HPETE]              (prostacyclin)
```

$5\text{-HPETE} \xleftarrow[\text{lipoxygenase}]{(a)}$ arachidonic acid, $aa \xrightarrow{(c)} 15\text{-HPETE} \xrightarrow{(f)} LX's$, $(b) \downarrow cyclo\text{-}oxygenase$

NOTE: HPETE = *hydroperoxyeicosatetraenoic acid (isomers!).
Further guidance, with a diagram, appears on p. 63 of Vol. 18*
('Bioanalysis of Drugs, Especially Anti-inflammatory and Cardio-
Vascular', ed. E. Reid *et al.*; Plenum, 1988), *which has arts. on
eicosanoid assay.*

#ncE-1

A Note on

DETECTION OF Na$^+$ AND Ca^{2+} MOVEMENTS BY CO-MEASURING O$_2$ CONSUMPTION AND ^{45}Ca^{2+} UPTAKE IN RAT CORTICAL SYNAPTOSOMES

D. Wermelskirchen[⊗], [†]J. Urenjak, B. Wilffert
and F. Tegtmeier

Janssen Research Foundation [†]Royal College of Surgeons,
4040 Neuss 21, Germany   London WC2A 3PN, U.K.

The pathophysiological changes in the CNS induced by O$_2$ deficiencies include an increase in intracellular Na$^+$ and Ca^{2+}. Since rat brain synaptosomes serve well for studying basic functions of nerve cells, many attempts have been made to investigate pathophysiological alterations of Na$^+$ and Ca^{2+} movements in this model. However, whereas Ca^{2+} movements can be determined by means of relatively simple ^{45}Ca^{2+} uptake meaurements, investigation of Na$^+$ movements requires intracellular ion-sensitive microelectrode techniques or the use of high-energy-emitting ^{22}Na. Therefore, an alternative method for assessing intracellular Na$^+$ concentrations was developed, based on recent findings that operation of the Na$^+$/K$^+$-ATPase draws substantially on intracellular ATP [1]. It was proposed that an increase in intracellular Na$^+$ concentration enhances the Na$^+$/K$^+$-ATPase activity, leading to an increase in the intracellular ADP level which in turn stimulates mitochondrial respiration [2]. Thus a stimulation of mitochondrial respiration should mainly reflect an increase in intracellular Na$^+$ concentration. It was therefore proposed that Ca^{2+} and Na$^+$ uptake in isolated rat-brain synaptosomes could be determined by measuring ^{45}Ca^{2+} uptake and O$_2$ consumption. The chosen stimulus was veratridine, which was shown to provoke both Na$^+$ entry and Ca^{2+} entry [3]. Furthermore, the effect of the two anti-ischaemic drugs flunarizine and R 56865* on veratridine-induced Na$^+$ and Ca^{2+} entry was investigated.

METHODS

Synaptosomes were prepared from rat brain essentially according to Nicholls [4]. They were suspended in Tyrode solution, containing (mM) 125 NaCl, 3.5 KCl, 1.2 CaCl$_2$, 1.2 MgCl$_2$, 25 NaHCO$_3$, 0.4 NaH$_2$PO$_4$ and 10 glucose, pH 7.4, and equilibrated with 95% O$_2$/5% CO$_2$. Standardized suspensions (1 mg synaptosomal protein/ml) were pre-incubated for 20 min at 36° in the absence or presence of the compound under study.

⊗addressee for any correspondence
*R 56865 *denotes*   *N*-{1-[4-(4-fluorophenoxy)butyl]-4-piperidinyl}-
N-methyl-2-benzothiazolamine; FCCP, carbonylcyanide-*p*-trifluoro-
methoxyphenylhydrazone. *Other abbreviations overleaf.*

Respiration measurements entailed polarographic recording of O_2 consumption at 30°. The following compounds were introduced at 4 min intervals: pyruvate, veratridine and, to test the basal and maximal mitochondrial respiration respectively, oligomycin and FCCP. O_2 consumption was expressed in nmol O_2/min per mg protein. Extra O_2 consumption caused by veratridine was calculated from O_2 consumption in the presence of veratridine (state 3) *minus* basal O_2 consumption in the presence of oligomycin (state 4), and is indicated as 'state 3 − 4' in Fig. 1 and Table 1.

$^{45}Ca^{2+}$ **uptake measurements** were initiated by adding $^{45}Ca^{2+}$-containing Tyrode solution (final radioactivity 1 µCi/ml) in the absence or presence of veratridine (to 10 µM). Incubations were terminated after 3 min by rapid addition of 5 ml ice-cold Tyrode solution and rapid filtration through a cellulose nitrate filter (pore size 0.45 µm). The filters were washed once with 5 ml ice-cold Tyrode solution. The remaining $^{45}Ca^{2+}$ was determined by liquid scintillation counting and expressed as dpm/mg protein.

All results are expressed as mean ±S.E.M. Statistical differences were calculated with the Wilcoxon-Mann-Whitney U-Test.

RESULTS

In the following summary, **T** and **F** indicate that Table 1 and Fig. 1 respectively should be consulted for concentrations used and for test values to which those below relate. The values are expressed as above for O_2 ('consumption' implied) and as dpm/mg protein for $^{45}Ca^{2+}$. Enhanced values induced by veratridine are denoted by the prefix **+V**. Tetrodotoxin is abbreviated TTX, and omega-conotoxin ω-CTX. None of the compounds tested, except ouabain, affected the Na^+/K^+-ATPase activity.

O_2 (consumption)
- **+V**: with 10 µM veratridine, O_2 rose from 10 ±0.3 to 25 ±1 (n = 34);
- subsequent oligomycin (4 mg/ml) gave a basal value of 8 ±0.2;
- FCCP raised the rate to 41 ±1;
- **(T)** TTX and ouabain gave maximal reductions of 87 ±2% and 78 ±1% (n = 4, 8) respectively;
- Na o-vanadate (2-100 µM) had no effect on the ouabain-resistant **+V**-O_2;
- pre-treatment with 10 µM BAY K 8644 or A23187 could not further raise **+V**-O_2;
- **(T)** ω-CTX or nitrendipine (Ca^{2+} entry blockers) had no effect on **+V**-O_2;
- **(F)** flunarizine and R 56865 suppressed **+V**-O_2 to 38 ±1.6% & 62 ±2% respectively (n = 6).

Table 1. Effect of veratridine after pre-treatment with certain agents (see opposite; values are % of effect of veratridine alone, **+V**), on O_2 consumption and $^{45}Ca^{2+}$ uptake.

	O_2, state 3—4	$^{45}Ca^{2+}$ uptake
TTX, 1 μM (max. effect; 0.1-10 tried*)	13 ±2% †	4 ±4% †
Ouabain, 1 mM (*ditto*; down to 1 μM tried*)	22 ±1% †	–
ω-CTX, 0.5 μM	106 ±4%	81 ±11%
nitrendipine, 10 μM	94 ±7%	95 ±17%

*effect dose-dependent   †$P < 0.05$

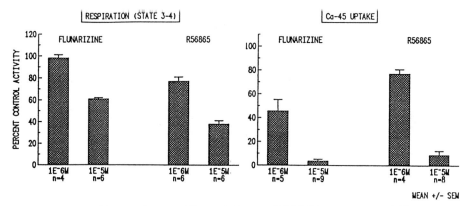

Fig. 1. Effect of 10 μM flunarizine or R 56865 on veratridine-induced O_2 consumption and $^{45}Ca^{2+}$ uptake. *$P < 0.05$.

$^{45}Ca^{2+}$ *uptake*

- **+V:** uptake rose from 20,000 ±1,800 to 59,000 ±5,000 (n = 7, 15), representing a Ca^{2+} influx of 20 nmol Ca^{2+}/mg protein;
- (**T**) TTX depressed **+V** uptake to 4 ±3 (n = 12);
- (**T**) ω-CTX or nitrendipine had no significant effect on the **+V** uptake, whereas
- (**F**) flunarizine and R 56865 completely abolished the **+V** uptake.

DISCUSSION

Na^+ and Ca^{2+} influxes occur simultaneously in response to veratridine, which is now shown to stimulate synaptosomal respiration. This increase in O_2 consumption could be blocked by the Na^+-channel blocker TTX, but not by the Ca^{2+}-entry blockers nitrendipine and ω-CTX. Furthermore, the Ca^{2+} agonist BAY K 8644 and the Ca^{2+} ionophore A23187 exhibited no effect on veratridine-induced O_2 consumption. Inhibition of the Na^+/K^+-ATPase with ouabain strongly reduced veratridine-induced O_2 consumption, and the residual level was insensitive to Na o-vanadate, this result excluding a possible contribution from other ATPase systems (e.g. Ca^{2+}-ATPase).

The detection of the oligomycin-induced basal respiration and the FCCP-induced maximal respiration enabled us to monitor the functional state of the mitochondria. The effect of the compounds tested on the Na^+/K^+-ATPase activity was investigated in a separate series of experiments. Thereby the possibility that the tested compounds interfere with either mitochondrial function or Na^+/K^+-ATPase activity was excluded. We therefore suggest that the veratridine-induced increase in O_2 consumption mainly represents Na^+ influx.

$^{45}Ca^{2+}$-uptake measurements demonstrated veratridine-induced Ca^{2+} uptake. Similarly to the respiration measurements this induced uptake was inhibited by the Na^+-channel blocker TTX and was insensitive to the Ca^{2+}-entry blockers nitrendipine and ω-CTX.

It was assumed that veratridine-induced ion movements are comparable to those induced by ischaemia. In accord with this assumption, two anti-ischaemic drugs, namely flunarizine and R 56865, were demonstrated to inhibit veratridine-induced Na^+ and Ca^{2+} movements.

In conclusion, the veratridine-induced increase in intracellular Na^+ concentration is reflected by a corresponding increase in synaptosomal O_2 consumption. The veratridine-induced increase in Ca^{2+} influx could be detected by $^{45}Ca^{2+}$ measurements. Accordingly, co-measurement of veratridine-induced O_2 consumption and $^{45}Ca^{2+}$ uptake in isolated rat-brain synaptosomes can serve as a suitable and easily accessible model for the investigation of putative anti-ischaemic drugs.

Acknowledgements

The authors thank Dr. J. Gleitz for helpful discussion. The skilled technical assistance of A. Beile, S. Khan, U. Nebel and A. Wirth enabled us to conduct the present study.

References

1. Astrup, J. (1982) *J. Neurosurg. 56*, 482-497.
2. Erecinska, M. & Dagani, F. (1990) *J. Gen. Physiol. 95*, 591-616.
3. Wada, A., Izumi, F., Yanagihara, N. & Kobayashi, H. (1985) *Naunyn-Schmiedeberg's Arch. Pharmacol. 328*, 273-278.
4. Nicholls, D.G. (1978) *Biochem. J. 170*, 511-522.

#ncE-2

A Note on

STUDIES ON EXTRACELLULAR OVERFLOW OF TRANSMITTER AMINO ACIDS DURING CEREBRAL ISCHAEMIA AND SPREADING DEPRESSION

D. Scheller, U. Heister, K. Dengler and F. Tegtmeier

Janssen Research Foundation,
Raiffeisenstrasse 8, D-4040 Neuss 21, Germany

The microdialysis (MD*) technique has become widely applied since the probes became commercially available [1]. The main component of a MD probe is a cylindrical, semipermeable membrane. It is internally perfused with an artificial bio-fluid and can be implanted in any organ or body fluid. The technique has been used mostly for its original purpose [2] to record extracellular transmitter changes within the brain.

Due to the implantation of the MD probe within the ECS, measurement entails some uncertainties.- (1) The measured transmitter changes are averaged over a certain brain area, which is large in relation to functional units (groups of neurons). (2) A small proportion of the transmitters being released into the synaptic cleft escapes the efficient re-uptake processes in detectable amounts [1]. (3) Accurate calculation of the true extracellular concentrations from the dialysis fractions by means of RR is hampered because of a difference between *in vitro* and *in vivo* in the RR of a MD probe [3-5]. (4) The time course of detectable changes is rather slow [2] compared to functional events.

MD measurements have commonly been related to stimulus-induced changes of transmitters and their metabolism, but not to electrical activity. In order to correlate transmitter changes with functional events, the time resolution of the MD equipment should be as high as possible.

It was the aim of this study to follow up changes in excitatory transmitters, asp(artate) and glu(tamate), in relation to electrophysiological recordings. For that purpose, a MD probe was implanted within the cerebral cortex in combination with a ME to register the DC signal, which enables EEG activities to be measured down to 0 Hz [6]. The analytical procedure was optimized according to Westerberg & co-authors [7].

---

*Abbreviations.- 2-APV, D-2-amino-7-phosphono-valerate; (a)CSF, (artificial) cerebrospinal fluid; EAA, excitotoxic amino acids; ECS, extracellular space; NMDA, *N*-methyl-D-aspartate; MD, micro dialysis; ME, micro-electrode; (c)sd, (cortical) spreading depression. RR: *see text* (*in vitro*-based relative recovery [1]).

To illustrate the usefulness of the technique, the electrical and biochemical changes during spreading depression
(**sd**) and cerebral ischaemia were monitored. Marked electrophysiological changes occur comparatively slowly (seconds to minutes)
in both situations. The former [8] is a spontaneously
occurring or experimentally inducible local and transient
depression of electrical activity (EEG) of the cerebral cortex,
which spreads across the whole hemisphere concerned; **sd** can
be identified as a propagating negative DC shift of ~20 mV
which returns to normal spontaneously within ~30 sec. During
ischaemia a similar DC shift occurs with a certain latency
in respect of the onset of the insult, returning rapidly
to normal if reperfusion is initiated immediately after DC
deflection. The negative DC shift in ischaemia as well
as in **sd** has been related to the synchronous depolarization
of the brain cells [6].

Release of asp and glu during ischaemia has been shown
to be related to the degree of ischaemic damage [9] but
not to cellular functions such as membrane depolarization.
However, whether these EAA's are released during **sd** has not
so far been investigated.

MATERIALS AND METHODS

Male Wistar rats (wt. ~250 g) were anaesthetized with
urethane (1.55 g/kg) and tracheotomized. The heads were
fixed within a stereotaxic frame. Body temperature was
maintained at 37° by use of a water jacket. On the right
hemisphere, a burr hole was drilled in the parietal bone
above the motor-cortex, and the dura mater was incised.
A MD probe (2 mm length, 0.5 mm o.d.; Carnegie Medicin,
Stockholm) was implanted to its full extent. Within a
distance of 100–200 μm a ME (for measuring DC) was inserted
at a depth of 1000 μm, the reference electrode being at
the animal's nose. The MD equipment was perfused with
aCSF at a flow-rate of 2 μl/min.

Fractions (0.5, 1 or 4 μl) were collected, and amino
acids were determined by HPLC with fluorescence detection
after automated pre-column derivatization with o-phthaldialde-
hyde. HPLC conditions were as follows: 60 × 4 mm Nucleosil C-18
(5 μm) column; eluents: **A**, 87.5% (by vol.) 0.1 M Na-acetate
buffer (pH 6.72)/10% methanol/2.5% tetrahydrofuran; **B**: methanol
alone. **A/B** ratios in the linear gradient with steps (times
are in min from start) were: 100/0 (2, start gradient), 90/10
(8), 80/20 (12), 50/50 (13), 0/100 (14); finally 100/0
(15); flow-rate 1.2 ml/min.

At 1–1.5 h after probe insertion a single **sd** was elicited,
switching perfusate from aCSF to aCSF-K$^+$ (NaCl replaced by

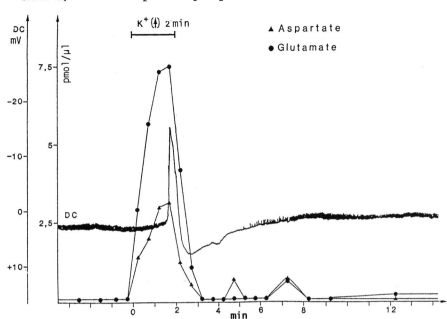

Fig. 1. A single **sd** induced by local application of K⁺.
Note the sequence of events of a representative experiment:
EAA's increased prior to the DC shift. The maximal dialysate
concentration was reached concomitantly with the DC shift
by asp and glu. EAA's returned very quickly to baseline levels.

KCl; K⁺ 128 mM) for 2 min or, alternatively, to NMDA-containing
aCSF (10 nM) for 30 sec and then back again to aCSF. After
sd had occurred, recovery periods of 40-60 min were permitted.
This procedure could be repeated several times using the
same animal. Finally, complete cerebral ischaemia was induced
by cardiac arrest (0.3 ml satd. $MgCl_2$ introduced i.v.).

'Relative recovery' (RR) was determined *in vitro* [cf. 1]
using 100 µM amino acid solution in aCSF (c_s); flow rate was
2 µl/min and perfusate (c_p) consisted of aCSF too (RR = c_s/c_p).

RESULTS

RR, determined to characterize the MD probes, was 4.2 ±1.6%
(±S.D.) for asp and 3.8 ±1.4% for glu; it was independent
of concentration but dependent on flow rate (data not shown).
The lowest time resolution was 0.5 min.

Local application of K⁺ caused the membrane potential
to partly depolarize. With a latency of 81.6 ±20.16 sec
(n = 19), **sd** occurred (Fig. 1). DC returned to basal levels
within ~30 sec. Prior to **sd**, asp and glu started to increase,
to a maximum near-concomitant with DC deflection. Recovery
occurred spontaneously and was complete in 6 min or less.

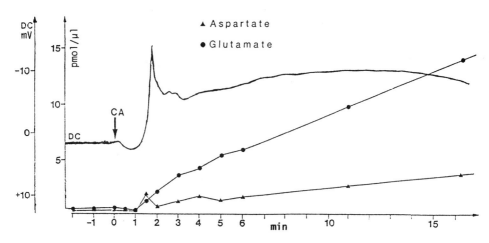

Fig. 2. Terminal DC negativation induced by a complete, irreversible ischaemia. Note the sequence of events.- After onset of ischaemia, DC turned slightly positive. EAA's did not change. Concomitantly asp and, in fewer cases, glu showed a small, transient increase, returning to baseline shortly thereafter; a continuous increase subsequently occurred.

At an average time of 110 ±21.0 sec (n = 9) after onset of ischaemia, DC shifted to the negative ('terminal DC negativation' [6]; Fig. 2). Subsequently asp and especially glu increased dramatically. A small but transient increase concomitantly with the DC negativation was seen in 6 out of 9 experiments with asp and in 4 out of 9 experiments with glu.

Fig. 3 demonstrates that asp and glu remained unchanged when the receptor agonist NMDA was applied, although **sd** occurred with a latency of 32.3 ±14.68 sec (n = 19).

DISCUSSION

The extracellular changes in asp and glu during cerebral ischaemia and **sd** were monitored with a time resolution of 1 min. Both amino acids increased during K^+ application and during ischaemia; but the time courses of these variations were different in relation to the DC deflection. EAA's increased prior to the DC shift in **sd** whereas in ischaemia the massive increase occurred subsequent to DC deflection albeit a slight and transient elevation could be observed concomitantly with the DC shift. Because the latter elevation did not occur regularly, it does not seem to play a causal role in the generation of terminal DC negativation.

However, EAA's seem to be essential for the generation of **sd**.- (1) NMDA-induced **sd** without affecting extracellular EAA's, which indicates that activation of the NMDA receptor

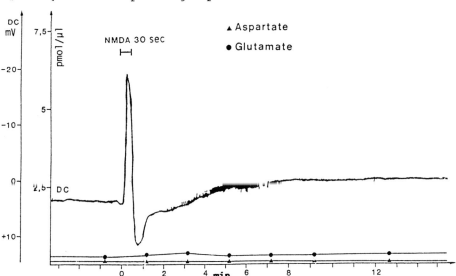

Fig. 3. A single **sd** induced by local application of NMDA.
Note that asp and glu remain unchanged.

can elicit **sd.** (2) EAA's – the endogenous agonists of the
NMDA receptor [10] – were released due to K^+ application
and increased prior to **sd.** (3) Local application of the NMDA
antagonist 2-APV suppressed the K^+ as well as the NMDA-induced
sd [11].

We therefore conclude that EAA's released by the K^+
stimulus activated the NMDA receptor and triggered the generation of **sd.** Evidently the mechanisms generating the DC
shift during ischaemia and **sd** are different although the
events look similar with respect to the electrical recording.
However, despite the optimization of the analytical procedure,
the time resolution of MD is poor compared to the ME recording,
although it is even worse for, e.g., dopamine (20 min).
This limits the applicability of the method to rather slow
processes.

As synaptic events occur in <1 sec and most of the
released transmitters are rapidly taken up, only a small
fraction will escape the re-uptake and diffuse into the
ECS.   Whereas diffusion of, e.g., asp or glu in solution
is in the μm/sec range (~1 mm/15 min [12]), it is ~2.5 times
slower within the brain [13], due to a lengthened diffusion
path around the cell bodies.   Glial re-uptake processes
or even extracellular catabolic events are provided with
enough time to further reduce the amount of the diffusing
compound. Accordingly, the changes in the transmitter overflow
[1] as detected by MD only indirectly reflect the quantity

of the synaptically released compound. We therefore assume that re-uptake from ECS and extracellular inactivating processes have to be compromised to make small transmitter changes (such as supposedly occur physiologically) detectable. This indicates that a very pronounced transmitter release had to occur to become detectable as in **sd** or ischaemia. Whether such marked changes take place physiologically and are detectable by MD remains to be established.

Moreover, the implantation of the MD probe itself induces **sd** [3, 14, 15] and some slight morphological damage [16]. Besides, **sd** induces a long-lasting reduction of cerebral blood flow [17, 18]. This affects cellular function around the probe for ~1 h or even longer (unpublished observations). A certain dead volume around the probe might remain [5], complicating the reliable estimation of levels of extracellular substances.

Despite the foregoing uncertainties, the results illustrate the successful use of the MD technique in respect of detection of the sequence of pathophysiological events:- in combination with the ME technique it was possible to elucidate the causal relationship between EAA release and DC shifts. The time resolution could be improved to ~1 min, which sufficed with respect to the investigated insults. The results obtained so far in our laboratory prove the MD technique to be a valuable tool for pathophysiological and pharmacological studies.

References

1. Ungerstedt, U. (1986) *Current Separations 7*, 43-46.
2. Benveniste, H. (1989) *J. Neurochem. 52*, 1667-1679.
3. Benveniste, H., Hansen, A. & Ottosen, N.S. (1989) *J. Neurochem. 52*, 1741-1750.
4. Lindefors, N., Amberg, G. & Ungerstedt, U. (1989) *J. Pharmacol. Meth. 22*, 141-156.
5. Scheller, D. & Kolb, J. (1991) *J. Neurosci. Meth.* (submitted).
6. Caspers, H., Speckmann, E-J. & Lehmenkühler, A. (1987) *Rev. Physiol. Biochem. Pharmacol. 106*, 127-178.
7. Westerberg, E., Kehr, J., Ungerstedt, U. & Wieloch, T. (1988) *Neurosci. Res. Comm. 3*, 151-158.
8. Leao, A.A.P. (1944) *J. Neurophysiol. 8*, 359-390.
9. Benveniste, H., Drejer, J., Schousboe, A. & Diemer, N.H. (1984) *J. Neurochem. 43*, 1369-1374.
10. Collinbridge, G.L. & Lester, R.A.J. (1990) *Pharmacol. Revs. 40*, 143-210.
11. Scheller, D., Dengler, K. & Heister, U. (1990) *Pfl. Arch. Eur. J. Physiol. Suppl. 1 to Vol. 415*, R80.

12. Macey, R.I. (1980) in *Membrane Physiology* (Andreoli, T.E.,
    Hoffman, J.F. & Fanestil, D.D., eds.), Plenum, New York,
    pp. 125-146.
13. Nicholson, C. & Phillips, J.M. (1981) *J. Physiol. 321*, 225-257.
14. Yergey, J.P. & Heyes, M.P. (1990) *J. Cerebr. Blood Flow
    Metab. 10*, 143-146.
15. Lauritzen, M., Hansen, A.J., Dronberg, D. & Wieloch, T.
    (1990) *as for* 14., 115-122.
16. Benveniste, H. & Diemer, N.H. (1987) *Acta Neuropathol.
    (Berl.) 74*, 234-238.
17. Lauritzen, M. (1987) *Acta Neurol. Scand. 76 Suppl. 111*, 9-40.
10. Benveniste, H., Drojer, J., Schoueboe, A & Diemer, N H
    (1987) *J. Neurochem. 49*, 729-734.

Comments on #E-1: J.W.C.M. Jansen - FIBRINOLYSIS REGULATION

M. Maley asked whether the endothelial cells in the study were from permanent cell lines or from primary short-term cultures. Reply.- The animal cells were from aorta, up to 4th passage - beyond which abnormalities arise, manifested in urokinase production. Human cells were from similar cultures of umbilical vein samples.

Comments on #E-2: S. Kellie - TARGETS FOR TYROSINE KINASES

Question by R.W. Bonser.- Adhesion plaques are reportedly enriched in PKC, yet you were unable to show any threonine or serine phosphorylation of integrin in your transformed cells. Since the integrins are substrates for PKC, can you comment on why integrin phosphorylation occurs only on tyrosine residues? Reply.- Whilst all the cytoskeletal proteins discussed are substrates for PKC *in vitro*, the situation is more complex within cells. Some cells, such as teratocarcinoma cells, respond to PMA by integrin phosphorylation whereas other cells do not. We have found serine phosphorylation of the α-chain, perhaps due to PKC, but not of the β-chain. These cells have not been treated with PMA to stimulate PKC, which may be necessary for optimal integrin phosphorylation.

P.A. Kirkham asked whether genistein, a specific TK inhibitor,* had been tried, to show that its inhibition of integrin tyrosine phosphorylation by pp60 does not lead to fibrinonectin loss. Reply.- Yes, pre-exposing the cells to it for up to 1 h at 37°; but we had difficulty with its solubility and did not obtain conclusive results. Answer to a query on the phosphorylation of vinculin and integrin: stoichiometry is in general low, 1-10% of maximum; but the interpretation may be complicated because of protein isoforms. Comment by Bruzzone.- Connexin 43, the heart gap-junction protein, should be added to the list of substrates for pp60^{v-src}. As for the other substrates you listed, connexin 43 is a membrane protein which is localized in the gap-junction plaques that comprise sites of cell-to-cell contact.

Relevant refs. noted by Senior Editor:
Linassier, C., *et al.* (1990) *Biochem. Pharmacol. 39*, 187-193.- 'Mechanisms of action in NIH-3T3 cells of **genistein,** an inhibitor of EGF receptor tyrosine kinase activity'; mitogenesis by thrombin also blocked, unrelated to TK. With isolated kinases including TK and PKC, blockage by **erbstatin** has been studied: Bishop, W.R., *et al.* (1990) *Biochem. Pharmacol. 40*, 2129-2135 (EGF receptor relevance).

Comments on #E-3: R. Bruzzone - GAP JUNCTION SIGNALLING ROLE

G. **Milligan asked** whether in a single pore there is only one connexin-gene product, or whether pores may be formed by the interaction of heterologous connexins. **Reply.-** It is not clear whether heterologous connexins will interact with each other to form one-half of a gap junction. Gap junctions can, however, be formed between cells expressing different connexin gene products. **Kellie asked** (1) whether the endogenous gap-junction proteins in the oocyte have been studied; (2) whether connexins are glycosylated. **Answers:** (1) two gap junction proteins have been cloned from oocytes, but their relationships with other connexins are unknown; (2) one side may be glycosylated, but this awaits proper study. **In reply to D. Scheller:** decoupling can be caused by pH change or, at very high concentrations, by Ca^{2+}; but Ca^{2+} has also been shown to be a cell-to-cell messenger through the gap junction. P-O. **Berggren,** concerning blockage of cell-to-cell communication by stimuli: one would really expect cells to benefit from being synchronized under stimulatory conditions, feasible only if they are coupled. **Bruzzone's response.-** Although, intuitively, one might reckon that synchronized cells will function better, there are several possible explanations of the increased pancreatic secretion following blockade of cell-to-cell coupling.-
(1) Molecules with intracellular messenger function oscillate under both resting and stimulated conditions in several cell types, including acinar cells. Oscillatory signals generated by one cell would be damped by the extensive coupling present between cells of an acinus, so depressing enzyme secretion rate.
(2) Uncoupling may trigger a 'stress' reaction of cells which are normally coupled, followed by secretion of digestive enzymes whose intracellular activation would be disastrous.
(3) Not all secretory cells show the same behaviour. In fact the B-cells are very poorly coupled, whereas insulin secretagogues such as glucose and glibenclamide induce the formation of gap junctions and increase the coupled territories. Furthermore, heptanol-induced coupling inhibits both glucose- and carbachol-induced insulin release.

Comments on #E-4: G.M.W. Cook - INTERACTIONS IN NEUROGENESIS

Replies to questions.- (1), from G.J. **Barritt** concerning matrix-cell interactions: extracellular-matrix glycoproteins may interact with a receptor on the nerve-cell surface and induce a cell change; (2), from H. **LeVine:** the Ab's to our PNA eluate are not blocking Ab's when applied to a cell system, being Ab's against protein epitopes; (3), from R. **Heath:** our system would suit well for trying retinoic acid, reported to be a neuronal morphogen; (4), from K.D. **Brown:** work is in hand to label the glycoproteins and look for binding to possible 'receptors' in the growth cone.

Comments on **#ncE-1:** D. Wermelskirchen - Na⁺ and Ca²⁺ MOVEMENTS
     #ncE-2: D. Scheller – SPREADING–DEPRESSION STUDIES

    Jansen asked **Wermelskirchen** what happens with N_2/CO_2 to get true ischaemic conditions? **Reply.-** Synaptosomes are not amenable to such conditions. There are indications that under real ischaemic conditions Na⁺–channels change; hence we use veratridine to simulate the ischaemic event. **Replies to other queries.-** We don't think there is a direct stimulation of respiration by Ca²⁺: respiration is unaffected by agents that alter $[Ca^{2+}]_i$ - BAY K 8644, A23187, Na-*o*-vanadate. The veratridine-induced ⁴⁵Ca²⁺ uptake was undiminished by Ca²⁺-antagonists but was completely blocked by the Na⁺-channel blocker TTX as well as by flunarizine and R 56865; we don't yet know whether the inhibition is caused by blockade of a Na⁺-channel or, subsequently, of Na⁺/Ca²⁺ exchange.

    Remarks by N.J. Birch.- In your use of O_2 uptake as a measure of Na⁺ transport, you should always try to take account of confounding factors which may be affecting O_2 utilization but not Na⁺ transport. For instance it is not widely recognized that A23187 has a binding constant ~10 times less than Ca²⁺ though $[Mg^{2+}]_i$ is 100-10,000 times higher than $[Ca^{2+}]_i$, this agent having been shown by you to slightly reduce O_2 uptake. In consequence, A23187 may deplete cell Mg²⁺ because Mg²⁺ is more able to compete for A23187 binding sites. Mg²⁺ is essential for the functioning of glycolysis and the Krebs cycle. Thus Mg²⁺ depletion may lead to reduced O_2 consumption as a result of reduced O_2 utilization rather than reduced Na⁺ transport.

    Scheller, answering **P-O. Berggren** who asked about damage associated with probe insertion: (1) traumatic insertion induces SD with effects persisting for ~2 h on cerebral blood flow and metabolism; (2) a ~200-300 µm zone of compressed cells is observable around the probe for ~4 h, their survival being problematical; (3) for long-lasting reduction in local cerebral blood flow and glucose consumption, see the Benveniste refs. **Replies to J.W.C.M. Jansen.-** Amongst neurotransmitters not so far measured, dopamine warrants study. Some literature points to minimization of damage at 3-4° (we work at 37°), and there may be effects if the glucose level is subnormal rather than normal as in our studies. **Reply to Heath.-** The spreading of the depression is thought to be caused by massive release of EAA's, affecting adjacent neurones/synapses. There is also a 'K⁺ hypothesis': the produced K⁺ is taken up by glial cells, causing depolarization which affects neighbouring neurones which in turn results in further depolarization. This is an active work-area.

SOME STUDIES ON NEURAL MATERIAL, noted by Senior Editor

Snyder, S.H. (1991) *Nature 350,* 195, citing   C. Hymen *et al.* (230-235) and alluding to **B. Knusel** *et al.* (1991) *Proc. Nat. Acad. Sci. 88,* 961-965.- Selectivity towards particular neurones is exhibited by **NGF**, neurotropin-3 (**NT-3**), and the recently discovered brain-derived neurotrophic factor (**BDNF**) which increases the abundance of **dopamine**-secreting cells.

Giambalvo, C.T. (1989) *Biochem. Pharmacol. 38,* 4445-4454.- The effect of drugs such as sulpiride which deplete striatal **dopamine** is associated with **PKC changes** (fall in cytosol, rise in membranes) for which an intact dopamine store is requisite.

Perez-Polo, J.R., *et al.* (1990) *Int. J. Clin. Pharm. Res. 10,* 15-26.- **NGF** in relation to aging in the CNS.

Conn, P.M., ed. (1988) *Neuroendocrine Peptide Methodology* (arts. from *Meths. Enzymol., 103, 124 & 168*), Acad. Press, 949 pp.

Gilbert, J.A., *et al.* (1989) *Biochem. Pharmacol. 38,* 3377-3382.- Analogues of **neurotensin(8-13)**, the potent C-terminal portion of neurotensin, were tested for stimulation of **cGMP** formation and (neuroblastoma clone) and for binding to neurotensin **receptors** (human-brain membranes, and intact clone cells).

Hasegawa, T. (1990) *Biochem. Pharmacol. 40,* 1463-1467. Brain cortex: **lipid peroxides** (locomotor depressants) stimulated synaptosomal **GTPase** activity similarly to adenosine; same receptor.

Rodriguez, R., *et al.* (1989) *Biochem. Pharmacol. 38,* 3219-3222.- Ca^{2+}/Mg^{2+}-**ATPase** in synaptosomes is modulated by **cAMP**.

Niles, L.P. & Hashemi, F.S. (1990) *Biochem. Pharmacol. 40,* 2701-2705.- Forskolin-stimulated **adenylate cyclase** activity in brain is inhibited by melatonin or analogues or by diazepam.

Llano, J., *et al.* (1987) *Pflügers Arch. 409,* 499-506.- $\Delta\psi$ (MP); $[Ca^{2+}]_e$.

Smith, R.E., *et al.* (1989) *J. Chromatog. Sci. 27,* 491-495.- Using cortex, hippocampus and striatum (rat), procedures were developed for **assay of InsP's** (and phosphocreatine etc.), with checking of the effects of post-sacrifice ischaemia. Homogenization in water followed by $CHCl_3$/methanol extraction was preferable to TCA, formic acid or PCA. Separation was by anion chromatography using conductivity detection (chemical suppression).

Hughes, P.J. & Drummond, A.H. (1987) *Biochem. J. 248,* 463-470, & Drummond (1987) *Trends Pharmacol. Sci. 8,* 129-133 (cf. #B-7, this vol.).- HPLC on pituitary tumour cells showed various **InsP's** derived from $InsP_3$ as it is recycled back to inositol (a peculiarity of brain); this restoration of inositol is blocked by Li^+ (crucial).

Ross, C.A..... Snyder, S.H. (1989) *Nature 339,* 468-470.- $InsP_3$ **receptor** is in rough and smooth e.r. of cerebellar Purkinje neurons.

Biochem. Pharmacol. arts. (rat brain).- On **InsP's**: Li, X. & Jope, R.S. (1989) *38,* 2781-2787 (S-containing amino acids inhibit PI hydrolysis in slices); (1990) *40:* Alexander, S.P.H., 1793-1799 ($[Ca^{2+}]_i$ influences); Brammer, M.J. & Weaver, K., 1901-1906 ($InsP_1$'s). On **ion channels** and intraterminal Ca^{2+}: Clark, J.M. & Brooks, M.W. (1989) *38,* 2233-2245. On **GTPase**: Ravindra, R. & Aronstam, R.S. (1990) *40,* 457-463 (tubulin role studied with Ab's).

Vogel, S.S., Chin, G.J., Schwartz, J.H. & Reise, T.S. (1991) *Proc. Nat. Acad. Sci. 88*, 1775-1778.-'Pertussis toxin-sensitive **G-proteins** are transported toward synaptic terminals by fast axonal transport'.

Cohen, J., *et al.* (1987) *Dev. Biol. 122*, 407-418.- 'The role of laminin and the laminin/fibronectin receptor complex in the **outgrowth** of retinal ganglion cell **axons**'. (Cf. art. #E-4.)

Earlier entries on neuroblastoma cells complement these neural items (see Index). The following items, overlapping with pp. 387-388, centre on **INTER-CELL PHENOMENA**, *also* **PANCREATIC ACINAR CELLS** *(see p. 434 for another entry) - cf. art. #E-3, also #C-1*

Willems, P.H.G.M., *et al.* (1987) *Biochim. Biophys. Acta 928*, 179-185.- PT-sensitive **G-proteins** mediate Ptd-InsP$_2$ formation after cholecystokinin binding to receptor on **acinar cells**; InsP/ Ca^{2+} pathway and enzyme secretion influenced by cAMP too.

Meyer, T. (1991) *Cell 64*, 675-680.- 'Cell signalling by second messenger waves'; a means for intra- and inter-cell **communication** and cell-response synchronization. e.g. in *Dictyostelium* (brain also?) - maybe involving Ca^{2+} waves (and gap junctions) or cAMP waves. *In same issue:* Proudfoot, N., 671-674.- '**Poly(A)** signals' (mRNA).

Benchimol, S., *et al.* (1989) *Cell 57*, 327-334.- **CEA**, abundant in embryonic and cancerous colon, mediates intercellular **adhesion**.

van Mourik, J.A., *et al.* (1990) *Biochem. Pharmacol. 39*, 233-239.- Commentary: 'Pathophysiological significance of **integrin** expression by vascular endothelial cells' (also platelets); integrins face outwards, mediating anchorage to extracellular matrix proteins, e.g. fibronectin and fibrinogen.

Takeichi, M. (1990) *Annu. Rev. Biochem. 59*, 237-252.- '**Cadherins:** a molecular family important in selective cell-cell adhesion' (and in **morphogenesis**?); Ca^{2+}-dependent association with some specific proteins and actin-based cytoskeletal elements which may regulate cadherins. Role in a **neurite outgrowth** system: Matsunaga, M., *et al.* (1988) *Nature 334*, 62-64.

Merritt, J.E. & Rink, T.J. (1987) *J. Biol. Chem. 262*, 17362-17369.- Role of **gap junctions**; see also Rink in Vol. 19, this series.

Petersen, O.H. & Gallagher, D.V. (1988) *Annu. Rev. Physiol. 50*, 65-80.- 'Electrophysiology of pancreatic and salivary **acinar cells**'; the Ca^{2+}- and voltage-activated **K$^+$ channel** is cardinal to secretory processes. *In same vol.:* Solthoff, S.P. & Cantley, L.C., 207-223.- '**Mitogens** and ion fluxes'.

Brown, E.M. (1991) *Physiol. Rev. 71*, 371-411.-'**Extracellular Ca^{2+}** sensing, regulation of parathyroid cell function, and role of Ca^{2+}'.

Brockenbrough, C.R. & Korc, M. (1987) *Cancer Res. 47*, 1805-1810.- Ca^{2+}- and PKC-independent regulation of secretion in isolated **acini**.

Adler, G., *et al.* (1991) *Gastroenterology 100*, 537-543.- Stimulation of **pancreatic secretion** in **humans**; cholecystokinin/cholinergics.

A Forum presentation for which no publication text was gained:

KENNETH D. BROWN, CHRISTOPHER J. LITTLEWOOD & ANTHONY N. CORPS
Department of Biochemistry, AFRC Institute of Animal Physiology and Genetics Research, Babraham, Cambridge CB2 4AT.

Quiescent cultures of Swiss 3T3 (mouse embryo-derived) cells can be stimulated to proliferate by a variety of mitogens, some of which have the capacity to stimulate phosphoinositide hydrolysis, activate protein kinase C and elevate intracellular Ca^{2+}. These include regulatory peptides such as bombesin and vasopressin, as well as polypeptide growth factors such as platelet-derived growth factor (PDGF) and fibroblast growth factor (FGF). However, the time courses of the production of inositol phosphates, and the elevation of intracellular Ca^{2+} concentration, are markedly different when cells stimulated by bombesin or PDGF are compared [1,2]. Indeed, recent results suggest that receptor coupling to phospholipase C, as well as the feedback control of this coupling, are fundamentally different between receptors for the peptide and polypeptide mitogens [1-3].

It is now clear that no single, unique signal is responsible for the stimulation of cell proliferation; instead, a complex network of interacting signal transduction pathways appears to be involved. Although activation of the inositol lipid pathway is often correlated with the stimulation of cell proliferation, a causal relationship has not been unequivocally established. Strategies and methods currently in use to try to resolve this problem [were] reviewed [in the Forum talk].

1. Blakeley, D.M., Corps, A.N. & Brown, K.D. (1989) Biochem. J. **258**, 177-185.
2. Corps, A.N., Cheek, T.R., Moreton, R.B., Berridge, M.J. & Brown, K.D. (1989) Cell Regulation 1, 75-86.
3. Meisenhelder, J., Suh, P.-G., Rhee, S.G. & Hunter, T. (1989) Cell 57, 1109-1122.

Comments on the above presentation

M. Maley asked Brown whether, in the studies on heterogeneity of Ca^{2+} levels in response to peptides, the cultures were synchronized. **Reply.-** Yes, by serum-deprivation, rather than serum-free, conditions; traces of serum advantageously provide attachment factors. **Reply to G.J. Barritt.-** We have measured $Ins(1,4,5)P_3$ (as distinct from total $InsP_3$) only over 2-3 min, not over longer periods. **Reply to J. Pfeilschifter. —** We have no explanation of our finding that PDGF-stimulated $InsP_3$ formation was attenuated in PKC-downregulated cells. **R. Bruzzone remarked,** concerning the temporal dissociation between PDGF-induced $InsP_3$ formation and $[Ca^{2+}]_i$ transients (prolonged $InsP_3$ formation), that it would be interesting to know the role of $InsP_3$ under these conditions. **P-0. Berggren commented** on the possibility that lowering of the Ca^{2+} outward transport capacity might account for the longer duration of the $[Ca^{2+}]_i$ increase in Brown's PKC-downregulated cells, and also alluded to a report (in *Science*) that the PDGF effect on DNA synthesis is unrelated to an increase in $[Ca^{2+}]_i$ or $InsP_3$.

FURTHER LITERATURE noted by Senior Editor,
especially on CELL PROLIFERATION/DIFFERENTIATION & GROWTH FACTORS

Taylor, C.W. (1989) *NATO ASI Ser. E 156*, 187-200.- The role
of phosphoinositate hydrolysis and ensuing **messenger generation**
in mitogen-induced cell proliferation.

Vincentini, L.M. (1986) *Life Sci. 38*, 2269-2276.- Review of
'putative **signals** for control of cell growth' (especially mitogen-induced): turnover of InsP's; $[Ca^{2+}]_i$; pH.

O'Brian, C.A. & Ward, N.E. (1989) *Biochem. Pharmacol. 38*, 1737-
1742 *(cf. 39, 49-57).*- Naphthalenesulphonamide inhibitors of **PKC**
(a key enzyme in growth regulation) may act by affecting the
ATP-binding site and/or the phospholipid cofactor.

Morris, H.R., Taylor, G.W., Masento, S.M., Jermyn, K.A. & Kay, R.R.
(1987) *Nature 328*, 811-814.- **DIF-1** (a **morphogen**: differentiation-
inducing factor), purified by HPLC, is a phenylhexanone; identified with aid of EI-MS, which superseded bioassay in *Dictyostelium*.
- DIF, passing from cytoplasm of one cell to adjoining cell, can
'tell' it whether to become a stalk cell or a spore-forming cell.

Arts. (1989) in *Cell 56*, 495-506 (Wong, S.T., *et al.*) & 691-700
(Brachmann, R., *et al.*).- **TGF-α**, which normally is snipped off the
outside of its cell of origin (animal) for entry (receptor) into a
non-adjoining cell where it causes transient Ca^{2+} **mobilization**
demonstrable by staining, has been rendered non-cleavable and
capable of influencing adjoining cells; model for **differentiation**.

Olsen, R., *et al.* (1990) *Biochem. Pharmacol. 39*, 968-972.- PKC
(inessential for **fibroblast** proliferation) not involved in the
inhibition by **staurosporine** of PDGF-induced proliferation and of
$[Ca^{2+}]_i$ transients; PDGF receptor autophosphorylation is blocked,
as is the **TK** involved in $[Ca^{2+}]_i$-release cascades.

Tavaré, J.M. & Holmes, C.H. (1989) *Cell.Signalling 1*, 55-64.- **EGF**
receptors (also insulin receptors) studied in developing human
placenta: they showed differential expression in different cells,
reflecting proliferative capacity and differentiation status.

Hayashi, T., *et al.* (1989) *Chromatographia 27*, 574-580.- **Assay of**
hEGF's in biological fluids by RP-HPLC; affinity precolumn to 'trap'.

Nieto, K.F. & Frankenberger, W.T. (1988) *J. Liq. Chromatog. 11*,
2907-2925.- Ion-suppression RP-HPLC for **assay of cytokinins**.

Seuwen, K. & Pouysségur, J. (1990) *Biochem. Pharmacol. 39*,
985-990.- Commentary: 'Serotonin as a growth factor'. As
for other neuropeptides including bombesin, BK and VP, also
thrombin, serotonin has diverse activities including mitogenesis,
depending on the target cell. Besides causing, when released
from platelets, EDRF (NO) liberation in vascular endothelial
cells, thereby affecting the surrounding smooth-muscle cells,
serotonin can stimulate proliferation of the latter in synergy
with PDGF and insulin. This proliferative action may involve
receptor interaction with G-proteins which control PL-C or
adenylate cyclase activity, whereas 'classical' growth factors,
e.g. EGF, PDGF, FGF and IGF_1, act through receptor TK's.

CONTINUATION of supplementary material, including:
ENDOTHELINS; EICOSANOIDS; TK's/CANCER-RELATED OBSERVATIONS

Yanagisawa, M. & Masaki, T. (1989) *Biochem. Pharmacol. 38*, 1877-1883.- Commentary: '**Endothelin**, a novel endothelium-derived peptide', now purified from porcine cells (21 residues; acidic); its precursor has a pre-pro-form as for some other peptide hormones. Its effects, besides vasoconstriction, include contractility and proliferation (it produces proto-oncogenes, e.g. in fibroblasts). It may stimulate Ca^{2+} influx *via* channels, and/or mobilize stored Ca^{2+}; PL-C may be stimulated. It may have a homeostatic as distinct from an 'emergency' role. Its sources and targets may include neurons, as for the kindred agent **sarafotoxin**. Both agents stimulate **InsP** production.- Sharif, N.A. & Whiting, R.L.(1990) *Biochem. Pharmacol. 40*, 1928-1931 (fibrosarcoma cell line). See also 2713-2717 (Takayasu-Okishio, M., *et al.*): **TXA$_2$** raised (by **PAF** also).

Kaever, V., Pfannkuche, H-J., Wessel, K. & Resch, K. (1990) *Biochem. Pharmacol. 39*, 1313-1319.- In macrophages, elevated $[Ca^{2+}]_i$ favours PKC-mediated synthesis of **LT** rather than **PG**.

Luscinkas, F.W., *et al.* (1990) *Biochem. Pharmacol. 39*, 359-365.- **LT-A$_4$** and epoxytetraenes (lipoxin precursors) caused $[Ca^{2+}]_i$ mobilization in human neutrophils.

Kainoh, M., *et al.* (1991) *Biochem. Pharmacol. 41*, 1135-1140.- **cAMP** rose in vascular endothelial cells treated with beraprost sodium (prostacyclin analogue), with enhanced expression of **thrombomodulin**.

Sraer, J., *et al.* (1989) *Biochem. Pharmacol. 38*, 1947-1954.- 'Dual effects of cyclosporine A on **arachidonate** metabolism by peritoneal macrophages': PL stimulation in resting cells (PG and 12-HETE rose) and, greater in stimulated cells, blockage of **thromboxane synthase**.

Kaplan, D.R., Martin-Zanca, D. & Parada, L.F. (1991) *Nature 350*, 158-160.- '**Tyrosine** phosphorylation and tyrosine kinase activity of the *trk* proto-oncogene product induced by **NGF**'.

Burke, Jr., T.R., *et al.* (1991) *Biochem. Pharmacol. 41*, R17-R20.- Generally there was no inhibition of EGF-receptor **TK** by opiates, disproving the idea that their antineoplastic action involves TK's.

Hsu, I.C., *et al.* (1991) *Nature 350*, 427-428.- 'Mutational hotspot in the **p53 gene** in human hepatocellular carcinomas': aflatoxin was reckoned to be the carcinogen that caused a precise mutation.

Nigg, E.A. (1990) *Adv. Cancer Res. 90*, 271-310.- 'Mechanisms of signal transduction to the cell **nucleus**'; primarily by shuttling of regulatory proteins, e.g. **PK**'s, for nucleus/cytoplasm integration.

Waterfield, M., ed. (1989) *Br. Med. Bull. 45*, 317-604.- '**Growth factors**'; diverse aspects incl. **gene induction** (Woodgett, J.R., 529-540).

Arts. (1989/90) in *Adv. Prostaglandin Thromboxane Leukotriene Res.- 19*: Vanderhoek, J.Y., *et al.* (78-81): Ca^{2+} only incidentally involved in **5'-lipoxygenase** activation by 5'-HETE in PT-18 cells. *20*: Raz, A., *et al.* (22-27): fibroblast/monocyte **prostanoid** regulation by **IL-1, endotoxin** & **glucocorticoids**. *20*: Crooke, S.T., *et al.* (127-137): transduction for **LTD$_4$** (*via* Ptd-InsP$_2$; different for **LTB$_4$**). *20*: Lewis, R.A. (170-178): eicosanoids and cytokines in **immune regulation**.

#F

LOCATION AND TRANSIT OF PROTEINS (BESIDES 'PKC')

#F-1

STRESS SIGNALS

C.A. Pasternak

Division of Biochemistry,
Department of Cellular & Molecular Sciences
St. George's Hospital Medical School,
Cranmer Terrace, London SW17 0RE,   U.K.

Environmental stress (e.g. hyperthermia, virus infection, toxic chemicals) causes cells to synthesize a class of proteins ('heat-shock proteins', hsp's) that function to ameliorate the deleterious consequences of stress.   Another response of cells is to increase their uptake of glucose, through translocation of the glucose transporter protein from an intracellular site to the p.m.*.   Experimental strategies employed to investigate these effects are now considered.

Stress is the illness of the age. We are all subject to it and are well aware of some of the causes such as death of spouse or divorce, financial worry or loss of a job, a thousand-and-one anxieties in our daily lives. Excessive stress leads to every kind of disease, from insomnia and back-ache through heart disease and stroke to cancer. The stress stimulus is not only mental but can be physical also - as in post-operative trauma, violent exercise or sepsis following an infectious episode.

Interestingly, isolated cells respond to virus infection and other forms of 'environmental' stress such as hyperthermia[†] or the presence of toxic chemicals in certain characteristic ways. This occurs whether the cells are taken from a human being, a fly or even a microbe such as *E. coli*. One response is to synthesize a group of proteins known as 'heat-shock proteins' or 'stress proteins', which function to protect the cell against a second stress stimulus such as hyperthermia, toxic chemicals or virus infection [1, 2]. A separate response, as studied by us, is increased capacity to take up glucose [3]. This, too, may protect the cell against further damage, and failure to increase glucose uptake may underlie the tissue damage often associated with sepsis and post-operative trauma.

---

[*]*Abbreviations.* - Ab, antibody; e.m., electron microscopy/micrograph; hsp, heat-shock protein; p.m., plasma membrane.
[†]Arts. in Vol. 19 (1989; eds. & publisher as now) - J.G. Hofmann; P.K. Wierenga - are pertinent. For toxic agents see Vol. 17.-*Ed.*

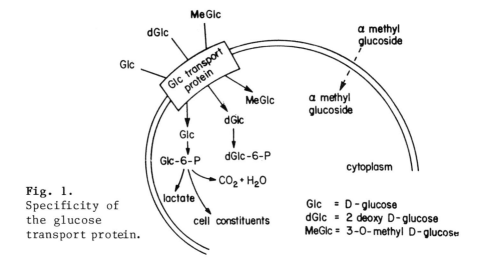

Fig. 1.
Specificity of
the glucose
transport protein.

Glc = D-glucose
dGlc = 2 deoxy D-glucose
MeGlc = 3-O-methyl D-glucose

Our original observations were with virus-infected cells [4]. We compared the uptake by infected and non-infected BHK cells of three glucose derivatives (key in Fig. 1): dGlc, which is transported by facilitated diffusion (using the specific glucose carrier protein) and then phosphorylated but not metabolized further; MeGlc, which is transported like dGlc but is not phosphorylated; and α-methyl-glucoside, which is not transported by a facilitated carrier mechanism but enters cells by simple diffusion only [5] (Fig. 1). The results were clear-cut. Virus infection increased the uptake of dGlc and MeGlc, but that of α-methyl-glucoside was unaffected [6]. Hence it is clear that the target for this type of stress is the glucose transport protein that catalyzes facilitated uptake into cells.

Since the effect, as for insulin-stimulated glucose uptake, is predominantly on V_{max} and not on K_m (Table 1), it is possible that the mechanism whereby the increase is brought about is the same for stress as for insulin [7]. In the latter case it has been shown that most of the increase can be accounted for by the movement of glucose transporter molecules from an intracellular (inactive) site to the p.m., rather than by the activation of existing transporter molecules at the p.m. [8]. What are the strategies by which such translocation might be revealed in the case of cellular stress?

Strategies for studying translocation of membrane proteins

Two types of strategy have been employed in the case of insulin-stimulated translocation of the glucose transporter protein (and in principle could also apply to cellular stress):

Table 1.   2-Deoxyglucose uptake: V_{max} (nmol/mg protein per min) and K_m (mM).   The values, for BHK cells, are mean ±S.D.

(Data already published [7].)

Treatment	Apparent V_{max}	Apparent K_m
Control	*12.7 ±1.4*	*3.33 ±0.49*
Semliki Forest virus (SFV)	48.3 ±2.7	2.13 ±0.26
Arsenite	60.5 ±0.6	2.45 ±0.4
Insulin	49.6 ±2.9	2.62 ±0.22
Heat shock	40.4 ±0.9	2.40 ±0.08

1) Qualitative: immunofluorescence.
2) Quantitative:
   (a) subcellular fractionation
   (b) surface (and total) labelling - i) sodium [^{3}H]borohydride
                          ii) Holman reagent.

1) **Qualitative** (or at best semi-quantitative) assessment of the distribution of the glucose transporter within cells can be made by staining permeabilized cells with Ab (e.g. raised in rabbits) to the transporter molecule, and then adding a second Ab (e.g. anti-rabbit IgG raised in goats) that is labelled in some other way, so as to make it detectable by visual or other means. Thus, for insulin-stimulated trans-location of the glucose transporter in differentiated 3T3-L1 (adipocyte-like) cells this has been achieved by using gold-labelled protein A to bind to the second Ab and examining ultrathin cell sections by transmission-e.m. [9].

For insulin- and stress-stimulated translocation of the transporter in BHK cells we have used fluorescein-labelled second Ab and examined the cells by UV light microscopy. This has the advantage over the gold method that an entire cell can be seen [10]; on the other hand it is less amenable to semi-quantitative assessment of the amount of transporter present at different intracellular sites. Instead of using first Ab raised against the transporter molecule, we used Ab raised against a synthetic peptide corresponding to residues 477-492 (the C-terminal end) of the molecule [11]; somewhat surprisingly, Ab raised against such a relatively short peptide is quite specific to the glucose transporter (e.g. see [9]).

The results of the gold-labelling e.m. analysis [9] and the fluorescent UV-light analysis were similar. In each case insulin-stimulated cells showed a movement of transporter from the perinuclear, Golgi-rich region to a more diffuse distribution compatible with insertion into the p.m. With the latter approach [10], arsenite, heat-shock or virus infection (vesicular stomatitis or Semliki Forest virus) showed a response similar to that elicited by insulin.

2) **Quantitative** assessment of the amount of glucose transporter moving between an intracellular site and the p.m.
may be obtained by subcellular fractionation of cells into
intracellular and p.m. vesicles, and measurement of the amount
of glucose transporter associated with each fraction. For
assaying the transporter, quantitative immuno-precipitation
of binding of cytochalasin B (a specific inhibitor of the
transport molecule) may be used.

The approach works well for adipocytes [8], but is unlikely
to be readily applicable to cells such as BHK cells, since
it is much more difficult to isolate p.m. free of intracellular
membranes from cells grown in monolayer culture.   There is
another difficulty. In the fractionation of insulin-stimulated adipocytes, there was a ~5-fold increase in cytochalasin
binding sites on p.m. vesicles, accompanied by an equivalent
loss from low-density microsomal vesicles [12]. Since insulin
stimulates glucose uptake by adipocytes >10-fold, the modest
2- to 3-fold stimulation of uptake seen in BHK cells [7]
is unlikely to be detectable by this technique.

An alternative approach is to label the glucose transporter
at the p.m. of intact cells, and to compare the amount
of label bound by control and by insulin-stimulated cells.
So as to be sure that one is labelling only the glucose
transporter situated at the p.m., the labelling agent should
not, of course, enter cells. Two labels have been used:
[^{3}H]borohydride in the case of 3T3-L1 cells [13], and a
^{3}H-labelled derivative of bis-mannose, ASA-BMPA, which specifically binds to the glucose transporter [14], for adipocytes
[15].   Insulin-treated cells show an increased labelling
with both agents.

In principle this approach can be extended to determining
the total amount of glucose-transporter protein by using
it with permeabilized or disrupted cells.   Unfortunately,
attempts to use the [^{3}H]borohydride method for BHK cells
have so far been unsuccessful (unpublished experiments with
C.C. Widnell & S.A. Baldwin).   Labelling with [^{3}H]ATB-BMPA
(a compound similar to ASA-BMPA) gave more promising results
(unpublished experiments with G.D. Clark & G.D. Holman).
Moreover, with adipocytes, ~3 times more [^{3}H]ATB-BMPA is demons-
trably bound to cells under conditions in which glucose uptake
has been stimulated ~4-fold by arsenite [16].

Specificity of stress-induced membrane protein translocation

In order to determine to what extent membrane protein
translocation in response to cellular stress is specific
to the glucose transporter molecule, the effect on other
membrane proteins may be measured. We have studied the

Table 2.   Effect of stress on receptor cycling in BHK cells.
Treatment options (37°): 2 µM insulin for 1 h; 10% serum for 6 h;
200 µM arsenite for 1 h; 87 µM phenylarsine oxide (PAO) for 3 min;
infected with Semliki Forest virus (SFV) for 4 h, then 30 min
with, if so indicated (+ PAO), 87 µM PAO. The cells were then
exposed to [^{125}I]-transferrin or -α_2-macroglobulin for 10 min at
37°, and isotope taken up measured. Values are % of control
uptake.   *C.C. Widnell & C.A. Pasternak, unpublished data.*

Treatment ⊗	2-Deoxyglucose	Transferrin	α-Macroglobulin
Control	*100*	*100*	*100*
Insulin	155	118	99
Serum (FCS)	250	105	87
Arsenite	325	110	80
PAO	100	59	18
SFV-infec.	227	102	108
SFV-infec. + PAO	135	56	17

⊗five *alternative* treatments, the 5th having a '+ PAO' variation

effect of cellular stress on two p.m. receptors: that for
transferrin and that for α_2-macroglobulin, each ligand being
^{125}I-labelled so that by measuring its uptake into BHK cells
[17] an estimate can be made of the total number of receptor
molecules that are recruited. The ^{125}I uptake will be increased
if receptors are internalized more effectively as a result
of stress, but decreased if internalization is decreased
as a result of stress.

     The results (Table 2) are clear: despite a 2- to 3-fold
stimulation of glucose uptake by SFV infection or arsenite,
and a 1.5- to 2.5-fold stimulation of glucose uptake by
insulin or serum, the uptakes for both ^{125}I-ligands were
essentially the same as in control cells. On the other
hand (Table 2), the uptake of each was dramatically affected
by adding PAO, which has no effect on glucose uptake on
its own and which reduces uptake by SFV-infected cells [7].
The reason for this effect, observed also with 3T3-L1 cells
[18], is not at present known. Taken together, the results
of Table 2 show that translocation of the glucose transporter
protein stimulated by stress (and insulin) is a specific
response of cells, and cannot be explained by an effect
on membrane protein recycling in general.

     Similar results were obtained on measuring fluid-phase
endocytosis, which likewise involves internalization of membrane
proteins. No difference in [^{14}C]sucrose uptake was observed
between control BHK cells and cells exposed to heat-shock,
arsenite or insulin [7]. This finding is again compatible
with the effect on translocation of the glucose transporter,
being specific to that particular membrane protein rather
than an effect on membrane recycling in general.

Acknowledgements

I am grateful to colleagues for permission to cite unpublished results and to NATO and the Cell Surface Research Fund for financial support.

References

1. Craig, E.A. (1985) *Crit. Rev. Biochem. 18*, 239-280.
2. Lindquist, S. (1986) *Annu. Rev. Biochem. 55*,1151-1191.
3. Warren, A.P., James, M.H., Menzies, D.E., Widnell, C.C., Whittaker-Dowling, P.A. & Pasternak, C.A. (1986) *J. Cell. Physiol. 128*, 383-388.
4. Gray, M.A., Micklem, K.J., Brown, F. & Pasternak, C.A. (1983) *J. Gen. Virol. 64*, 1149-1156.
5. Wohlheuter, R.M. & Plagemann, P.G.W. (1980) *Int. Rev. Cytol. 64*, 171-240.
6. Gray, M.A., James, M.H., Booth, J.C. & Pasternak, C.A. (1986) *Arch. Virol. 87*, 37-48.
7. Warren, A.P. & Pasternak, C.A. (1989) *J. Cell.Physiol. 138*, 323-328.
8. Simpson, I.A. & Cushman, S.W. (1986) *Annu. Rev. Biochem. 55*, 1059-1089.
9. Blok, J., Gibbs, E.M., Lienhard, G.E., Slot, J.W. & Geuze, H.J. (1988) *J. Cell Biol. 106*, 69-76.
10. Widnell, C.C., Baldwin, S.A., Martin, S. & Pasternak, C.A. (1990) *FASEB J. 4*, 1634-1637.
11. Davies, A., Meeram, K., Cairns, M.T. & Baldwin, S.A. (1987) *J. Biol. Chem. 262*, 9347-9352.
12. Simpson, I.A., Yver, D.R., Hissin, P.J., Wardzala, L.J., Karnieli, E., Salans, L.B. & Cushman, S.W. (1983) *Biochim. Biophys. Acta 763*, 393-407.
13. Calderhead, D.M. & Lienhard, G.E. (1988) *J. Biol. Chem. 263*, 12171-12174.
14. Holman, G.D., Karim, A.R. & Karim, B. (1988) *Biochim. Biophys. Acta 946*, 75-84.
15. Holman, W., Parkar, B.A. & Midgley, P.J.W. (1986) *Biochim. Biophys. Acta 855*, 115-126.
16. Barnett, P. (1990) *Biochemistry project*, Univ. of Bath.
17. Simon, K.O., Gardamone, J.J., Whitaker-Dowling, P.A., Youngner, J.S. & Widnell, C.C. (1990) *Virology 177*, 375-379.
18. Frost, S.C., Lane, M.D. & Gibbs, E.M. (1989) *J. Cell. Physiol. 141*, 467-474.

#F-2

MECHANISMS OF INTRACELLULAR TARGETING AND TRANSLOCATION ACROSS MEMBRANES OF NEWLY-SYNTHESIZED PROTEINS

Mingyue He, Andrew Robinson, Ian Adcock, Olwyn Westwood and Brian Austen[†]

Department of Surgery, St. George's Hospital Medical School, Cranmer Terrace, Tooting, London SW17 0RE, U.K.

How newly synthesized secretory proteins are translocated across the e.r.* membrane has become clarified from the development of in vitro translation/translocation systems. Thereby two targeting mechanisms have been demonstrated. One involves the participation of the SRP ribonucleoprotein particle and its receptor in the membrane. Besides targeting, the role of SRP may be to retain precursor protein nascent chains in an unfolded state. Small precursor proteins bypass SRP by a route which may be better understood when a small protein precursor, prepromelittin, becomes available in bulk, expressed from E. coli. Both SRP-dependent and SRP-independent precursors may be recognized by a second receptor, a 43 kDa protein, integrated into the membrane. Targeting and translocation require energy in the form of GTP.

INTRACELLULAR PROTEIN TARGETING PATHWAYS

Proteins are synthesized in the cytoplasm of cells, then many are targeted to intracellular destinations by virtue of small stretches of signal or transit polypeptide sequences located in their initially-translated forms. These sequences are recognized at their target membranes by specific receptor systems, interaction with which activates a translocation machinery that allows the protein to pass into or across the membrane.

Digestive enzymes of the pancreas are initially synthesized on ribosomes attached to membranes in the rough e.r.*, and are targeted to and through the membrane by their amino-terminal hydrophobic signal sequences which bring about translocation across to the lumen; here precursor sequences are cleaved off by signal peptidase. In this article we describe how identification and isolation of some of the receptor systems and translocation machinery have been carried out. There is increasing evidence from the literature that acute pancreatitis

---

[†]addressee for any correspondence

**Abbreviations.*- Ab, antibody; CT, cholera toxin; e.r., endoplasmic reticulum; PAGE, polyacrylamide gel electrophoresis; SRP, signal recognition particle; TCA, trichloroacetic acid; IPTG, isopropyl-β-thiogalactopyranoside.

is caused by protein mis-targeting of lysosomal enzymes [1], which are likewise targeted through the e.r.; hence understanding these receptor-mediated processes may indicate novel therapeutic approaches for this disease.

IN VITRO SECRETORY-PROTEIN TRANSLATION AND PROCESSING

The development of mRNA-directed translation in a cell-lysate to which membranes can be added has allowed identification of the molecular entities which regulate the targeting and translocation of proteins to and across the membrane.

Cell-free systems may be prepared from lysates of a number of cell types including wheat-germ, reticulocytes and yeast. They contain the ribosomes and additional cofactors required for protein translation, but need to be supplemented with buffers, energy sources and amino acids, including [^{35}S]-methionine so that the protein may be visualized by fluorography after separation on SDS-PAGE. The mRNA for the protein of interest may be extracted from tissues which synthesize a lot of the protein. An alternative is to take the cDNA coding for the protein of interest and ligate into a plasmid downstream of a T7 or SP6 promoter, and then generate the mRNA by *in vitro* transcription, using purified RNA polymerase.

Vesiculated portions of the e.r. are isolated in the microsomal fraction by homogenizing pancreas or yeast spheroblasts in isotonic buffers, then separating them from other membranous components by centrifugation on sucrose density gradients [2]. Microsomes are added to a lysate wherein the mRNA of a secretory protein is being translated, either during or after translation, to observe the processing of the polypeptide that occurs upon crossing to the lumen of the vesicles, which generates a loss in molecular weight.

Results obtained by adding microsomes at different times after initiation to a lysate translating pancreatic mRNA are shown in Fig. 1. In this experiment, the re-initiation inhibitor edeine was added after 2 min so that nascent chains of varying length could be assessed for susceptibility to processing. A decrease in mol. wt. due to signal peptide cleavage was observed for pre-amylase, pre-trypsinogen and preprophospholipase when microsomes were added at zero time or 1.5 min, but if added at 4.5 min only pre-amylase and preprophospholipase were processed. No processing is observed when microsomes are added at 10 min.

It can be calculated from the elongation rates of these proteins that whereas pre-trypsinogen and pre-amylase can be processed only up to a stage where the nascent chains

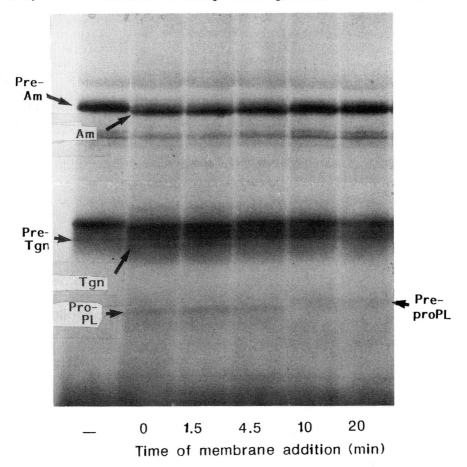

Fig. 1. *In vitro* translocation and processing of pancreatic
proteins translated from pancreatic mRNA (canine, 66 µg/ml)
in a nuclease-treated rabbit reticulocyte lysate (147 µl) with
[^{35}S]cysteine and placental ribonuclease inhibitor (91 u./ml).
The system was synchronized by adding edeine to 10 µM after
2 min at 30°. One aliquot was kept free of membranes (—, *on
left*) while further aliquots (18 µl) were added to EDTA-stripped
microsomes (final concn. 5 A.U.$_{260}$/ml) at the times shown, and
protein synthesis allowed to proceed at 30° for 1 h. Radiolabelled
proteins were separated by SDS-PAGE (10 to 17.5% gradient) and
detected by autoradiography. Am, amylase; Tgn, trypsinogen;
PL, phospholipase.

are ~80 and ~140 residues in length respectively, preprophospholipase (M_r 14,000), which is elongated at 40 residues/min,
is completed before the capacity for processsing is lost.
In keeping with these findings, it has now been established
that the translocation of many precursor proteins is initiated
co-translationally, but there is no obligatory coupling of

elongation to translocation and, as with preprophospholipase
in the present study, small proteins such as prepromelittin
and the frog prepropeptide GLa can be translocated post-
translationally [3, 4]. This poses the question, considered
in the final section of this article, how these smaller
proteins are recognized at the e.r. membrane if they bypass
the SRP-docking protein recognition system, which operates
only while the nascent chain is still attached to the ribosome
(see below).

ROLE OF SIGNAL RECOGNITION PARTICLE (SRP)

An 11 S ribonucleoprotein particle containing 6 proteins
of 72, 68, 54, 19, 14 & 9 kDa bound to a 7 S RNA species
has been isolated from salt-extracts of pancreatic rough
microsomes. This particle restores the ability of salt-stripped
pancreatic microsomal vesicles to translocate precursor
secretory proteins *in vitro* [5].   Moreover, it has been
found to halt the elongation of some secretory proteins (Fig. 2).
Elongation resumes in the presence of a component of microsomal
membranes known as docking protein, which consists of two
subunits of M_r 72,000 and 30,000. It is thought that the
SRP binds to signal sequences on nascent chains as they emerge
from the ribosome when the nascent chain is ~80 residues
in length, halting elongation until SRP contacts docking
protein at the membrane, at which point elongation resumes
and the nascent chain translocates across the e.r. membrane.

Sequencing of the larger subunit of docking protein
reveals a consensus GTP-binding sequence in its C-terminal
domain [6].   Also the 54 kDa subunit of SRP possesses a
GTP-binding domain as well as a signal-sequence binding domain
[7], and SRP has been found to bind to GTP-agarose (Fig. 3).

Because of the participation of GTP-binding proteins
in the targeting and translocation of secretory proteins,
we have examined the effect of toxin-catalyzed ADP-ribosylation.
Many GTP-binding proteins, including transducin and the regulatory subunits of adenylate cyclase, are ADP-ribosylated in
the presence of microbial toxins. We found that microsomes
ADP-ribosylated with CT were less active in the co-translational
processing of preprolactin, but that signal peptidase activity
is unchanged [8]. However, when membranes or deoxycholate
extracts were incubated with ^{32}P-labelled NAD and CT, the
major protein to be labelled was at the 22 kDa gel position,
whereas labelling was absent from the docking protein and
SRP subunits. The 22 kDa protein is nonribosomal and conserved
across species, and may form part of the translocation complex.

Fig. 2. Effect of SRP on translation of preprolactin. Bovine pituitary mRNA was translated in a wheat-germ lysate (final vol. 20 µl) in the presence of [^{35}S]-methionine and in the absence (track 1) and presence of increasing amounts of SRP (0.05 µg/ml solution) for 60 min at 25°. Translation products were analyzed by SDS-PAGE and fluorography (**a**), and TCA-precipitable cpm are shown (**b**, as % translation of pPL).

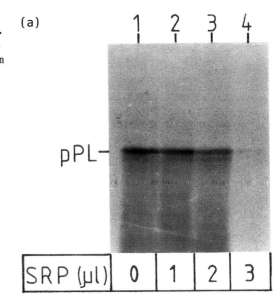

(a)

pPL—

| SRP (µl) | 0 | 1 | 2 | 3 |

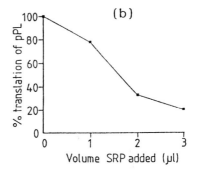

(b)

Bottom right:
Fig. 3 Binding of SRP to GTP-agarose. Wheat-germ microsomes (25 A.U.$_{260}$/ml) were extracted with 0.5 M K-acetate at 0°, centrifuged at 120,000 **g** for 2.5 h, and the supernatant applied to a column of aminopentyl-agarose in 0.5 M K-acetate. Elution was with 1 M K-acetate containing 0.01% (w/v) Nikkol (Nikko Chem[1]. Co., Japan; a polyoxyethylene detergent). The eluate (**P**) was applied to GTP-agarose in the same buffer, washed (**W**) and eluted with 2 mM GTP. **P,W** and fractions, after TCA (10%) and SDS-PAGE, were Western blotted with antisera to pancreatic SRP sub-unit, 54 kDa. GTP-eluted fractions had heaviest staining at 54 kDa.

SRP
54kDa

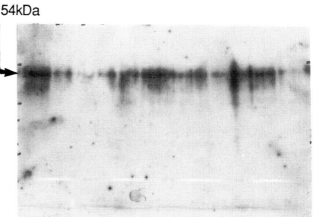

P W 1 2 3 4 5 6 7 8 9 10

Column fractions

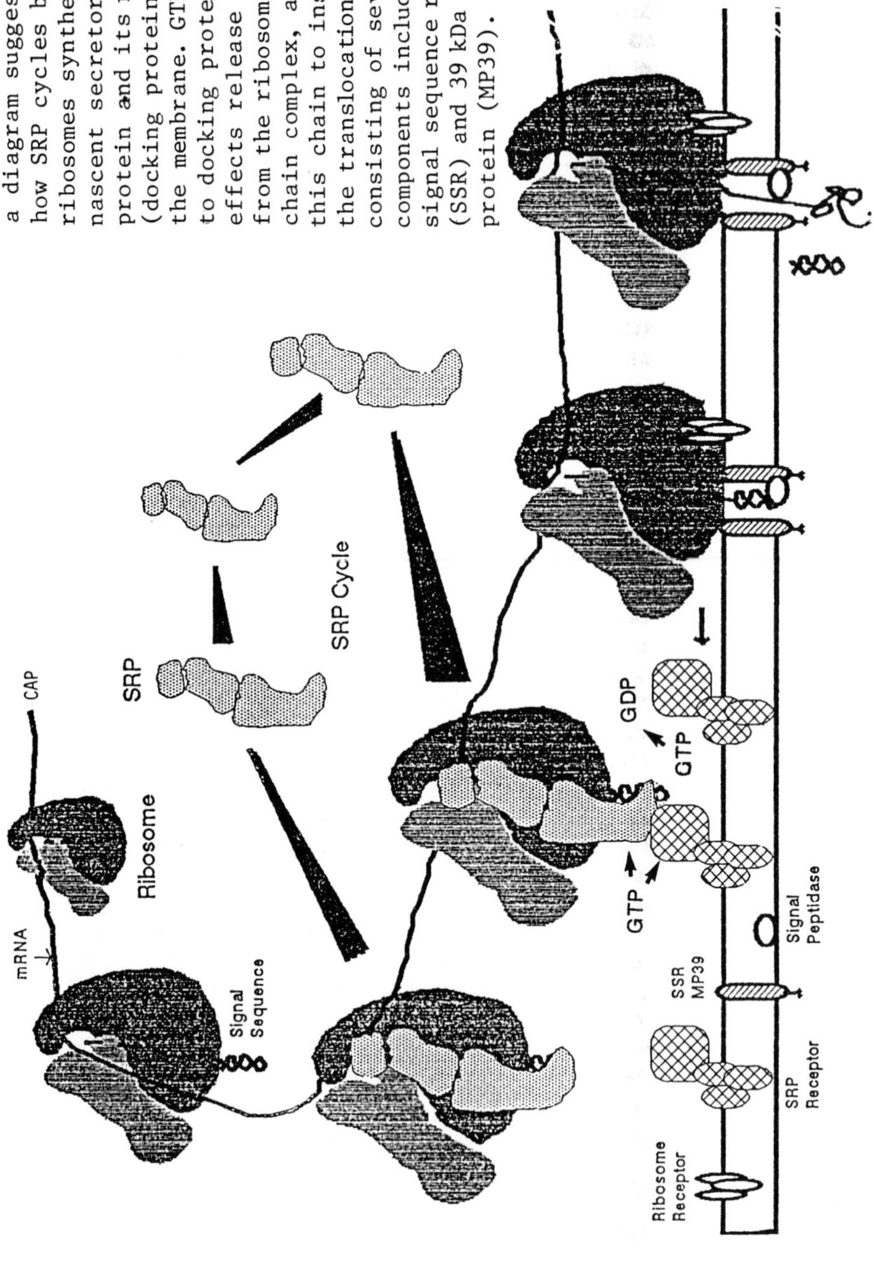

Fig. 4. The SRP cycle: a diagram suggesting how SRP cycles between ribosomes synthesizing nascent secretory protein and its receptor (docking protein) in the membrane. GTP binding to docking protein effects release of SRP from the ribosome-nascent chain complex, allowing this chain to insert into the translocation complex consisting of several components including a signal sequence receptor (SSR) and 39 kDa glyco-protein (MP39).

EXPRESSION OF PREPROMELITTIN AS A FUSED PROTEIN

The specific localization of docking protein to e.r. membranes adequately explains how nascent secretory proteins select these membranes rather than mitochondrial or other intracellular membranes. Small precursor proteins circumvent SRP and are translocated post-translationally. So as to study the physicochemical properties of these pre-proteins and the way they interact with the e.r. membrane, we are endeavouring to produce large quantities of prepromelittin as a fused protein in *E. coli*, cleavable after purification.

A cDNA encoding honeybee prepromelittin was isolated from a plasmid pBM13 (kindly provided by Dr. R. Zimmermann) by Pst1 digestion, and subcloned into the Pst1 site of pUC19 to give pUCM/A. So as to place Ile-Glu-Gly-Arg, a sequence cleaved by the protease factor Xa, directly upstream to the prepromelittin gene, synthetic oligonucleotides containing the corresponding coding sequence were inserted into the Xbal-Hincll site of pUCM/A to give pUCHxM.

An expression plasmid encoding a fusion protein with the factor Xa site between the N-terminal 655 residues of β-galactosidase and prepromelittin under control of the lac promoter was constructed by isolating the DNA coding for the factor Xa site and prepromelittin as an Xbal-Pst1 fragment from pUCFxM and inserting it into the Xbal-Pst1 site in a plasmid pBD1 containing the coding sequence for β-galactosidase. Low levels of expression of this protein were obtained in *E. coli*, and the signal sequence was recognized by the bacterial translocation machinery, leading to ~10% secretion of the expressed protein to the periplasm, with concomitant cleavage of the signal sequence. Because of the instability of this fusion protein, a second expression plasmid containing a factor Xa cleavage site between bacteriophage gene 10 and prepromelittin under control of a T7 promoter was constructed for expression in a strain in which T7 polymerase is under control of the lac promoter. Under induction with IPTG over 20 h this fusion protein is expressed and, because of its insolubility, is stable (Fig. 5), and can readily be purified as a precipitate after extraction with detergent. After cleavage with factor Xa, we hope to determine the 3-D structure of prepromelittin and use it to explore the SRP-independent translocation machinery in the e.r. membrane.

IDENTIFICATION OF A SIGNAL SEQUENCE RECEPTOR IN THE MEMBRANE

For receptor investigation on the e.r. membrane, we have chemically synthesized the signal peptide
Lys-Lys-Ser-Ala-Leu-Leu-Ala-Leu-Met-Tyr-Val-Cys-Pro-Gly-Lys-Ala-Asp-Lys-Glu
and converted it to a photoreactive peptide by S-alkylation

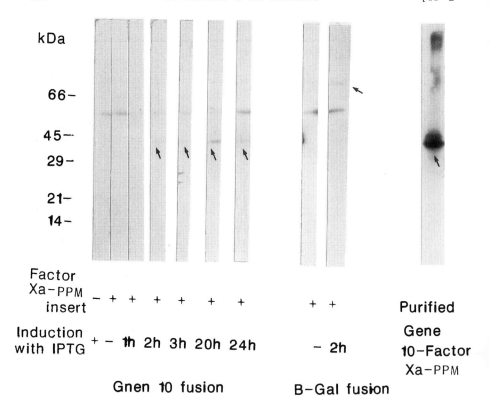

Fig. 5. Expression of prepromelittin as a fused protein:
Western blot of an SDS-PAGE gel of extracts of transformed
E. coli using an Ab to melittin, showing induction of a
fusion protein containing the bacteriophage protein gene 10,
a factor Xa cleavage site, and prepromelittin (PPM). Purification was by precipitation after lysozyme and Triton X-100
treatment. Inducing agent: IPTG (see footnote on title page).

with p-azidophenacyl bromide in slight excess. After purification of the ligand, the photoreacted peptide inhibited the
translocation of preprolactin into microsomes. The major
protein cross-linked in microsomes incubated with radio-
iodinated peptide ran at 45 kDa on SDS-PAGE [9] (Fig. 6,
track A). The protein subunits of SRP labelled with Bolton-
Hunter reagent are shown for comparison in Track B. It
is clear that only weak cross-linking to the 68 and 54 kDa
subunits of SRP occurs. Track C shows the results of cross-
linking the peptide to isolated SRP.

Cross-linking to the 45 kDa protein was inhibited by
the underivatized synthetic peptide, and also by a signal
peptide from ovalbumin at Kd ~10^{-7} M indicating that binding
to the receptor prior to cross-linkage was saturable and
specific.

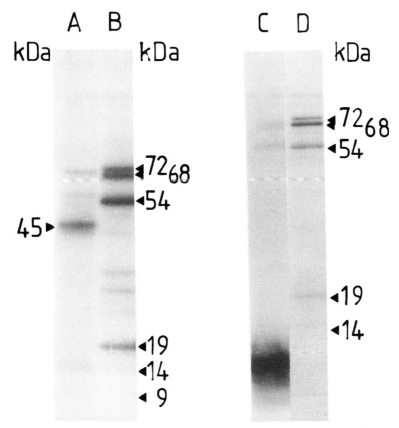

Fig. 6. Cross-linking with a photoreactive signal peptide. Tracks on the gel: **A**, proteins in pancreatic rough microsomes cross-linked under UV light to the peptide; **B**, protein sub-units of SRP labelled with Bolton-Hunter reagent; **C**, proteins in purified SRP cross-linked to the peptide; **D**, Coomassie Blue staining of SRP subunits.

The adduct which contained the cross-linked M_r 2,000 peptide, giving a total M_r of 45,000, was extracted from the membranes by concentrations of the non-ionic detergents octyl glucoside and NP-40 that exceed their critical micelle concentration, but was not extracted by 0.1 M Na_2CO_3, showing that it is integrated into the membrane. Trypsin and chymotrypsin both cleaved away a M_r 15,000 cytoplasmic domain. The cross-linked protein could be generated by photoreaction of rough but not smooth microsomes, as would be expected if the protein is functional in translocation. A similar protein of M_r 50,000 is produced by cross-linking the signal peptide to yeast microsomal membranes.

After solubilization with NP-40, the protein could be partially purified by ion-exchange chromatogrpahy. The labelled protein has been eluted from an SDS-PAGE gel, and used to immunize rabbits. The antisera will be used to investigate the function of the protein in an *in vitro* translocation assay. However, indications are that this protein forms an important receptor system integrated into the e.r. membrane, and may be part of a translocation complex comprising several integral membrane proteins, one of which - a glycoprotein of M_r 39,000 - has been cross-linked from a photoreactive nascent secretory chain [10, 11].

Acknowledgements

We gratefully acknowledge the support of the Medical Research Council, Amersham International plc, the Science & Engineering Research Council and the Wellcome Trust.

References

1. Saito, I., Hashimoto, S., Saluja, A., Steer, M.L. & Meldolesi, J. (1987) *Am. J. Physiol. 253,* G517-G526.
2. Kaderbhai, M. & Austen, B.M. (1984) *Biochem. J. 217,* 145-157.
3. Zimmermann, R. & Mollay, C. (1986) *J. Biol. Chem. 261,* 12889-12895.
4. Schlenstedt, G., & Zimmermann, R. (1987)*EMBO J. 11,* 3553-3557.
5. Walter, P. & Blobel, G. (1980) *Proc. Nat. Acad. Sci. 77,* 7112-7116.
6. Connolly, T. & Gilmore, R. (1989) *Cell 57,* 599-610.
7. Romisch, K., Webb, J., Herz, J., Prehn, S., Frank, R., Vingron, M. & Dobberstein, B. (1989) *Nature 340,* 478-481.
8. Austen, B.M. & Robinson, A. (1987) *FEBS Lett. 218,* 63-67.
9. Robinson, A., Kaderbhai, M.A. & Austen, B.M. (1987) *Biochem. J. 242,* 767-777.
10. Kreig, U.K., Johnson, A.E. & Walter, P. (1989) *J. Cell Biol. 109,* 2033-2043.
11. Wiedmann, M., Kurzchalia, T.V., Bielka, H. & Rapoport, T.A. (1987) *J. Cell Biol. 104,* 201-208.

#F-3

DIFFERENTIAL ROUTING OF MEMBRANE PROTEINS AND CELL SURFACE-BOUND LIGANDS ON HEPATOCYTE ENDOCYTIC PATHWAYS

Jorge H. Perez, Barbara M. Mullock and J. Paul Luzio

Department of Clinical Biochemistry, University of Cambridge, Addenbrooke's Hospital, Hills Road, Cambridge CB2 2QR, U.K.

Ligand transfer via *hepatocyte endocytic pathways has been studied (1) by subcellular fractionation of ipRL* and (ii) by immunohistochemistry and IRMA. After single-pass injection of ^{125}I-radiolabelled pIgA into ipRL, ~½ was delivered (as sIgA) to bile and ~½ to lysosomes. Approach (i) showed that radiolabelled protein was present sequentially in sinusoidal p.m., light endosomes, dense endosomes, very dense endosomes and lysosomes with a similar time course to that found in vivo. Microtubule disruption with nocodazole (20 µM for 1 h) resulted in inhibition of the transfer of a first dose of pIgA to bile but not to lysosomes. When a second dose was given to the same ipRL after recovery from microtubule disruption, it was transferred normally to bile and lysosomes; however, the pIgA given earlier continued to be delivered to lysosomes but not to bile. It was concluded that following endocytosis pIgA is taken to an endosomal compartment where it is 'sorted' for delivery to lysosomes or, requiring microtubule integrity, to bile. Details are given of techniques used.*

Indirect evidence that the blood-to-bile pathway operates in human liver came from approach (ii), by immunohistochemistry of frozen sections using a mAb to human secretory component and by 2-site immunoradiometric assay of free secretory component (FSC) in human bile. In human liver, as in rat, anti-secretory component Ab was found to stain hepatocyte as well as bile-canalicular membranes, and FSC was found in large amounts in bile. These findings support the presence of the blood-to-bile transhepatocytic pathway in humans.

The hepatocyte is a polarized cell separating blood from bile. Many ligands bind to receptors on the blood sinusoidal surface of the hepatocyte and are internalized by receptor-mediated endocytosis. After delivery to a common,

---

**Abbreviations.-* Ab, antibody (mAb, monoclonal); ALD, alcoholic liver disease; alkP, alkaline phosphatase; ALT, alanine aminotransferase; ASF, asialofetuin; FSC, free secretory component; ipRL, isolated perfused rat liver; IRMA, immunoradiometric assay; i.v., intravenous; 5'N, 5'-nucleotidase; p.m., plasma membrane; TCA, trichloroacetic acid; pIgA is polymeric, sIgA is secretory immunoglobulin A.

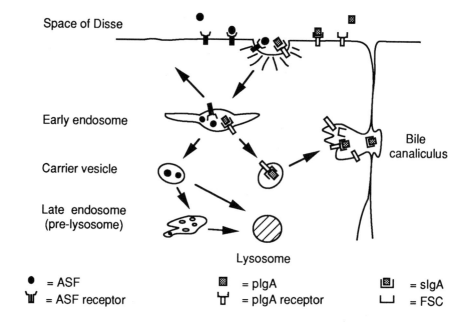

Space of Disse

Early endosome

Carrier vesicle

Late endosome
(pre-lysosome)

Bile
canaliculus

Lysosome

● = ASF   ▨ = pIgA   ▣ = sIgA
Ψ = ASF receptor Ϯ = pIgA receptor ⊔ = FSC

Fig. 1. Hepatocyte endocytic pathways for pIgA and ASF (schematic).

early (peripheral) endocytic compartment, some ligands including asialoglycoproteins are sorted for delivery to lysosomes along the degradative pathway, some are returned back to the cell surface whereas at least one, pIgA, is excreted into bile by transcytosis [1, 2] (Fig. 1).

In a number of mammalian species including rats and rabbits it has been shown that hepatocytes synthesize the pIgA receptor which is used to transfer pIgA and pIgA-immune complexes across the cell from blood to bile [3, 4]. Electron-microscopic and subcellular fractionation studies on rat liver have given much insight into the intracellular transcytic pathway, whereby pIgA bound to the blood-sinusoidal cell-surface receptor is internalized by a coated-pit mechanism into an early endocytic compartment. Here a sorting proccess takes place, resulting in much of the pIgA and its receptor being transferred across the cell to the bile-canalicular p.m., where specific proteolysis of the receptor occurs, releasing sIgA and FSC into bile [4-9]. Transcytosis and proteolytic cleavage of the receptor are constitutive and occur whether or not pIgA is bound. The molecular cloning of the rabbit pIgA receptor [10] and subsequent molecular cut-and-paste experiments [11] have indicated the importance of the cytoplasmic tail in determining the correct intra-cellular membrane traffic pathway for this molecule.

The passage of asialofetuin (ASF) through the endocytic pathway in hepatocytes is also well described [12-15]. It requires binding to asialoglycoprotein receptors [13, 16, 17], internalization *via* coated pits [7, 12], appearance in an early, peripheral endosome compartment(s) [12-14 & 18] where the ligand becomes dissociated from receptor [13, 19], and finally digestion in lysosomes [14] (& see T. Berg's survey and protocols in Vol. 17, this series -*Ed.*).

The present article describes methods used to study the pathways followed by ASF and by pIgA following their uptake by the hepatocyte, the role of microtubules in the route-guidance of these ligands, and data which suggest that in humans, as in rats, the pIgA transcytic pathway operates in hepatocytes.

ISOLATED PERFUSED RAT LIVER (ipRL)

A problem inherent in the study of timed events in organs *in situ* is the occurrence of recycling of the excess ligand not taken up in the first pass, such that a continuous supply of ligand becomes presented for uptake, and the overall picture is complex. That this situation applies to pIgA was supported by Perez *et al.* [4], in whose experiments the uptake of label by rat liver *in vivo* was monitored continuously following i.v. injection of ^{125}I-pIgA, and the radiolabel was increasingly seen in the liver, maximally at 15 min. This problem may be overcome by the use of an ipRL preparation where the liver is removed from a donor rat with a minimal period of hypoxia, kept in a cabinet with ambient temperature constant at 37°, and its viability maintained by supplying it with physiological amounts of O_2, CO_2, electrolytes and nutrients.

Perfusate composition

The perfusate used for ipRL is a 40% (v/v) suspension of human red cells, freshly washed in 0.1 M NaCl/0.06 M NaHCO$_3$, in the following perfusion medium (Dulbecco's modified Eagle's medium):- bovine serum albumin, 30 g/L; glucose, 25 mM; penicillin, 50,000 u./L; streptomycin, 50,000 µg/L; hepatin, 1,000 u./L; and CaCl$_2$, 2.5 mM.

Perfusion apparatus (Fig. 2)

1. The isolated liver is placed on a fine plastic net, above a plastic funnel which collects the perfusate as it leaves the liver into a collecting reservoir.

2. From this reservoir a peristaltic pump (HRE 200; Watson-Marlow, Falmouth, U.K.) takes the perfusate *via* 12 G plastic tubing through an on-line blood filter, then through an oxygenator made up of 80 ft. of silastic medical-grade tubing (602-235;

Fig. 2.
An ipRL
system
(schematic).

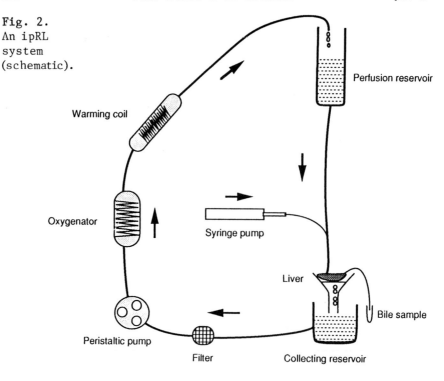

Dow Corning Corpn., Medical Products, Midlands, MI 48640)
of 0.058 inch i.d./0.077 inch o.d., arranged in 6 parallel
lines and enclosed in a 240 × 80 mm plastic cylinder wherein
it is equilibrated with 95% air/5% CO_2 flowing at 100 ml/min.

3. On leaving the oxygenator, the perfusate passes through
a warming coil set at 37° and then collects in a perfusion
reservoir placed directly above the liver preparation.

4. From this reservoir the perfusate enters the liver preparation by gravity. The perfusion flow is in direct proportion
to the pressure given by the height of the reservoir above
the liver preparation, which may be adjusted as required,
and is in inverse proportion to the vascular resistance of
the liver preparation, which cannot easily be adjusted. The
resistance from the tubing is constant. The liver preparation
and the perfusion apparatus are maintained in an enclosed
cabinet, 800 × 800 × 600 mm, with a constant ambient temperature
of 37° provided by an electric fan heater coupled to a
thermostat *via* a temperature probe.

5. Bile salts (2 mM sodium taurocholate in perfusion medium;
Sigma, Poole BH17 7NH, U.K.) are introduced as a continuous
infusion with a digital syringe pump (Treonic IP3; Vickers
Ltd., Basingstoke, U.K.) connected to the perfusing catheter
by a 3-way tap.

6. The system is primed with 400 ml of perfusate for enough time to bring it to 37°, the pH to 7.4, the pCO_2 to ~5 kPa and the pO_2 to ~12 kPa (~1 h; 288 Blood Gas System, Ciba-Corning Diagnostics Corpn., Medfield, MA 02052). During this time the pump is adjusted to deliver 1 ml/min per g of the expected liver weight to the perfusion reservoir and to establish what should be the height of the reservoir above the liver preparation so as to deliver this flow rate. This height measured down from the perfusate gives the zero tissue resistance or zero liver-perfusion pressure.

7. The measured values for the perfusate at this point should be Na^+ = 150 mM, K^+ = 3.5 mM (654 $Na^+/K^+/Li^+$ Analyzer; Ciba-Corning, as above), Ca^{2+} (ionized) = 1.14 mM (288 Blood Gas System; Ciba-Corning) and osmolality = 290 mosmol/kg (3D II Digimatic Osmometer; Advanced Insts. Inc., Highland Ave., Needham Heights, MA 02194). The apparatus is now ready for perfusing a liver.

Operative procedure

1. The donor rat (suitably male Wistar, 150-200 g) is anaesthetized with Hyponorm (fentanyl citrate 0.315 mg/ml and fluanisone; Janssen Pharmaceutical Ltd., Grove, Oxford OX12 0DQ, U.K.), 0.5 ml/kg intramuscularly.

2. Mid-line and transverse 25 mm abdominal incisions are made under the anaesthesia; the bile duct, portal vein and inferior vena cava are identified after displacing the intestinal loops to the animal's left side.

3. The operative field is maintained exposed with small retractors applied to the abdominal wall and adjacent viscera.

4. The distal end of the bile duct is occluded with mosquito forceps and the duct is allowed to distend (~5 min).

5. Meanwhile loose ties (3/0 silk) are placed around the distal end of the bile duct, the proximal end of the portal vein and around the inferior vena cava, above the right renal vein.

6. A small incision in the wall of the bile duct (by now distended) is made at the distal end with fine ophthalmic scissors.

7. A cannula is introduced into the bile duct (plastic tubing of 0.61 mm o.d., 0.20 mm i.d., 250 mm long; Portex Ltd., Hythe CT21 6JL, U.K.) with its tip placed 10 mm from the origin of the common bile duct, then tied in place, free bile flow is observed immediately.

8. After clamping the distal end of the portal vein with mosquito forceps, a small incision is made in the vein wall. The perfusion cannula (plastic tubing 14G × 280 mm with a

16G × 30 mm blunt-end needle having a groove at 6 mm from the distal end to facilitate ligation) is introduced and tied in place, and the perfusion started immediately. The time from clamping the portal vein to the beginning of perfusion should be <15 sec so that tissue hypoxia is minimized.

9. The chest is cut open by a mid-sternal incision and the thoracic vena cava cut off above the diaphragm to drain the liver which must be uniformly perfused and not enlarged.

10. The inferior vena cava is tied and cut off below the ligature, effectively isolating the liver from the donor.

11. The liver is freed from all attached tissues and transferred with the bile duct cannula to its final position in the constant-temperature cabinet, placed on the diaphragm side making sure that there is good drainage of perfusion medium. Twisting the portal vein or the bile duct must be avoided. Once in the perfusion cabinet the liver is covered with a gauze soaked in normal saline (0.15 M NaCl) to prevent tissue drying.

12. The bile production should have continued uninterrupted; the perfusion pressure should be adjusted to 3-5 cm of perfusate height above the zero perfusion pressure. The sodium tauro-cholate infusion is started at 40 µmol/h and the distal end of the bile cannula is placed in a 1.5 ml plastic tube to collect bile.

13. Before any experiments are conducted, 20 min is allowed for the liver to achieve 'steady state'.

Parameters of viability and function of the perfused liver

The isolated liver remains viable for at least 3 h as judged by the following parameters.
- Macroscopic appearance. The liver colour remains uniform, of viable tissue appearance and non-oedematous.
- Phenol red excretion. Phenol red present in the perfusion medium at a concentration of 15 mg/L appears in bile within 3 min of the start of the perfusion and is continuously excreted throughout the experimental period.
- Portal vein pressure. This is estimated as the height of the perfusate column above the level observed when the perfusion medium is flowing freely, before being connected to the portal vein. Portal vein pressure is initially ~10 cm but within 15 min decreases to between 3 and 5 cm; a steady pressure rise is seen in those livers with poor viability.
- Oxygen consumption. This is calculated from the difference in the pO_2 content of the perfusion medium before entering and after leaving the liver, the perfusate haemoglobin concentration and the volume of perfusate introduced per min. O_2 consumption is ~3 µmol/min per g liver.

- Bile flow: averages 1.4 ml/min per g liver.
- Release of hepatocyte marker enzymes (see initial abbreviations list), which may be measured by a centrifugal analyzer (IL Multistat II; Instrumentation Lab., Lexington, MA 02173) in bile and perfusate (sampled from liver output) using kit reagents: alkP (Boehringer Mannheim Diagnostica, Bell Lane, Lewes, E. Sussex BN7 1LG), ALT (Beckman Insts., Fullerton, CA 92634) and 5'N (Sigma).

> Enzyme content (u./L) in perfusion medium (and in normal rat serum and bile, and iPRL bile): alkP, 13-31 (serum, 246; bile, 20; iPRL bile, 17-22); ALT, 8-91 (serum, 22; bile, 5; iPRL bile, 27-28). Perfusate ALT show a moderate rise towards the higher values indicated, reflecting progressive hepatocyte damage; that alkP is low compared with serum is to be expected since perfusion-medium alkP lacks the bone component of the alkP present in normal serum from young rats. In iPRL bile the value for 5'N is 150-170.

- Histological appearance. After 3 h of perfusion, samples from perfused livers may be taken for histological examination and should show preservation of the liver architecture and little cell damage.
- Ligand transport from perfusate to bile. A single bolus injection of radiolabelled pIgA given to the iPRL is transported into bile with a time course similar to that seen in bile-duct-cannulated rats (Fig. 3).
- Distribution of organelle marker enzymes after subcellular fractionation (Ficoll and Nycodenz density gradients) of livers perfused for 2 h resembles that for fractions from livers of freshly killed rats in respect of β-hexosaminidase, succinate dehydrogenase (B-Hex, SDH) and latent 5'N (Fig. 4).

The effluent perfusate may be pooled for re-oxygenation and recycling, or left to drain and discarded after a single pass through the liver: any excess of radiolabelled ligand not taken up in a single pass can thus be discarded, allowing the presentation of a pulse dose of radiolabelled protein to the hepatocytes.

DIFFERENTIAL ROUTING OF pIgA AND ASF BY THE RAT HEPATOCYTE

The differential handling of ligands by rat liver may be illustrated by study of ligand distribution between liver, bile and perfusate following the administration of radiolabelled ligand to an iPRL. In Fig. 5 the handling of ^{125}I-ASF and of ^{125}I-pIgA are illustrated. For ASF, which normally follows the degradative pathway, 93% of the radiolabel at 2 h after single-pass administration was found in the perfusate and was >80% TCA-soluble, while little label was found in bile or in liver homogenate. By contrast, when radiolabelled pIgA was given, 62% of the label was found in the perfusate:

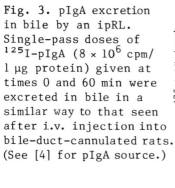

Fig. 3. pIgA excretion in bile by an ipRL. Single-pass doses of ^{125}I-pIgA (8×10^6 cpm/1 µg protein) given at times 0 and 60 min were excreted in bile in a similar way to that seen after i.v. injection into bile-duct-cannulated rats. (See [4] for pIgA source.)

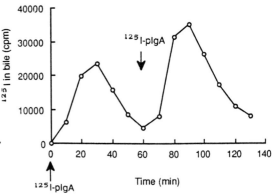

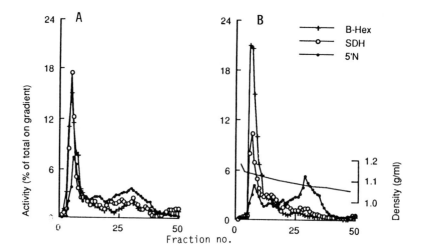

Fig. 4.   Comparison of (A) ipRL and (B) liver *in situ* for marker enzyme distribution among subcellular fractions from Ficoll density gradients. Assay methods as in [6].

20% of this was TCA-soluble, suggesting return of the intact protein from the liver into the perfusate (retroendocytosis); 20% was found in bile (transcytosis) and 18% remained in the liver homogenate.

SEQUENTIAL pIgA TRANSFER THROUGH ENDOCYTIC CELL COMPARTMENTS

The time sequence of events in ^{125}I-pIgA movement across the rat hepatocyte was examined by subcellular fractionation of ipRL at specific times after ligand administration (Fig. 6). After single-pass administration, ^{125}I-pIgA was seen consecutively in the light, dense and very dense endosomal compartments and the lysosomes.   These findings are in contrast to those reported by Branch & co-authors [18] for experiments *in situ*

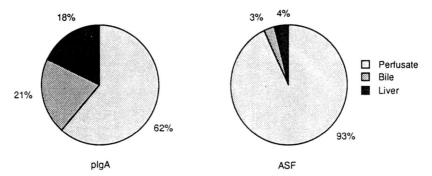

Fig. 5. ^{125}I distribution 2 h after 2 single-pass doses of radiolabelled ligands (see text).

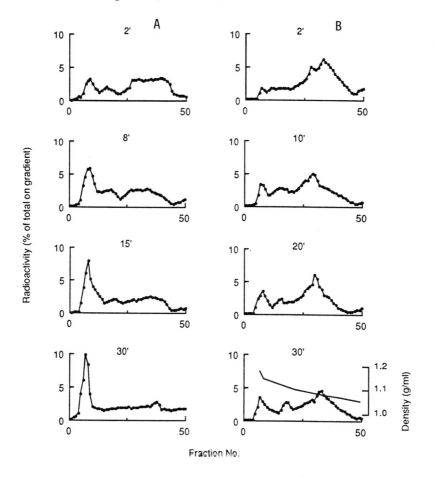

Fig. 6. Comparison of (A) ipRL and (B) liver *in situ* for sequential distribution of ^{125}I-pIgA among subcellular fractions from Ficoll gradients onto which post-mitochondrial fractions were loaded.

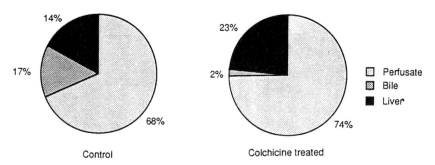

Fig. 7. Distribution of [125]I 3 h after a single-pass injection of radiolabelled pIgA.

where subcellular fractionation of livers was carried out at specific times after giving rats single i.v. doses of radiolabelled ligand: a constant peak of radiolabel was seen in the light endosomal fraction as well as movement of ligand through the denser compartments. In ipRL this peak was only transient, suggesting that this compartment in livers *in situ* is continuously supplied with ligand from the circulation.

MICROTUBULE INVOLVEMENT IN pIgA TRANSPORT

Treatment of ipRL with colchicine (1.6 μM, for 1 h), which inhibits microtubule polymerization [20], resulted in a dose-dependent inhibition of the transport of pIgA into bile. This inhibition of biliary excretion of pIgA was associated with a correspondingly increased presence of radiolabel in the liver tissue (Fig. 7). To establish the intracellular site at which microtubules are involved in the transport of pIgA into bile, two possibilities may be considered. Firstly, if the inhibition occurred before the point of pIgA sorting into bile and liver fractions, the label should accumulate in an early compartment before the block, and following removal of the block the ligand should be delivered to the bile and to lysosomes. An alternative possibility would be for the inhibition to occur at a place between the ligand sorting and bile: the ligand would continue its passage to the dense endosomes, very dense endosomes and lysosomes and no ligand would be delivered to the bile after recovery from microtubule disruption.

Nocodazole which disrupts microtubular function in a similar way to colchicine has a short duration of action [20]. A single-pass dose of ^{125}I-pIgA was given to an ipRL treated with nocodazole (20 μM, for 1 h). The transport of ^{125}I-pIgA into bile was inhibited as expected, but did not recover when microtubule function was restored, 90 min after removal of nocodazole from the perfusion medium. This liver was, however, capable of normally transporting a second single-pass dose of pIgA, this time labelled with ^{131}I, and given 30 min after nocodazole removal from the medium. Sub-cellular fractionation of this ipRL at 8 min after the injection of the second dose of pIgA showed that the ^{125}I-pIgA given early (which was not delivered to bile) was present during nocodazole treatment in the very dense endosome and lysosome compartments, while the ^{131}I-pIgA given late (which was excreted into bile) was present after nocodazole removal in the light and dense endosomal compartments [4]. These findings are consistent with microtubular involvement in delivery of ^{125}I-pIgA from the light or dense endosome into bile, at a point beyond sorting into the transcytic rather than the lysosomal pathway.

TRANSCYTOSIS OF pIgA BY HUMAN HEPATOCYTES

It is widely thought that the pIgA receptor is not present on human hepatocytes, largely because of the difficul-ties in detecting the receptor immunohistochemically in liver sections, lack of Northern blotting of the mRNA and the absence of large amounts of FSC in bile [21-28]. Human bile contains sIgA, and measurements of bile/blood ratios for pIgA [see 24] and sIgA [29] have indicated that in man, like other species, pIgA is actively transported from blood to bile. Moreover, the primary sequences of the rabbit [10], rat [30] and human [26, 31] pIgA receptors show considerable homology, particularly in the cytoplasmic tail, the region known to be incorporated for correct intracellular targeting and transcytosis [11]. Transport in the absence of the pIgA receptor on hepatocytes has been explained by postulating that in humans pIgA is transferred solely *via* the bile-duct lining cells, in which receptor is easily detected. This explanation has been largely accepted despite some early reports of immunohistological localization of the receptor on human hepatocytes [32, 33], evidence that pIgA is bound to the surface of normal human isolated hepatocytes [34], reports that patients with chronic parenchymal liver disease exhibit a proportionately higher increase in serum pIgA than in total serum IgA [32, 35], evidence that the presence of circulating IgA-containing immune complexes is directly related to severity of liver parenchymal damage [36], and evidence that in acute hepatitis the persistence of hepatocytes seems necessary for maintaining the high serum sIgA [37].

Most recently three new lines of evidence have provided further support for the human liver pIgA transcytic pathway being similar to that in rat. Firstly, a gamma camera-scinti-graphy study revealed the liver as the major organ mediating the removal of macromolecular, chemically cross-linked, ^{125}I-labelled pIgA in humans [38].

Secondly, using one of a set of mouse mAb's to FSC prepared in this laboratory and gentle fixation conditions it was possible to demonstrate immunohistologically the presence of the pIgA receptor on hepatocyte p.m., as well as on bile-duct lining cells, in all sections examined from 8 normal and 3 abnormal human livers [39].

Thirdly, in contrast to earlier studies [24, 27, 40] but in agreement with the findings from rodents, we have recently shown that the FSC content of human bile is large and equivalent to the sIgA concentration, provided that precautions are taken to prevent proteolysis in the collected samples. In bile samples from bile-duct drains in patients who had undergone liver transplantation, FSC mean concentration was 2.7 mg/L (range 0.7-33.3; n = 8), while in corresponding samples from patients who had undergone cholecystectomy the mean was 0.063 mg/L (range 0.010-0.165; n = 3). This lower concentration of FSC in bile from cholecystectomy patients was partly due to proteolytic activity of these biles towards FSC, since it was shown that they could degrade ^{125}I-FSC *in vitro* [41].

FSC serum concentration is raised in patients with liver disease

In human serum both sIgA and FSC are present, but the large excess of both pIgA and IgM ensures that the concentration of FSC is low [41, 42]. Using mAb's directed towards different epitopes, a 2-site IRMA for FSC has been established in which a 100-fold excess of sIgA does not interfere. A 2-site IRMA has also been developed using an anti-FSC mAb as a capture Ab with a radiolabelled, commercially available anti-α-chain Ab as signalling reagent. These assays have been used to establish reference ranges for serum sIgA and FSC [41], similar to previously published values [24, 25, 43-46]. FSC and sIgA concentrations were measured in the serum of patients with liver disease, and in control serum wherein FSC averaged 3.4 µg/L (range 2-11; n = 38); FSC was significantly higher (P <0.001) in liver-disease patients, averaging 14.4 µg/L (range 2-100; n = 41). For sIgA the mean serum concentration was 11.8 mg/L (range 2-30; n = 38) in controls and significantly higher ($P = 0.013$) in serum from liver-disease patients (mean 17.5 µg/L; range 3-45; n = 36) [41].

ROLE OF THE pIgA TRANSCYTIC PATHWAY IN LIVER-DISEASE AETIOLOGY

The demonstration of the pIgA transcytic pathway in humans has important implications in the understanding of the pathogenesis of liver disease and of conditions resulting from the failure of clearance of pIgA and pIgA-containing immune complexes by the liver. An important clinical situation where it may be hypothesized that abnormalities in liver transcytosis of pIgA are relevant is chronic liver disease, particularly alcoholic (ALD), in which liver pIgA is thought to be passively deposited and there is often IgA nephropathy [47]. In rats, ethanol has been shown to interfere with intracellular hepatocyte membrane-traffic pathways including transcytosis [48], and subcellular fractionation studies of samples of human liver from patients with ALD have shown abnormalities consistent with cytoskeleton and membrane-traffic disruption [49, 50].

We propose the following course of events for the production of hepatocellular damage in ALD.- (1) Acetaldehyde, a major metabolite of ethanol, may impare microtubule function by inhibiting tubulin polymerization [51]. (2) Microtubule dysfunction would result in impairment of pIgA transcytosis, but not binding to the pIgA receptor on the hepatocyte surface. (3) pIgA on the hepatocyte surface may activate complement [52], leading to the formation of membrane-attack complexes [53] and thus hepatocellular damage. Besides, (4) failure of pIgA removal by the liver would result in an increased concentration of pIgA in the circulation and possibly deposition of pIgA at distant sites, resulting in damage to the involved tissue.

Acknowledgements

We thank the Medical Research Council and the East Anglian Regional Health Authority for their financial support.

References

1. Pastan, I. & Willingham, M.C., eds. (1985) *Endocytosis*, Plenum, New York, 326 pp.
2. Geuze, H.J., van der Donk, H.A., Simmons, C.F., Slot, J.W., Strous, J.G. & Schwartz, A.L. (1986) *Int. Rev. Exp. Path.* 29, 113–171.
3. Sztul, E.S., Howell, K.E. & Palade, G.E. (1985) *J. Cell Biol.* 100, 1255-1261; also 1248-1254 and (1983) 97, 1582-1591.
4. Perez, J.H., Branch, W.J., Smith, L., Mullock, B.M. & Luzio, J.P. (1988) *Biochem. J.* 251, 763-770.
5. Mullock, B.M., Luzio, J.P. & Hinton, R.H. (1983) *Biochem. J.* 214, 823-827.
6. Mullock, B.M., Hinton, R.H., Peppard, J.V., Slot, J.W. & Luzio, J.P. (1987) *Cell Biochem. Funct.* 5, 235-243.
7. Geuze, H.J., Slot, J.W., Strous, G.J.A.M., Peppard, J., von Figura, K., Hasilik, A. & Schwartz, A.L. (1984) *Cell* 37, 195-204.
8. Musil, L.S. & Baezinger, J.U. (1988) *J. Biol. Chem.* 263, 15799–15808.
9. Quintart, J., Baudhuin, P. & Courtoy, P.J. (1989) *Eur. J. Biochem.* 184, 567-574.
10. Mostov, K.E., Friedlander, M. & Blobel, G. (1984) *Nature* 308, 37-43.

11. Breitfield, P.P., Casanova, J.E., Simister, N.E.,
    Ross, S.A., McKinnon, W.C. & Mostov, K. (1989) *Curr.
    Opinion Cell. Biol. 1*, 617-623.
12. Wall, D.A., Wilson, G. & Hubbard, A.L. (1980) *Cell 21*, 79-93.
13. Geuze, H.J., Slot, J.W., Strous, G.J., Lodish, H.F. &
    Schwartz, A.L. (1983) *Cell 32*, 277-287.
14. Berg, T., Kindberg, G.M., Ford, T. & Blomhoff, R. (1985)
    Exp.Cell Res. 161, 285-296 (*& see* ref. 49: pp. 315-325 - *Ed.*)
15. Van Berkel, T.J.C., Kruijt, J.K., Narkes, L.,
    Nagelkerke, J.F., Spanjer, H. & Kempen, H.M. (1986) in
    Site-Specific Drug Delivery (Tomlinson, E. & Davis, S.S.,
    eds.), Wiley, Chichester, pp. 49-68.
16. Drickamer, K., Mamon, J.F., Binns, J. & Leung, O.J. (1984)
    J. Biol. Chem. 259, 770-778.
17. Steer, C.J. & Ashwell, J. (1986) *Prog. Liver Dis. 8*, 99-123.
18. Branch, W.J., Mullock, B.M. & Luzio, J.P. (1987) *Biochem.
    J. 244*, 311-315.
19. Mueller, S.C. & Hubbard, A.L. (1986) *J. Cell Biol. 102*,
    932-942.
20. Dustin, P. (1984) *Microtubules*, 2nd edn., Springer-V'lag, Berlin.
21. Brandtzaeg, P. (1985) *Scand. J. Immunol. 22*, 111-146.
22. Chandy, K.G., Hubscher, S.G., Elias, E., Berg, J.,
    Khan, M. & Burnett, D. (1983) *Clin. Exp. Immunol. 52*, 207-218.
23. Daniels, C.K. & Schmucker, D.L. (1987) *Hepatology 7*, 517-521.
24. Delacroix, D.L. & Vaerman, J.P. (1983) *Ann. N.Y. Acad. Sci.
    409*, 383-400.
25. Fukuda, Y., Nagura, H., Asai, J. & Satake, T. (1986) *Am. J.
    Gastroenterol. 81*, 315-324.
26. Krajci, P., Solberg, R., Sandberg, M., Oyen, O., Jahnsen, T.
    & Brandtzaeg, P. (1989) *Biochem. Biophys. Res.Comm. 158*, 783-789.
27. Nagura, H., Smith, P.D., Nakane, P.K. & Brown, W.R. (1981)
    J. Immunol. 126, 587-595.
28. Tomana, M., Kulhavy, R. & Mestecky, J. (1988)
    Gastroenterology 94, 762-770.
29. Mullock, B.M., Shaw, L.J., Fitzharris, B., Peppard, J.,
    Hamilton, M.J.R. & Simon, M.T. (1985) *Gut 26*, 500-509.
30. Banting, G., Brigitte, B., Braghetta, P., Luzio, J.P. &
    Stanley, K.K. (1989) *FEBS Lett. 254*, 177-183.
31. Eiffert, H., Quentin, E., Decker, J., Hillemier, S.,
    Hufschmidt, M., Klingmuller, D., Weber, M.H. & Hilschmann, N.
    (1984) *Hoppe-Zeyl. Z. Physiol. Chem. 365*, 1489-1495.
32. Foss-Bowman, C., Jones, A.L., Dejbakhsh, S. & Goldman, I.S.
    (1983) *Ann. N.Y. Acad. Sci. 409*, 822-823.
33. Hsu, S. & Hsu, P. (1980) *Gut 21*, 985-989.
34. Hopf, V., Brandtzaeg, P., Hutterof, T.H. &
    Meyer zum Buschenfelde, H.K. (1978) *Scand. J. Immunol. 8*, 543-549.
35. Kutteh, W.H., Prince, S.J., Phillips, J.O., Spenney, J.G.
    & Mestecky, J. 81-82) *Gastroenterology 82*, 184-193.
36. Van de Wiel, A., Valentijn, R.M. Schuurman, H.J., Daha, M.R.,
    Hene, R.J. & Kater, L. (1988) *Dig. Dis. Sci. 33*, 679-684.

37. Delacroix, D.L., Reynaert, M., Pauwels, S., Geubel, A.P. & Vaerman, J.P. (1982) *Dig. Dis. Sci. 27*, 333-339.

38. Rifai, A., Schena, F.P., Montinaro, V., Mele, M., D'Addabbo, A., Nitti, L. & Pezzullo, C. (1989) *Lab. Invest. 61*, 381-388.

39. Perez, J.H., Wight, D.G.D., Wyatt, J.I., van Schaik, M., Mullock B.M. & Luzio, J.P. (1989) *Immunology 68*, 474-478.

40. Orlans, E., Peppard, J.V., Payne, A.W.R., Fitzharris, B.M., Mullock, B.M., Hinton, R.H. & Hall, J.G. (1983) *Ann. N.Y. Acad. Sci. 409*, 411-426.

41. Perez, J.H., van Schaik, M., Mullock, B.M., Bailyes, E.M., Price, C.P. & Luzio, J.P. (1991) *Clin. Chim. Acta*, in press.

42. Kvale, D. & Brantzaeg, P. (1988) *J. Immunol. Meth. 113*, 279-281.

43. Homburger, H., Casey, M., Jacob, G. & Klee, G. (1984) *Am. J. Clin. Path. 81*, 569-574.

44. Kvale, D., & Brandtzaeg, P. (1986) *J. Immunol. Meth. 86*, 107-114.

45. Vincent, C. & Revillard, J.P. (1988) *J. Immunol. Meth. 113*, 283-285.

46. Wood, G.M., Trejdosiewicz, L.K. & Losowsky, M.S. (1987) *J. Immunol. Meth. 97*, 269-274.

47. Clarkson, A.R., Woodroffe, A.J., Bannister, K.M., Lomax-Smith, J.D. & Aarons, I. (1984) *Clin. Nephrol. 21*, 7-14.

48. Okanoue, T., Kondo, I., Ihrig, T.J. & French, S.W. (1984) *Hepatology 4*, 253-260.

49. Peters, T.J. (1987) in *Cells, Membranes and Disease* [Vol. 17, this series] (Reid, E., Cook, G.M.W. & Luzio, J.P., eds.), Plenum, New York, pp. 25-34.

50. Tuma, D.J., Casey, C.A. & Sorrell, M.F. (1990) *Alcohol & Alcoholism 25*, 117-125.

51. Jennett, R.B., Sorrell, M.F., Saffari, F.A., Ockner, J.L. & Tuma, D.J. (1989) *Hepatology 9*, 57-62.

52. Hiemstra, P.S., Biewenga, J., Gorter, A., Stuurman, M.E., Faber, A., van Es, L.E. & Daha, M. (1988) *Mol. Immunol. 25*, 527-533.

53. Luzio, J.P., Arafaine, A., Richardson, P.J., Daw, R.A., Sewry, C.A., Morgan, B.P. & Campbell, A.K. (1987) *as for* 49., pp. 199-209.

#F-4

SUBCELLULAR DISTRIBUTION OF TRIMERIC AND
LOW MOLECULAR WEIGHT G-PROTEINS IN LIVER

W. Howard Evans and Nawab Ali

National Institute for Medical Research,
Mill Hill, London NW7 1AA, U.K.

We describe the sub-cellular distribution in rat-liver homogenates of various G (GTP-binding) proteins. The α- and β-subunits of the trimeric G-proteins involved mainly in signal transduction were shown, using Ab's* and Western blotting, to be present mainly at the sinusoidal p.m. and in endosomes, but the highest relative amounts were present at the bile-canalicular p.m. A similar distribution among the fractions of GTP-binding 23 kDa ras oncoproteins was demonstrated by Western blotting. In contrast, however, certain Ab's to ras oncoproteins indicated that only 'early' and 'light' endosomes were rich in these proteins. These endosomally located 28 kDa polypeptides emerge as members of an expanding ras polypeptide family, and may correspond to the rab polypeptides shown to be present in secretory and endocytic pathways in cultured cells and postulated to be involved in regulating intracellular vesicular trafficking.

G-proteins are emerging as ubiquitous membrane-associated components of cells [1]. The number identified is proliferating, and two major categories have been investigated in detail. The trimeric G-proteins function in the transduction of external signals involving receptors and regulated channels at the p.m.* In contrast, the recently identified low-M_r G-proteins appear to be more widely distributed in the cell. For example, the GTP-binding *ras* oncoproteins are present in most mammalian tissues, and act by regulating normal cell proliferation and differentiation as well as possessing trans-forming activity [2]. An increasing number of such low-M_r proteins (18-30 kDa), many associated with transformation, are being identified, and diverse roles in secretory and endocytic functions are being unravelled [3, 4].

Herein we describe the subcellular distribution of G-proteins in rat liver by exploiting the availability of relevant Ab's to the proteins and the availability of fractions of high purity derived from the hepatocyte's p.m. functional domains and the endocytic compartment. The results show two different

---

**Abbreviations.-* Ab, antibody; p.m., plasma membrane(s).
M_r signifies (apparent) mol. wt. G-proteins feature in sect. #A, and past work by the Evans group was described in earlier Vols. (ed. E. Reid *et al.*) - 13 & 17 (Plenum) and 19 (Roy. Soc. Chem.).

sets of distribution patterns of G-proteins. One set is distributed mainly between the surface domains and the endocytic compartment, whereas the other is predominantly in the endocytic compartment.

METHODS

Subcellular fractionation of liver homogenates to yield characterized domain-specified p.m.'s, lysosomes, Golgi and endosomes is shown in Scheme 1. The isolation of p.m., first described by Wisher & Evans [5], has been modified further, involving the introduction of additional sonication and Nycodenz density-gradient fractionation steps. The procedure yields bile-canalicular p.m., >100-fold purified *vs.* homogenate levels as judged by marker enzymes [6]. This procedure also furnishes lateral p.m. [6]. The endosome fractions were isolated as already described and are characterized, according to the time elapsing after the uptake from the circulation of various ligands, as 'early' (2-5 min) or 'late' (10-20 min); a further 'heavy' or receptor-enriched endosome fraction that may feature in receptor recycling was also recovered [7].* In this procedure, Golgi fractions ('endosome-depleted') were also prepared [8]. Lysosomes were isolated by a standard procedure [9], but metrizamide was replaced by Nycodenz.

Ab's and Western blotting.- Ab's to the C-terminal amino acid sequence of the α-subunit of inhibitory G-proteins were obtained from Dr. G. Milligan (cf. #A-4; Univ. of Glasgow), as was an antiserum specifically recognizing the β-subunit raised against a mixture of holomeric pertussis toxin-sensitive G-proteins purified from bovine brain [10, 11]. Two Ab's recognising 28 kDa proteins in endosome fractions were used - Y-13259 rat monoclonal Ab to a conserved amino acid region of *ras*, and a mouse monoclonal Ab (E_{546}) of unknown epitope specificity obtained from Dr. Roger Grand (Univ. of Birmingham, U.K.). The Ab's recognising a 23 kDa protein in the fractions were a mouse monoclonal Ab (RAS/10) of unknown specificity and mouse monoclonal 6B7 generated to the effector domain of *ras* protein, both obtained from Dr. Grand [12].

Membrane fractions (50 μg protein) were resolved by electrophoresis in 12.5% or gradient (9-18%) SDS-polyacrylamide gels, with electrophoretic transfer to nitrocellulose sheet (0.1 μm pore size; Schleicher & Schull) [13]. With rabbit Ab's, ^{125}I-protein A was used directly, whereas with rodent Ab's a second Ab (rabbit anti-rat or mouse IgG) was used before adding ^{125}I-protein A. Labelled antigens were detected by autoradiography and were scanned using a densitometer to quantify labelling.

Note by Ed.- In *Investigation of Membrane-located Receptors*, Vol. 13 of this series, Evans cites ligands used, and key findings.

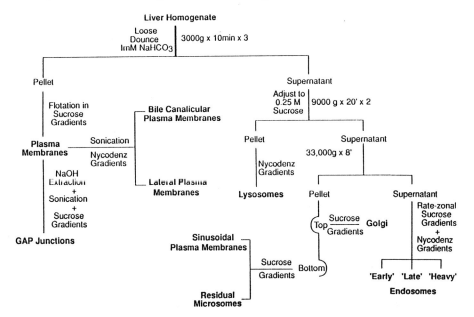

Scheme 1. Schematic representation of methods used to prepare the various liver subcellular fractions (amplified in METHODS).

RESULTS

The distribution in liver subcellular fractions of various G-proteins was assessed by Ab binding after gel separation (Fig. 1). As anticipated, both α- and β-subunits of the inhibitory trimeric G-protein were located at the sinusoidal p.m. of hepatocytes, reflecting the receptors functioning at this blood-facing domain in signal transduction. e.g. glucagon and epinephrine receptors [14]. However, less expected was the high relative distribution, on a protein basis, of the G-protein subunits at the canalicular p.m. where few receptors are located, and where no signalling function is envisaged [15]. A possible mode of transfer of the G-proteins across the hepatocyte was suggested by the detection of the subunits in 'early' and 'late' endosomes. This result suggests that proteins are endocytosed in concert with receptors, but unlike those receptors that are recycled back to the sinusoidal p.m. from the endocytic compartment, the G-proteins are transferred onwards to the bile-canalicular p.m. via 'early' and 'late' regions of the endocytic compartment.

The distribution in liver of the G-protein α- and β-subunits mirrored closely that of the inositol tris- and tetrakis-phosphatases, enzymes involved in signal transduction involving as yet unidentified G-proteins [16]. The distribution in liver was also similar to that of ectoenzymes such as 5'-nucleotidase [16], except that the components involved in signalling were located at the cytoplasmic face of the p.m.

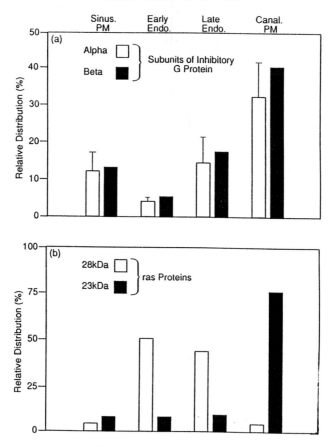

Fig. 1. Relative distribution of G-proteins in subcellular fractions. As amplified in METHODS, gel-separated proteins transferred to nitrocellulose were exposed to relevant Ab's and ^{125}I-protein A; autoradiographs were subjected to densitometry. **Endo,** endosomal; **Canal. & Sinus.** denote p.m. (**PM**) regions.

A number of GTP-binding proteins, M_r 18-30 kDa (low mol. wt. G-proteins), have been detected in tissues, especially brain [17]. In liver membranes these proteins were identified, after resolution in polyacrylamide gels and transfer to nitrocellulose supports, by their ability to bind a ^{35}S-labelled non-degradable GTP analogue [12]. It is noteworthy that the higher-M_r (trimeric) G-proteins described above, when resolved in polyacrylamide gels, do not bind GTP under these conditions. One specific type of G-protein is the *ras* oncoprotein associated with the inner face of the p.m. The subcellular distribution of these *ras* proteins was investigated using a range of Ab's generated to *ras* proteins. The conventional *ras* oncoprotein (M_r 23 kDa in liver membranes) showed a similar distribution in the fractions to that of the G-protein α- and β-subunits, the highest levels being recorded in bile-canalicular p.m. (Fig. 1b). However, with two of the Ab's examined, one of unknown epitope specificity and the other to a conserved amino acid sequence in *ras*, it was evident that antigens of 28 kDa were present almost exclusively in endosomes.

The demonstration of a discrete class of *ras*-related proteins in rat liver endosomes accords with emerging observations on transport-deficient yeast showing that low-M_r G-proteins are involved in regulating intracellular trafficking [18]. In addition, a role for low-M_r G-proteins in endocytosis has been reported [19]. It is likely that two of the Ab's to *ras* now employed have identified related members of this family, e.g. the *rab* proteins. The *ras* and *rab* proteins are closely related to the Sec 4p and Ypt 1p proteins studied in *Saccharomyces cerevisiae* [18]. Also, certain *rab* proteins have been shown to be present in Golgi membranes, whereas others are present in endosomes of cultured cells [20].

These results add, in a liver context, to the emerging view that a number of low-M_r GTP-binding proteins are involved in controlling, through GTP hydrolysis, the intricacies of intracellular vesicular trafficking. The 28 kDa polypeptide now identified using Ab's to *ras* oncoproteins appears to be specific to 'early' and 'late' regions of the liver's endocytic compartment [21]. A further novel 100 kDa putative liver G-protein has also been located to the 'heavy' or receptor-enriched endosomes, where it may function in the return of endocytosed receptors to the p.m. [22]. It is clear that the functional difference of the various categories of G-proteins is reflected in their location to specific subcellular compartments as shown by subcellular fractionation studies.

Acknowledgements

Nawab Ali thanks the Wellcome Trust for a Fellowship. We are grateful to Drs. G. Milligan and R. Grand for providing us with Ab's, and to Mrs. Lydia Pearson for preparing the manuscript.

References

1. Taylor, C.W. (1990) *Biochem. J. 272*, 1-13.
2. Macara, I.G. (1989) *Physiol. Rev. 69*, 797-820.
3. Bourne, H.R. (1988) *Cell 53*, 669-671.
4. Bourne, H.R., Sanders, D.A. & McCormick, F. (1990) *Nature 348*, 125-131.
5. Wisher, M.H. & Evans, W.H. (1975) *Biochem. J. 146*, 374-388.

6. Ali, N., Aligue, R. & Evans, W.H. (1990) *Biochem. J. 271*,
    185-192.
7. Evans, W.H. & Flint, N. (1985) *Biochem. J. 232*, 25-32.
8. Evans, W.H. (1985) *Meths. Enzymol. 109*, 246-257.
9. Wattiaux, R., Wattiaux-de-Coninck, S., Ronveaux-Dupal, M.F.
    & Dubois, F. (1978) *J. Cell Biol. 78*, 349-368.
10. Ali, N., Milligan, G. & Evans, W.H. (1989) *Biochem. J.
    261*, 905-912.
11. Ali, N., Milligan, G. & Evans, W.H. (1989) *Mol. Cell.
    Biochem. 91*, 75-84.
12. Ali, N. & Evans, W.H. (1990) *Biochem. J. 271*, 179-183.
13. Burnette, W.N. (1981) *Anal. Biochem. 112*, 195-203.
14. Evans, W.H. (1990) *Biochim. Biophys. Acta 604*, 27-64.
15. Enrich, C., Tabona, P. & Evans, W.H. (1990) *Biochem. J.
    271*, 171-178.
16. Shears, S.B., Evans, W.H., Kirk, C.J. & Michell, R.H.
    (1988) *Biochem. J. 256*, 363-369.
17. Burgoyne, R.D. (1989) *Trends Biochem. Sci. 14*, 388-390.
18. Goud, B., Salminen, A., Walworth, N.C. & Novick, P.J.
    (1988) *Cell 53*, 753-768.
19. Mayorga, L.S., Diaz, R. & Stahl, P.D. (1989) *Science
    244*, 1475-1477.
20. Goud, B., Zahraoui, A., Tavitan, A. & Saraste, J. (1990)
    Nature 345, 553-556.
21. Evans, W.H. & Enrich, C. (1989) *Biochem. Soc. Trans. 17*, 619-622.
22. Traub, L.M., Evans, W.H. & Sagi-Eisenberg, R. (1990)
    Biochem. J. 272, 453-458.

#ncF

NOTES and COMMENTS relating to

LOCATION AND TRANSIT OF PROTEINS (BESIDES 'PKC')

#ncF-1

A Note on

βVLDL AND LDL MAY FOLLOW DIFFERENT INTRACELLULAR PATHWAYS FOLLOWING ENDOCYTIC UPTAKE IN LIVER PARENCHYMAL CELLS

T. Berg, O. Gudmundsen and M.S. Nenseter

Department of Biology, Division of Molecular Cell Biology, and Institute for Nutrition Research, University of Oslo, Blindern, 0316 Oslo 3, Norway

The liver is the main organ for removing cholesterol-containing lipoproteins such as LDL and HDL from blood. Cholesterol feeding of experimental animals results in marked changes in plasma lipoproteins. Unlike normal VLDL and chylomicrons, the d <1.006 g/ml lipoproteins that accumulate in these animals are rich in cholesteryl esters, possess primarily apolipoproteins B and E, and display β-electrophoretic mobility [1]; they are referred to as βVLDL. They appear atherogenic, seemingly reflecting their ability to cause deposition of cholesteryl esters in macrophages, both *in vivo* and *in vitro* [1]. However, when βVLDL are injected into rats or rabbits most of the labelled apoproteins are removed by the liver. Separation of liver cells following injection of βVLDL has shown that the uptake takes place mainly in the parenchymal cells*.

The present aim was to follow the intracellular transport of endocytosed βVLDL in rat-liver parenchymal cells and to compare this process with the simultaneous transport of LDL and galactosylated albumin, which are internalized *via* the LDL receptor and the Gal-receptor respectively. Receptor-mediated endocytosis *via* the Gal-receptor has been studied in great detail and may therefore serve as a 'standard' pathway in the liver parenchymal cells.

All ligands were labelled with ^{125}I-tyramine cellobiose. The labelled degradation products from such ligands are trapped in the degradative organelle in which they are formed and may therefore serve as ideal markers for these organelles [2]. Studies were done both in the intact animal and in suspensions and primary cultures of rat liver parenchymal cells. The intracellular transport was followed by means of subcellular fractionations in density gradients and by differential centrifugation.

The results obtained so far indicate that LDL are transported differently from the other two ligands studied. By means of subcellular fractionation it could be shown that

*In Vol. 17 of this series (1987, ed. E. Reid *et al.*; Plenum), Berg *et al.* described pertinent concepts and techniques.- *Ed.*

degradation of apoβ in LDL was initiated in a lysosome that was less dense than that involved in degradation of galactosyl-ated albumin or βVLDL.   The degradation products from all three ligands accumulated eventually in the same dense secondary lysosome.

References

1. Mahley, R.W. (1983) *Arch. Path. Lab. Med. 107*, 393-399.
2. Nenseter, M.S., Wiik, T. & Berg, T. (1989) *Biol. Chem. Hoppe-Seyler 370*, 475-483.
   Nenseter, M.S., *et al.* (1989) *Biochem. J. 261*, 587-593.

Comments on #F-1: C.A. Pasternak - STRESS SIGNALS

H. Joost asked what is happening specifically in the long lag phase between cell infection and glucose-transporter translocation, and whether the infection affects transporter biosynthesis. **Reply.**- We assume that specific (heat-shock?) proteins have to be synthesized before transporter transloca-tion occurs. We have not investigated transporter synthesis; but the effects of the infection are independent of trans-scription and translation as shown with actinomycin and cycloheximide. **Answers to J.T. O'Flaherty.**- The transporter is bound in vesicles, other proteins in which await identification [but see Zorzano *et al.* (1989) *JBC 264*, 12358]. Our immuno-fluorescence approach can't cast light on your old observation that glucose transporter is located just beneath the p.m. Answer to A.K. Campbell: it is a good suggestion that we try heating the serum at 56° to see whether complement caused the effect. **Question from G.J. Barritt.**- Does the use of Na-[³H]-borohydride plus galactose oxidase label all cell-exterior proteins that have a carbohydrate moiety? **Reply.**- It labels all such proteins if they possess an aldehyde group and so are reducible by borohydride.

Comments on #F-2: B. Austen - MOVEMENT OF NEWLY FORMED PROTEINS

A.K. Campbell.- Have you measured the efficiency of your reticulocyte lysate and wheat-germ *in vitro* translation systems in terms of protein copies per mRNA? **Reply.**- Yes; ~3 protein molecules per mRNA molecule. **Reply to another ques-tion.**- Although diphtheria toxin is not a newly synthesized protein, its translocation across membranes as described by Madshus (#A-8) indeed shows some similarities to what we have found, involving docking, protein unfolding, exposing hydrophobic residues, and proteolysis.

Comments on #F-3: J.H. Perez *et al.* - HEPATOCYTE ENDOCYTIC ROUTES
           #F-4: W.H. Evans - G- AND *ras*-PROTEIN MOVEMENTS

Answer to A.P. Dawson.- Colchicine and nocodazole indeed impair bile flow, but only by ~20% - whilst inhibiting trans-port by ~80% - if they are titrated in. **Reply (by Perez) to Hilderson.**- Our demonstration of 5'N latency gives assurance that we are indeed getting endosomes in our Ficoll gradients. **Reply to question by Døskeland.**- The IgA secreted into the bile is not taken up from the gut and so recirculated; the secreted IgA is protected by a secretory component and therefore stays in the gut. The polymeric IgA taken up by the liver originates from lymphocytes in the gut.

Austen, to Evans.- The labelling of low-M_r GTP-binding proteins with GTPγS in Golgi membranes was strong, but there was little cross-reaction with anti-*ras* antisera. Could any of these proteins be involved in protein secretion? **Reply.**- The *ras*-related proteins are restricted to endosomes, but there are *ras*-related proteins in yeast that are clearly involved in protein secretion. **Reply to G.J. Barritt.**- The Ab's (other than Y13-259) used to detect '*ras*-like proteins' probably have a broad specificity and detect a variety of low-M_r G-proteins. **Point put by Pasternak (Evans** agreed): conceivably the presence of G-proteins on the bile-canalicular membrane prior to membrane fusion may be part of the mechanism by which secretion takes place. **Pasternak asked** whether, with thin (permeabilized) liver sections, the various G-proteins had been located *in situ.* **Reply.**- Yes! **O'Flaherty asked** whether the effects of ligands on the distribution of G_i-proteins had been studied (**Reply:** no!), and whether hepatic 'constitutional' as distinct from ligand-induced endocytosis had been studied. **Reply:** our studies on identifying 'early' and 'late' endocytic compartments were done with radio-iodinated ligands, but non-perfused and perfused liver can be regarded as equivalent; normal liver performs 'constitutional' endocytosis, e.g. trans-cytosis of poly-IgA [cf. #F-3 - *Ed.*].

SOME LITERATURE PERTINENT TO 'F' THEMES, *noted by Senior Editor*

Yamashiro, D.J. & Maxfield, F.R. (1988) *Trends Pharmacol. Sci.* **9**, 190-193.- 'Regulation of **endocytic processes** by pH'.

Cutler, D.F. (1988) *J. Cell Sci. 91*, 1-4.- Commentary: 'The role of transport signals and retention signals in **constitutive export** [of proteins] from animal cells'. Cf.(1985) *Cell 42*, 489-496.

Doxsey, S.J., *et al.* (1987) *Cell 50*, 453-463.- 'Inhibition of **endocytosis** by anti-clathrin antibodies'.

Keim, V. & Rohr, G. (1987) *Int. J. Pancreatol.* **2**, 117-126.- 'Evidence *in vivo* of asynchronous intracellular transport of rat pancreatic **secretory proteins**'.

Amir, S. & Shechter, Y. (1988) *Nature 336*, 528.- A polymyxin B-sensitive membrane phospholipid involved in mediating the effect of insulin on **glucose transport** (a diacylglycerol?).

Woychik, N.A. & Dimond, R.L. (1987) *J. Biol. Chem. 262*, 10008-10014.- Intracellular transport of **lysosomal protein** from the rough e.r. is prevented by a single mutation.

Hollenbeck, P. (1989) *Nature 338*, 294-295, citing an accompanying art. by J.M. Scholey *et al.* and work in the labs. of N. Hirokawa and J.T. Yang.- Transport systems (for organelles as well as small molecules) in **microtubules**: carrier molecules include kinesin and dynein (cf. the actin microfilament carrier, myosin).

Tasaka, K., *et al.* (1991) *Biochem. Pharmacol. 41*, 1031-1037.- **Microtubule** role in Ca^{2+} release from the e.r. and histamine release from rat peritoneal mast cells.

Subject Index

This Index, patterned on those in previous vols. to aid back-tracking, focuses on cellular constituents, phenomena and processes, non-comprehensively where text mentions are numerous (as with PKC). Index listing is minimal for use of techniques such as centrifugation ('Subcellular' is pertinent) or gel electrophoresis. Entries are lacking or sporadic for bioactive agents used as 'tools', e.g. drugs, secretagogues and vasopressin. Where there are minor entries for growth factors (e.g. TGF; conventional abbreviations used) or hormones, it may be the agent's *receptor* that features in the text. Eicosanoid interrelationships are outlined on p. 367.

For major citations the page is represented (e.g.) '25-', the *hyphen* signifying that ensuing pp. are relevant too.

[Entry continues overleaf

REMINDER: some entries for agonists, e.g. TGF, *actually
refer to receptors.*